SUBBAND IMAGE CODING

THE KLUWER INTERNATIONAL SERIES
IN ENGINEERING AND COMPUTER SCIENCE

VLSI, COMPUTER ARCHITECTURE AND
DIGITAL SIGNAL PROCESSING

Consulting Editor
Jonathan Allen

SUBBAND IMAGE CODING

by

Editor

John W. Woods
ECSE Department
Rensselaer Polytechnic Institute
Troy, New York 12180-3590

KLUWER ACADEMIC PUBLISHERS
Boston/Dordrecht/London

Distributors for North America:
Kluwer Academic Publishers
101 Philip Drive
Assinippi Park
Norwell, Massachusetts 02061 USA

Distributors for all others countries:
Kluwer Academic Publihsers Group
Distribution Centre
Post Office Box 322
3300 AH Dordrecht, THE NETHERLANDS

Library of Congress Cataloging-in-Publication Data

Subband image coding / editor, John W. Woods.
    p. cm. — (The Kluwer international series in engineering and
computer science)
    Includes bibliographical references and index.
    ISBN 0-7923-9093-8
    1. Image processing—Digital techniques. I. Woods, John W. (John
William), 1943–   . II. Series.
TA1632.S855 1991
621.36'7—dc20   90–46887
                                                             CIP

Printed on acid-free paper.

Printed in the United States of America

DIP LAB

Contents

To Margaret and Kathryn

Preface

This book concerns a new method of image data compression which well may supplant the well-established block-transform methods that have been state-of-the-art for the last 15 years. Subband image coding or SBC was first performed as such in 1985, and as the results became known at first through conference proceedings, and later through journal papers, the research community became excited about both the theoretical and practical aspects of this new approach. This excitement is continuing today, with many major research laboratories and research universities around the world investigating the subband approach to coding of color images, high resolution images, video- including video conferencing and advanced television, and the medical application of picture archiving systems. Much of the fruits of this work is summarized in the eight chapters of this book which were written by leading practitioners in this field.

The subband approach to image coding starts by passing the image through a two- or three-dimensional filter bank. The two-dimensional (2-D) case usually is hierarchical, consisting of two stages of four filters each. Thus the original image is split into 16 subband images, with each one decimated or subsampled by 4x4, resulting in a data conservation. The individual channel data is then quantized for digital transmission. In an attractive variation an octave-like approach, herein termed *subband pyramid*, is taken for the decomposition resulting in a total of just eleven subbands. What is remarkable about these new approaches is that the coding efficiency of such a small number of subbands surpasses that of the 16x16 discrete cosine transform (DCT) which has 256 coefficients or channels. Also the more regular and local structure of these shift-invariant filters offers some implementation advantages over the block structure of the DCT.[1]

The first chapter by Pearlman "Performance Bounds for Subband Coding" treats the information-theoretic justification for the subband coding approach. He demonstrates the asymptotic optimality of subband coding for the Gaussian source and variable width subbands. He presents several methods, with a range of computational intensities, for approximate implementation. Measures of memory reduction are used to show that subbands are easier to encode in a rate-distortion sense, in relation to the full band data. This communication theory chapter treats the subband filters as ideal with zero aliasing. The next three chapters are generally devoted to the synthesis problem for practical filters that approximate this ideal.

The second chapter by Vetterli "Multirate Filter Banks for Subband Coding" is devoted to the theory of subband filtering and filter banks. It first provides a useful summary of multirate signal processing, that includes block processing

---

[1] Actually, Chapters 2 and 4 show that block processing, of which the DCT is an example, can be included into the multirate signal processing approach of subband coding. It is just the extreme case where the length of the filters is equal to the decimation ratio.

as a special case. Real multirate filter banks are studied in both the polyphase and modulation domains. The emphasis is on perfect reconstruction FIR filter banks, including the elegant but restrictive paraunitary case. Design or synthesis of filters for multirate filter banks is considered in some detail, with attention to implementation complexity. In a later section, he treats the 2– and 3–D case with general sampling matrices, and includes quincunx[2] sampling as an example.

The third chapter by Smith "IIR Analysis/Synthesis Systems" emphasizes the use of recursive or IIR filters for generating the subbands. These filters are much more efficient than FIR filters for the equivalent filter magnitude specifications. Their non-causality, which precluded their use in 1–D time domain processing, is not a problem in the 2–D spatial domain. Design strategies and all-pass polyphase implementation structures are also provided. Smith presents coded results comparing FIR to IIR filters. He finds the IIR filters offer slightly less coding efficiency, but at an 8-to-1 reduction in multiplies per pixel versus an 16–tap QMF filter. Also presented here is the elegant symmetric extension method, which comes into play when the FIR subband filter's active memory overlaps the image data boundary.

The fourth chapter by Simoncelli and Adelson "Subband Transforms" considers a unification of the subband approach with the block-transform approach. Using the concepts of *scale and orientation*, *spatial localization*, and *orthogonality*, they compare and evaluate existing subband transforms, including QMF and DCT. Also included here is a novel orthogonalization of the Laplacian pyramid decomposition, a precurser to subband coding which emerged from computer vision research. The QMF approach is chosen based on its very good properties relative to the above criteria and a design method is presented. Extensions to 2–D hexagonal sampling QMF's are recommended for their additional directional selectivity, with only modest increases in required computation versus the conventional separable QMF.

The last four chapters in the book emphasize the application of subband coding to color images, video conferencing, high definition television, and medical image archiving, in that order. They implicitly make use of and combine results from the earlier chapter on communication/information theory and the chapters on multirate signal processing and subband filter design.

Chapter 5 by Westerink, Biemond, and Boekee "Subband Coding of Color Images" considers the coding of both monochrome and color still images. After a careful modeling of the probability densities of the subband signals, mean-square error optimal quantizers are designed for the generalized Gaussian density. The authors' optimal bit assignment algorithm is applied and extended to color

---

[2]   This is a type of interlaced sampling used in digital television, wherein a full frame is broken into two fields consisting of even and odd lines respectively, and only alternate samples in alternate lines are retained.

images using a weighted color-error criterion. Such a color error criteria, when used adaptively in subband coding, avoids the need for fixed sub-sampling of the chrominance components, eg. 4:2:2 in CCIR 601, which can be a problem for some highly saturated scenes. Also presented is a detailed comparison of SBC with DCT for the monochrome case, illustrating both the SNR and subjective superiority of the subband technique.

Chapter 6 by Gharavi "Subband Coding of Video Signals" presents current results on subband coding of moving images for video conferencing. *Intraframe*, or within frame coding is used for the first frame or after a scene change. Then *interframe*, or between frame coding is used for the general frame. Here, the block matching method is used to estimate motion vectors which are, in turn, utilized in a temporal DPCM loop that creates motion-compensated frame differences. These compensated frame differences are then subband coded. An additional feature of this chapter's approach is that runlength codes are used to transmit spatial blocks of significant higher subband energy. Also included is an experimental study of achievable back-to-back filter bank performance as a function of computational bit accuracy for different filter-tap lengths and number of subbands.

In the next Chapter 7 "Advanced Television Coding Using Exact Reconstruction Filters," Ansari and LeGall discuss research on the integration of the subband concept into new approaches for advanced and high definition television (HDTV). First analog approaches to advanced television are reviewed, along with the place of analog HDTV is this hierarchy. Then the topic of digital HDTV is taken up for transmission over optical fiber at 135 Mbps. Due to the high sample rates involved, only intraframe coding is considered using short perfect reconstruction filters. Conventional rectangular sampling grids are considered as well as non-rectangular ones involving a novel diamond prefilter prior to quincunx sub-sampling. The Chapter concludes with a brief description of the recent MIT proposal to the Federal Communications Commission for advanced broadcast television in the United States.

Finally in Chapter 8 "Medical Image Compression: Possible Applications of Subband Coding," Rompelman presents results of a study of SBC for medical image compression as used in picture archiving and communication systems (PACS). The Chapter starts out with a summary of the various medical imaging modalities and their nominal spatial and bit resolutions. The application area of PACS is then described. Both vector quantization (VQ) and multidimensional scalar quantization (SQ) are employed and results shown. The latter technique is favored because it allows for *progressive transmission* wherein a low resolution image is retrieved first, followed by smooth increases towards the full spatial resolution. A prototype board-level system was designed based on the Motorola 56001 DSP chip with a 68020 microprocessor host. Using 16–tap separable QMF filters and multidimensional SQ, the filtering time was about 2 seconds and coding time 1/2 second. Coded X-ray and CT images are shown.

An optional video tape of the still image (color and monochrome) and video coding results is being made available. In many cases the digital images were transferred directly to tape. It is hoped that this will provide an economical way to evaluate the image quality without intervening photographic and printing press degradations. Of course, it's the only way to really show the video coding results. The last page of the book contains an order form which may be copied.

John W. Woods
Troy, New York

Acknowledgement

I would like to thank all the authors of the chapters for their excellent contributions to this book. Thanks also to R. Cohen and T. Naveen who helped me with various LaTeX and Postscript matters, including obtaining prints of some of the coded image results. My work on image coding, including the editorship of this book, has been supported since 1986 by the National Science Foundation through Grant No. MIP-8519948. Additional support has been provided by the Center for Advanced Television Studies (CATS). Finally thanks are due to my wife Harriet, who has put up with one year of near nightly reports on "progress with the book...."

Chapter 1

Performance Bounds for Subband Coding

by: William A. Pearlman
   Electrical, Computer, and Systems Engineering Department
   Rensselaer Polytechnic Institute
   Troy, NY 12180

The purpose of this chapter is to present some tutorial material in information theory which is pertinent to the subject of subband coding and to develop this material further in order to gain insight into the superior performance realized in subband coding systems. Except for the well-known result that PCM coding of subbands outperforms direct PCM coding of the full-band original (see [7]), which is known to be generally inefficient, there have been hardly any theoretical analyses which have shown a mean-squared error improvement for subband coding. The superior results of subband coding systems are largely empirical in nature and they have indeed been impressive in coding of images. In this chapter these superior results are quantified in formulas, which have not yet appeared in print elsewhere.

The Chapter starts with an overview of information theory and its source coding branch called rate-distortion theory. Much of this material appears in standard information theory textbooks (e.g. [3]), but it is hoped that the tutorial treatment to follow will help the reader bridge theoretical concepts and practice more readily. The next Section 1.2 introduces the subband decomposition of a source with ideal filters and presents some relatively recent theory [9] on optimal coding of subbands. The main results here are the optimal distribution of rate among subbands and the fact that optimal performance can be obtained by coding subbands. Nowhere in this treatment or others in the following sections is there any restriction that the subbands must be of equal size. This is particularly important because of the current research activity in pyramid and other multi-resolution decompositions, which contain octave related subband bandwidths.

The subjects of PCM and DPCM coding comprise the next Section 1.3. The now standard treatments in [7,11] for equal size subbands are generalized to subbands of any sizes (as long as the ratio of the bandwidth to the lowest frequency is an integer) with similar conclusions. Also presented are more general rate allocation formulas and distortion formulas applicable to low rates. In the final section are presented derivations, many of them new, which show that the reduction of required rate for subband coding is due to the narrowing of the gap between effective finite order entropy and entropy rate in a subband decomposition [13], which can be interpreted loosely as a reduction of average memory between the full-band source and its subbands. The presentation illuminates some of the advantageous characteristics of subband coding. It is hoped that, along with generalizations of older coding results and the introductory tutorial review, this material proves useful to current and future users of subband coding.

1.1   Overview of Rate Distortion Theory

Concepts in information theory are of fundamental importance for understanding the theoretical advantages and limitations of subband coding. We present now a selective overview of information theory which will aid us in our objective of understanding the basis of subband coding techniques and pointing the way toward future research. The treatment will be somewhat informal in order to illuminate concepts, but will be founded in rigorous theory. For the sake of notational simplicity, one-dimensional (1-D) notation will be used to indicate a sequence in time or space. All formulas and results are applicable to two dimensions with obvious replacements of single indices and variables by double ones, and intervals replaced by regions. Whenever the corresponding two-dimensional (2-D) extension has a special characteristic, it will be noted.

1.1.1   Noiseless Source Coding

Consider an information source which emits a vector random variable of dimension N denoted by $\mathbf{U} = (U_1, U_2, \ldots, U_N)$ according to a probability law characterized by a probability mass function or probability density function

$q_{\mathbf{U}}(\mathbf{u})$, depending on whether the random vector takes on continuous or discrete values $\mathbf{u}$ in N-dimensional Euclidean space. Assume the source is stationary so that the probability function is the same for any N and length N vector emitted at any time. The source is said to be memoryless when the individual components of the vector, called the source letters, are statistically independent, i.e.,

$$
\begin{aligned}
q_{\mathbf{U}}(\mathbf{u}) &= q_{U_1}(u_1)\, q_{U_2}(u_2) \dots\, q_{U_N}(u_N) \\
&= q_U(u_1) q_U(u_2) \dots\, q_U(u_N) \ .
\end{aligned}
$$

The last equality which removes the dependence of the probability distribution on time follows from stationarity.

Suppose that the source is memoryless and discrete. It emits at any time values (letters) from a countable set (alphabet), i.e.,

$$
\begin{aligned}
u_i \ \varepsilon \ [a_1, a_2, \dots, a_K] & \quad K \quad \text{finite or infinite} \\
i &= \ 1, 2, \dots, N \ ,
\end{aligned}
$$

with respective probabilities $P(a_1), \dots, P(a_K)$. The entropy or average uncertainty of the source in bits per source letter is

$$
\frac{1}{N} H(\mathbf{U}) \equiv H(U) = \sum_{k=1}^{K} P(a_k) \ \log \ \frac{1}{P(a_k)} \ , \tag{1.1}
$$

where the base of the logarithm is 2 unless otherwise specified.

Through the non-negativity of $-\log P(a_k)$ for all k and the inequality $\log_e x \le x - 1$, it is easy to prove that

$$
H(U) \ge 0
$$

and

$$
H(U) \le \log \ K \ , \tag{1.2}
$$

with equality in the latter if and only if the probabilities $P(a_k)$ are equal.

When the source is not memoryless, one can show that

$$
H_N(\mathbf{U}) \equiv \frac{1}{N} H(\mathbf{U}) \ \le \ \frac{1}{N} \sum_{i=1}^{N} H(U_i) = H(U) \ , \tag{1.3}
$$

which means that the uncertainty per source letter is reduced when there is memory or dependence between the individual letters. Furthermore, as N tends toward infinity, $H_N(\mathbf{U})$ goes monotonically down to a limit $H_\infty(\mathbf{U})$. The following source coding theorem can now be stated:

Theorem 1. For any $\epsilon > 0, \delta > 0$, there exists N sufficiently large that a vector of N source letters can be put into one-to-one correspondence with binary sequences of length $L = N[H_\infty(\mathbf{U}) + \epsilon]$ except for a set of source sequences occurring with probability less than δ.   •

Note that when the source is memoryless $H_\infty(\mathbf{U}) = H(U)$. The ramification of this theorem is that we can select the $M = 2^{N[H_\infty(\mathbf{U})+\epsilon]}$ vectors from the source which occur with probability greater than $1 - \delta$, and index each of them with a unique binary codeword of length $L = N[H_\infty(\mathbf{U}) + \epsilon]$. If we transmit the binary index of one of these vectors to some destination where the same correspondences between the M indices and vectors are stored, then the original source vector is perfectly reconstructed. When the source emits a vector which is not among the M indexed ones, an erasure sequence is transmitted with no recovery possible at the destination. The probability of this *error event* is less than δ. The set of M binary sequences is called a code with rate in bits per source letter of $R = \frac{1}{N} \log_2 M = H_\infty(\mathbf{U}) + \epsilon$.

Consider now the case of a memoryless source. If one is willing to transmit binary codeword sequences of variable length, one can theoretically eliminate the error event associated with fixed length codeword sequences. In practice, however, one needs to utilize a fixed length buffer which may overflow or become empty with a finite probability when operating for a finite length of time. If we ignore the buffering problems by assuming an infinite buffer, the idea is to choose for $\mathbf{U} = \mathbf{u}$ a binary codeword sequence of length $L(\mathbf{u})$ such that

$$\log \frac{1}{q_\mathbf{U}(\mathbf{u})} \leq L(\mathbf{u}) < \log \frac{1}{q_\mathbf{U}(\mathbf{u})} + 1 . \tag{1.4}$$

Averaging this inequality, dividing by N and using the memoryless property of the source, we obtain

$$H(U) \leq \frac{\bar{L}}{N} < H(U) + \frac{1}{N} . \tag{1.5}$$

As N approaches infinity, the average number of binary digits per source letter $\bar{L}$ or rate of the code approaches $H(U)$, which parallels the result for fixed length codes.

When variable length codewords are strung together and transmitted, there is no assurance that they can be uniquely separated and decoded at the destination unless an extra mark digit is inserted between each codeword. As insertion of an extra digit for each codeword is uneconomical, one seeks codes (sets of codewords) which are uniquely decodable and of minimum average length. Huffman [1] has demonstrated a procedure whereby codewords are assigned optimally to terminal nodes of a tree and hence no one codeword can be the prefix of another, which guarantees unique decodability.

For sources with memory, the same conclusions can be drawn for variable length codes if we substitute $H_\infty(\mathbf{U})$ for $H(U)$ in (1.5) and construct vector Huffman codes. The general conclusion is that perfect reconstruction at the destination is possible both for fixed length and variable length codes if the code rate R is no less than the entropy rate, $H_\infty(\mathbf{U})$ or $H(U)$, when the source is memoryless. A converse coding theorem states that perfect reconstruction is impossible if the rate R of the code is strictly less than $H_\infty(\mathbf{U})$. Therefore, the minimum possible rate for achieving perfect reconstruction is the average uncertainty or entropy per source letter of the source. The objective of an efficient distortionless source coding scheme is to operate at a rate as close as possible, within complexity constraints, to the entropy rate of the source.

1.1.2 Continuous-Amplitude Sources

When the source letters are continuous in amplitude, the entropy rate $H_\infty(\mathbf{U})$ or $H(U)$ is infinite. It is useful, however, to define a quantity call differential entropy of a source, in which probability mass distributions are replaced by densities and integrals by sums. For a vector source $\mathbf{U}$ with joint probability density function $q_{\mathbf{u}(u)}$, the differential entropy is defined as

$$h(\mathbf{U}) \;=\; E\!\left[\log\frac{1}{q_{\mathbf{U}}(\mathbf{u})}\right] \;=\; \int q_{\mathbf{U}}(\mathbf{u})\,\log\frac{1}{q_{\mathbf{U}}(\mathbf{u})}\,d\mathbf{u}\;.$$

The differential entropy has two failings with regard to entropy: (1) it can be negative; and (2) it changes with a linear scaling of the vector or any of its components. It does, however, reflect a dispersive property of the distribution $q_{\mathbf{U}}(\mathbf{u})$ and is therefore sometimes called dispersion.

A memoryless source U with probability density function $q_U(u)$ has a differential entropy defined to be

$$h(U) = -\int q_U(u)\,\log q_U(u)\,du\;. \qquad (1.6)$$

When the source is Gaussian with probability density

$$q_U(u) = \frac{1}{\sqrt{2\pi\sigma^2}}\exp\!\left\{-\frac{u^2}{2\sigma^2}\right\}, \quad -\infty \;<\; u \;<\; \infty\;, \qquad (1.7)$$

the differential entropy is evaluated to be

$$h(U) = \frac{1}{2} \log \left(2\pi e \sigma^2\right) . \tag{1.8}$$

One can also show for any source U with variance σ^2,

$$h(U) \leq 1/2 \log \left(2\pi e \sigma^2\right) , \tag{1.9}$$

with equality if an only if the source is Gaussian. The Gaussian source has the largest dispersion of all sources having the same variance. A similar conclusion holds for vector sources with and without memory, but we shall defer the discussion of this subject until later when vector sources are studied in more detail.

1.1.3   Source Coding with a Fidelity Criterion

When the real-valued source letters are not discrete, but continuous amplitude, the entropy rate is generally infinite. Perfect reconstruction is therefore impossible with a finite code rate. In order to reproduce the source vector of length N at a remote point, a certain amount of distortion must be accepted. First, a measure of distortion is defined between the source vector realization $\mathbf{u}$ and its corresponding reproduction $\mathbf{v}$ and denoted by $d_N(\mathbf{u}, \mathbf{v})$. At the source is a list of M possible reproduction vectors $\{\mathbf{v}_1, \mathbf{v}_2, \ldots, \mathbf{v}_M\}$ of length N called the codebook or dictionary C. When $\mathbf{u}$ is emitted from the source, the codebook C is searched for the reproduction vector $\mathbf{v}_m$ that has the least distortion, i.e.,

$$d_N(\mathbf{u}, \mathbf{v}_m) \leq d_N(\mathbf{u}, \mathbf{v}_k) \text{ for all } k \neq m ,$$

and the binary index of m is transmitted to the destination. The rate R of the code in bits per source letter is therefore

$$R = \frac{1}{N} \log_2 M .$$

As the same codebook and indexing are stored at the destination in a table, the lookup in the table produces the same reproduction vector and same distortion found at the source through search if the channel is assumed to be distortion-free. Since the source vector is a random variable (of N dimensions), the average distortion obtained for a given codebook C can in principle be calculated with a known source probability distribution. Through the search of the codebook, the minimum distortion reproduction vector $\mathbf{v}_m$ is a function of the source vector $\mathbf{u}$ and is also a random variable. The average distortion per source letter of the code C is therefore

$$E\left[\frac{1}{N}\,d_N(\mathbf{U},\mathbf{v}_m(\mathbf{U}))|C\right] ,$$

where the expectation is calculated for the given source distribution $q_\mathbf{U}(\mathbf{u})$. The optimal code C is the one that produces the least average distortion for the same rate R, i.e.,

$$\min_C\; E\left[\frac{1}{N}\,d_N(\mathbf{U},\mathbf{v}_m(\mathbf{U}))|C\right] ,$$

where the minimization is over all codebooks C of rate R.

To find an optimal code through an exhaustive search through an infinite number of codes of size $M = 2^{RN}$ is truly an impossible task. Furthermore, optimal performance is obtained in the limit of large N, as it can be proved that as N approaches infinity, $E\left[\frac{1}{N}\,d_N(\mathbf{U},\mathbf{v}_m(\mathbf{U}))|C\right]$ approaches a limit in a monotone and non-increasing fashion. We must therefore resort to another approach to find the minimum average distortion of a code of rate R. This approach is through a theory called rate-distortion theory and actually yields a method of code construction which is "highly probable" to achieve optimal performance. We shall now introduce this theory to make the above statements more precise.

1.1.4 Rate-Distortion Theory

Consider a continuous-amplitude vector source ensemble $\mathbf{U}$ and a reproduction vector ensemble $\mathbf{V}$ and a stochastic relationship between them governed by the conditional probability density function $p_{\mathbf{V}/\mathbf{U}}(\mathbf{v}/\mathbf{u})$ of a reproduction vector $\mathbf{V} = \mathbf{v}$ given a source vector $\mathbf{U} = \mathbf{u}$. The conditional probability density function $p_{\mathbf{V}/\mathbf{U}}(\mathbf{v}/\mathbf{u})$ is called a test channel, which is depicted in Fig. 1.1.

This test channel description relating output to input, being probabilistic, is not coding, in which the equivalent channel is entirely deterministic. A theorem to be stated later will establish the relationship to coding. The average distortion per source letter can be calculated for the joint density function $q_\mathbf{U}(\mathbf{u})\,p_{\mathbf{V}/\mathbf{U}}/(\mathbf{v}/\mathbf{u})$ of the test channel joint input-output ensemble $\mathbf{UV}$ as

$$E\left[\frac{1}{N}d_N(\mathbf{U},\mathbf{V})\right] = \int\int\frac{1}{N}d_N(\mathbf{u},\mathbf{v})q_\mathbf{U}(\mathbf{u})p_{\mathbf{V}/\mathbf{U}}\,(\mathbf{v}/\mathbf{u})\,d\mathbf{u}d\mathbf{v} . \quad (1.10)$$

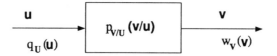

Figure 1.1: Test channel for the rate-distortion function.

The (average mutual) information per letter between the input and output ensembles is

$$\frac{1}{N}I_N(\mathbf{U};\mathbf{V}) = E\left[\frac{1}{N} \log \frac{p_{\mathbf{V}/\mathbf{U}}(\mathbf{v}/\mathbf{u})}{w_{\mathbf{V}}(\mathbf{v})}\right]$$

$$= \frac{1}{N} \int \int q_{\mathbf{U}}(\mathbf{u})p_{\mathbf{V}/\mathbf{U}}(\mathbf{v}/\mathbf{u}) \log \frac{p_{\mathbf{V}/\mathbf{U}}(\mathbf{v}/\mathbf{u})}{w_{\mathbf{V}}(\mathbf{v})} \, d\mathbf{u}d\mathbf{v}, \ (1.11)$$

where $w_{\mathbf{V}}(\mathbf{v}) = \int q_{\mathbf{U}}(\mathbf{u})p_{\mathbf{V}/\mathbf{U}}(\mathbf{v}/\mathbf{u}) \, d\mathbf{u}$ is the probability density function of the output vector random variable $\mathbf{V}$. The source vector probability distribution $q_{\mathbf{U}}(\mathbf{u})$ is considered to be known, so that the information $I_N(\mathbf{U};\mathbf{V})$ is a function of the test channel and is hence denoted as $I_N(p)$.

Consider now the problem of minimizing the average distortion per letter over all test channels $p_{\mathbf{V}/\mathbf{U}}(\mathbf{v}/\mathbf{u})$ giving an average mutual information per letter no greater than some fixed rate $R \geq 0$. The mathematical statement is

$$D_N(R) = \inf_p \left\{ E\left[\frac{1}{N} d_N(\mathbf{U},\mathbf{V})\right] : \frac{1}{N}I(p) \leq R\right\} . \qquad (1.12a)$$

The result is an average distortion $D_N(R)$ which is a function of R and depends on N. The function $D_N(R)$ is called the N-tuple distortion-rate function. The distortion-rate function $D(R)$ is

$$D(R) = \inf_N D_N(R) = \lim_{N\to\infty} D_N(R) . \qquad (1.12b)$$

The corresponding inverse functions, $R_N(D)$ and $R(D)$, are stated mathematically as

$$R_N(D) = \inf_p \left\{ \frac{1}{N} I_N(p) : E\left[\frac{1}{N}d_N(\mathbf{U},\mathbf{V}) \leq D\right]\right\}$$
$$R(D) = \inf_N R_N(D) = \lim_{N\to\infty} R_N(D) , \qquad (1.13)$$

and are called the N-tuple rate-distortion function and rate-distortion function, respectively.

We now restrict the distortion measure to be based on a single-letter distortion $d(u, v)$ between corresponding vector components according to

$$d_N(\mathbf{u}, \mathbf{v}) = \sum_{i=1}^{N} d(u_i, v_i) \,.$$

We now state the following theorems which connect the above functions to coding.

Theorem 2 (Distortion-Rate Theorem and Converse). For stationary and ergodic sources (under certain technical conditions) and a single-letter based distortion measure, for any $\epsilon > 0$ and N sufficiently large, there exists a code C of rate R such that

$$E\left[\frac{1}{N} \, d_N(\mathbf{U}, \mathbf{v}_m(\mathbf{U})) | C\right] \; < \; D(R) + \epsilon \,.$$

Furthermore, there exists no code of rate R such that

$$E\left[\frac{1}{N} \, d_N(\mathbf{U}, \mathbf{v}_m(\mathbf{U})) | C\right] \; < \; D(R) \,.$$

The distortion-rate function $D(R)$ is an unbeatable lower bound on average distortion for a code of rate R. The inverse statement of the above theorem is also common.

Theorem 3 (Rate-Distortion Theorem and Converse). For stationary and ergodic sources (under certain technical conditions) and a single-letter based distortion measure, given $D \geq 0$ and $\epsilon > 0$, for N sufficiently large there exists a code of rate R such that

$$R < R(D) + \epsilon \,.$$

Furthermore, there exists no code of rate

$$R < R(D) \,.$$

The rate-distortion function $R(D)$ is an unbeatable lower bound on rate for a code with average distortion D. The above two theorems are likewise valid for discrete-alphabet sources.

1.1.5   Rate-Distortion Functions

Memoryless Sources.   For memoryless sources and a single letter based distortion measure, the rate-distortion (or distortion-rate) function can be stated more simply since

$$E\left[\frac{1}{N} d_N(\mathbf{U}, \mathbf{V})\right] = E[d(U, V)] \ ,$$

where U and V are scalar input and output random variables of a scalar test channel $p_{V/U}(v/u)$ and average mutual information may be calculated for this test channel (instead of the vector one) as

$$I(p) = E\left[\log \frac{p_{V/U}(v/u)}{w_V(v)}\right] \ , \tag{1.14}$$

where $w_V(v) = \int q_U(u) p_{V/U}(v/u) \, du$ is the test channel output probability density. The rate-distortion function is

$$R(D) = \inf_p \ \{I(p) \ : \ E[d(U, V)] \le D\} \ . \tag{1.15}$$

In some cases $R(D)$ can be evaluated or bounded analytically. Otherwise there is a computational algorithm due to Blahut [2] for evaluating the points of $R(D)$. $R(D)$ is a monotone, non-increasing, convex $\bigcup$ (concave) function of D. Some important special cases, where exact formulas or analytical bounds are known, follow.

Case 1:   Gaussian source, squared error distortion measure.

$$
\begin{aligned}
q_U(u) &= \frac{1}{\sqrt{2\pi\sigma^2}} \exp\left\{-\frac{u^2}{2\sigma^2}\right\} \ , \quad -\infty < u < +\infty \\
d(u, v) &= (u - v)^2 \\
R(D) &= \max\left\{0, 1/2 \log \frac{\sigma^2}{D}\right\} \quad \text{bits/source letter or} \\
D(R) &= \sigma^2 \, 2^{-2R} \ , \quad R \ge 0 \ .
\end{aligned}
$$

Note the two extremal points $D = 0, R = \infty$ and $R = 0, D = \sigma^2$. The first one says that infinite rate is required for zero mean squared error, as expected. The second says that a mean squared error of σ^2 can be achieved for zero rate. This is done by setting any input u to zero, the mean value. The reception of all zeros is certain, so no information rate is required to convey it.

Given a source of unknown density and variance σ^2 and squared error distortion measure, the rate-distortion function is overbounded as:

$$R(D) \leq \max \left\{ 0, \frac{1}{2} \log \frac{\sigma^2}{D} \right\} . \tag{1.16}$$

The upper bound is the rate-distortion function of the Gaussian source. This means that the Gaussian source of the same variance is the most difficult to encode, since it requires the largest rate to obtain the same distortion. A lower bound to $R(D)$ for squared error distortion is

$$R(D) \geq h(U) - 1/2 \log (2\pi e D) , \tag{1.17}$$

where $h(U)$ is the differential entropy of the source.

Case 2: Bernoulli source, Hamming distortion measure

$$q_U(0) = q_U(1) = 1/2$$
$$d(u,v) = \begin{cases} 0 , u = v \\ 1 , u \neq v \end{cases} \qquad u, v = 0 \text{ or } 1$$

$$R(D) = \begin{cases} 1+ D \log D + (1 - D) \log (1 - D) & D \leq 1/2 \\ 0 & , D > 1/2 . \end{cases}$$

The average distortion here is equivalent to probability of error. Zero probability of error is obtainable for a rate of 1 bit per source symbol, which is the entropy $H(U)$ of the source. This is the case of no coding. Clearly, at the other extreme, a probability of error of 1/2 is achieved when all source symbols are mapped to one of the symbols 0 or 1, which again requires zero rate.

1.1.6   Optimal Code Construction

The Rate-Distortion or (Distortion-Rate) Theorem guarantees the existence of optimal codes but does not tell us how to construct these codes. However, there is a methodology in the proof of the theorems that will produce an optimal code with probability approaching one. Consider the memoryless source case and denote by $p_0(v/u)$ the test channel that achieves the rate-distortion function in (1.13). The output probability of this optimal test channel is

$$w_o(v) = \int q_U(u) \, p_o(v/u) du . \tag{1.18}$$

Choose variates v_{in} independently and at random from the probability distribution $w_o(v)$ until M vectors of length N are obtained ($n = 1, 2, \ldots, N, i = 1, 2, \ldots, M$). The reproduction code book is

$$
\begin{aligned}
\mathbf{v}_1 &= (v_{11}, v_{12}, \ldots, v_{1N}) \\
\mathbf{v}_2 &= (v_{21}, v_{22}, \ldots, v_{2N}) \\
&\vdots \\
\mathbf{v}_M &= (v_{M1}, v_{M2}, \ldots, v_{MN}) \, ,
\end{aligned}
$$

where the rate of the code is $R = N^{-1} \log_2 M$. For any $\epsilon > 0$ and sufficiently large N, the code will have average distortion no more than $D(R) + \epsilon$ with probability close to unity. As pointed out previously, for large N, the storage of $NM = N2^{RN}$ real variates and the $N2^{RN}$ multiplications and additions to find the minimum distortion $\mathbf{v}_m$ for a given source vector $\mathbf{u}$ is an enormous burden of storage and complexity. Fortunately, code books may be stored along codeword-dependent structures such as trees and trellises, which relieve some of the burdens of storage and complexity, while still maintaining optimality. The rate R_T of a code residing on a tree or trellis with branching factor α and β branches emanating from each node is $R_T = \beta^{-1} \log \alpha$. It will not be our task here to explain the workings of such codes, only to point out their existence.

1.1.7   Stationary Gaussian Sources

Let the vector source ensemble $\mathbf{U}$ be Gaussian distributed with zero mean and N-dimensional covariance matrix $\boldsymbol{\Phi}_N = E[\mathbf{UU}^t]$. This (Toeplitz) matrix is real, symmetric, and positive definite. There exist a set of N orthonormal eigenvectors $\mathbf{e}_1, \mathbf{e}_2, \ldots, \mathbf{e}_N$ and a corresponding set of N non-negative, not necessarily distinct, eigenvalues $\lambda_1, \lambda_2, \ldots, \lambda_N$ for $\boldsymbol{\Phi}_N$. A given source sequence $\mathbf{U} = \mathbf{u}$ can be represented with respect to the eigenvector basis by

$$
\mathbf{u} = \mathbf{Q}\, \tilde{\mathbf{u}} \quad , \quad \mathbf{Q} = [\mathbf{e}_1 | \mathbf{e}_2 | \ldots | \mathbf{e}_N] \, ,
$$

where $\mathbf{Q}$ is a unitary matrix ($\mathbf{Q}^{-1} = \mathbf{Q}^t$).

The vector $\tilde{\mathbf{u}}$ is called the transform of the source vector $\mathbf{u}$. The covariance matrix of the transform vector ensemble $\tilde{\mathbf{U}}$ is

$$
\begin{aligned}
\boldsymbol{\Lambda} \;=\; E[\tilde{\mathbf{U}}\tilde{\mathbf{U}}^t] \quad &=\quad E[\mathbf{Q}^{-1}\mathbf{U}\mathbf{U}^t\mathbf{Q}] \\
=\; \mathbf{Q}^{-1}\boldsymbol{\Phi}_N\mathbf{Q} \quad &=\quad
\begin{pmatrix}
\lambda_1 & & & \\
& \lambda_2 & & 0 \\
& & \ddots & \\
0 & & & \lambda_N
\end{pmatrix}.
\end{aligned}
\tag{1.19}
$$

The components of $\tilde{\mathbf{U}}$ are uncorrelated with variances equal to the eigenvalues. Moreover, since the source vector ensemble is Gaussian, the ensemble $\tilde{\mathbf{U}}$ is also Gaussian with independent components.

When we consider the problem of source encoding $\mathbf{U}$ with the single-letter squared error distortion measure $d(u,v) = (u-v)^2$, it is equivalent to encode $\tilde{\mathbf{U}}$ with the same distortion measure $d(\tilde{u},\tilde{v}) = |\tilde{u} - \tilde{v}|^2$, because average mutual information and average squared error are preserved in the unitary transformation of $\mathbf{U}$ to $\tilde{\mathbf{U}}$. The N-tuple rate-distortion (or distortion-rate) function is solved for the transform ensemble $\tilde{\mathbf{U}}$ and is found to be expressed parametrically by

$$
R_N = \frac{1}{N}\sum_{n=1}^{N}\max[0, 1/2\ \log\ \frac{\lambda_n}{\theta}]\ ,
\tag{1.20a}
$$

$$
D_N = \frac{1}{N}\sum_{n=1}^{N}\min[\theta, \lambda_n]\ .
\tag{1.20b}
$$

As N approaches infinity, the eigenvalues approach in a pointwise fashion a sampling of the discrete-time source power spectral density (psd) $S_U(\omega) = \sum_{n=-\infty}^{\infty}\phi(n)e^{-j\omega n}$, where $\{\phi(n)\}$ is the autocorrelation sequence, and the formulas above tend toward the limit of the rate-distortion (or distortion-rate) function:

$$
R_\theta = \frac{1}{2\pi}\int_{-\pi}^{\pi}\max\left[0, 1/2\ \log\ \frac{S_U(\omega)}{\theta}\right]d\omega
\tag{1.21a}
$$

$$
D_\theta = \frac{1}{2\pi}\int_{-\pi}^{\pi}\min\left[\theta,\ S_U(\omega)\right]d\omega\ .
\tag{1.21b}
$$

Interpretation of Formulas and Code Constructions

The interpretation of the R_N, D_N in (1.20) above is facilitated by defining the distortion sequence

$$
e_n = \min\left\{\theta, \lambda_n\right\}\ ,
\tag{1.22a}
$$

and the rate sequence

$$r_n = \max\left\{0, 1/2\log\frac{\lambda_n}{\theta}\right\}. \tag{1.22b}$$

For those components of $\widetilde{\mathbf{U}}, \widetilde{U}_n$, such that $\lambda_n > \theta$, the average distortion $e_n = \theta$ and the rate is $r_n = 1/2\ \log\frac{\lambda_n}{\theta}$. These components are scalar Gaussian variates of variance λ_n to be encoded with distortion θ with the minimal possible rate $1/2\ \log\frac{\lambda_n}{\theta}$. For every component whose variance does not exceed θ, the rate $r_n = 0$ and $\widetilde{U}_n$ is set to zero, its mean value, to give distortion equal to its variance λ_n. The collection of such components is hence called the stop band. A similar interpretation may be applied to the limiting N case where the frequency axis is continuous.

A coding theorem [3, Sec. 9.7] gives a code construction by which the above rate-distortion bounds may be achieved with high probability of success for sufficiently large N. It utilizes the random coding procedure described earlier with two modifications. First, the given source vector $\mathbf{u}$ is transformed to $\widetilde{\mathbf{u}}$ by $\mathbf{u} = \mathbf{Q}\ \widetilde{\mathbf{u}}$. Secondly, the codebook consists of transform reproduction vectors whose components are drawn independently and at random from the test channel output distributions of the components $\widetilde{V}_n$ given by

$$w_o(\widetilde{v}_n) = \frac{1}{\sqrt{2\pi\lambda_n}}\ \exp\left\{-\frac{\widetilde{v}_n^2}{2(\lambda_n - \theta)}\right\}, \tag{1.23}$$

where $\lambda_n \geq \theta$ for any $n = 1, 2, \ldots, N$. The nth component of the reproduction vector $\widetilde{\mathbf{v}}_m, m = 1, 2, \ldots, M$ is drawn from $w_o(\widetilde{v}_n)$, if $\lambda_n > \theta$. Otherwise the component is set to zero.

The codebook construction above creates an unstructured list to be searched for the $\widetilde{\mathbf{v}}_m$ which has the lowest mean squared error with respect to the given transform source vector $\widetilde{\mathbf{u}}$.

In [4] and [5], it has been have shown how to construct an optimal code by depositing randomly selected reproduction words along the paths of a tree or trellis, whose branching factor (for the tree only) and/or number of letters per branch vary with level in the structure. Although optimality of the resulting code is guaranteed only for large N, the tree or trellis structure provides an opportunity for savings in storage and computation, especially when selective search procedures, which give nearly optimal performance, are invoked.

The formulas in (1.22) may be regarded as the rate r_n and mean-squared error e_n to be assigned to the nth transform component in an optimal coding scheme achieving an overall rate R_N and distortion D_N in (1.20). These formulas are also guidelines for sub-optimal coding schemes. For example, each component is often independently quantized with a rate determined from (1.22b).

The distortion of (1.22a) can not be achieved, however, unless $r_n = 0$ when such quantization is an optimal scheme. The rate distribution formula, however, can be modified somewhat to take into account the distortion-versus-rate characteristic of the particular quantization scheme. The result is still fairly close to (1.22b) for most cases, because of the exponential dependence of distortion with rate. Many schemes of quantization, however, restrict the rate r_n to be an integer number of bits. Such a confinement to integer rates imparts yet another deviation from optimality. In practice, block, tree, or trellis codes can achieve to almost any required accuracy, the rate-distribution formula in (1.22b) and distortions much closer to (1.22a) than those of quantization.

An optimal codebook for a stationary Gaussian source with squared error distortion can be constructed on a regular tree with reproduction words for the original source word **u** [6]. The reproduction sequences along the tree must possess the statistics of the test channel output achieving the rate-distortion function. These statistics are Gaussian with psd,

$$S_V(\omega) = \max \left\{ 0, S_U(\omega) - \theta \right\}, \tag{1.24}$$

where θ is the parameter value solving eqn (1.21a) for the desired rate $R = R_T$ of the tree. The way to construct such a tree is to first populate a subsidiary tree of rate R_T with independent, Gaussian, unit variance letters. The sequence along every path through the subsidiary (or "white") tree is fed to a causal filter $H(\omega) = \sqrt{S_V(\omega)}$ and then deposited along the corresponding path through the final code tree (the "colored tree"). The latter tree is the one searched for the best reproduction word (in the squared error sense) for the given source word **u**. The corresponding path is then released to the channel and received at the destination, wherein the same white tree and filter reside. Since the path maps of the white and colored tree are in unique correspondence, perfect reception of the path map allows tracing through the white tree to find the white sequence to feed to the filter, whose output is the same reproduction word found by search at the source. The first use of this kind of system in coding of images is in [17].

We have treated primarily optimal methods for finite N to show that the N- tuple rate-distortion bounds can be approached. These searched stochastic code methods are not generally well known and are not used in practice for images. They have been much more prevalent in the speech coding literature since their introduction by Anderson and Bodie in 1974 [18] (see also [19]). The stochastic block code procedure just described, however, is currently used in CELP (code excited linear prediction) speech coding systems [20]. The common block techniques in image coding are forms of vector quantization, which is suboptimal for finite N and optimal only in the limit of large N. Vector quantization for large N is computationally not feasible at this time.

1.1.8   Spectral Flatness Measure

The rate-distortion function formulas in (1.20) and (1.21) may be further interpreted in light of spectral flatness measures. The spectral flatness measure (SFM) of the psd $S_U(\omega)$ is defined to be [7]:

$$\gamma_U^2 \;=\; \frac{\exp\left\{\frac{1}{2\pi}\int\limits_{-\pi}^{\pi}\log_e S_U(\omega)d\omega\right\}}{\frac{1}{2\pi}\int\limits_{-\pi}^{\pi} S_U(\omega)d\omega}\;. \tag{1.25}$$

The analogous definition for an eigenvalue spectrum $\{\lambda_n\}_{n=1}^{N}$ is

$$\gamma_{N,U}^2 \;=\; \frac{\left(\pi_{n=1}^{N}\lambda_n\right)^{1/N}}{\frac{1}{N}\sum\limits_{n=1}^{N}\lambda_n}\;, \tag{1.26a}$$

and

$$\lim_{N\to\infty}\gamma_{N,U}^2 \;=\; \gamma_U^2\;. \tag{1.26b}$$

The SFM in (1.26a) for the eigenvalue spectrum is the ratio of the geometric to arithmetic means of the eigenvalues. From the non-negativity of the eigenvalues and the geometric-to-arithmetic mean inequality [8],

$$0 \;\le\; \gamma_{N,U}^2 \;\le\; 1\;, \tag{1.27}$$

with equality if and only if the eigenvalues are equal. A flat eigenvalue spectrum gives the maximum SFM of unity. Likewise, in view of (1.26b), the SFM in (1.25) of the psd is a maximum of unity if and only if it is a constant for all frequencies, i.e., is flat.

The numerators of the SFM's in (1.25) and (1.26) also have interesting interpretations in terms of entropy. The differential entropy rate $h_N(\mathbf{U})$ of a Gaussian source with covariance matrix $\mathbf{\Phi}_N$ and eigenvalue spectrum $\{\lambda_n\}_{n=1}^{N}$ is

$$h_N(\mathbf{U}) \;=\; \frac{1}{2}\log_e 2\pi e \mid \mathbf{\Phi}_N \mid^{1/N} \;=\; \frac{1}{2}\log_e 2\pi e\left(\pi_{n=1}^{N}\lambda_n\right)^{1/N}\quad\text{nats/letter.} \tag{1.28}$$

In the limit as N approaches infinity,

$$h_\infty(\mathbf{U}) = \lim_{N \to \infty} h_N(\mathbf{U})$$

$$= \frac{1}{2} \log_e 2\pi e \left(\exp\left\{ \frac{1}{2\pi} \int_{-\pi}^{\pi} \log_e S_U(\omega) d\omega \right\} \right).$$

The quantity

$$Q_U = \exp\left\{ \frac{1}{2\pi} \int_{-\pi}^{\pi} \log_e S_U(w) d\omega \right\}, \qquad (1.30)$$

is called the entropy power of the source. Its eigen-spectrum equivalent is

$$Q_U^{(N)} = \left(\pi_{n=1}^N \lambda_n \right)^{1/N}. \qquad (1.31)$$

Bear in mind that the Gaussian source has the largest differential entropy rate of any source with the same psd or eigenvalue spectrum. Note that the flatness measures are entropy powers normalized by the source variance.

The spectral flatness measure plays a role in the rate-distortion function formulas in (1.20) and (1.21). Suppose that $\lambda_n \geq \theta$ for all n so that $D_N = \theta$ in (1.20b). Then the minimal rate R_N in (1.20a) can be expressed in terms of the SFM in (1.26a) as

$$R_N = \frac{1}{2} \log(\gamma_{N,U}^2 \sigma_U^2 / D_N),$$

and, similarly, in the limit of large N,

$$R_\theta = 1/2 \log(\gamma_U^2 \sigma_U^2 / D_\theta),$$

when $\theta \leq \min_\omega S_U(\omega)$. These expressions imply that to achieve a given signal-to-noise ratio, the minimal rate increases with the flatness of the psd.

1.2 Theory of Optimal Coding of Subbands

We turn now to the decomposition of the source sequence into subbands and show that these subbands can be coded such that the reconstruction is optimal. The filter decomposition assumed is ideal in every aspect. The following Chapters 2, 3, and 4 treat various techniques of filtering and sampling which are required to achieve a given decomposition and their limitations with respect to this ideal.

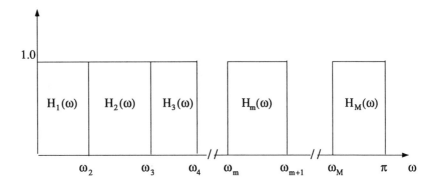

Figure 1.2: Subband filter transfer functions.

1.2.1 Subband Decompositions of a Source

Consider now a subband decomposition, wherein the source sequence is fed to a bank of bandpass filters which are contiguous in frequency so that the set of output signals can be recombined without degradation to produce the original signal. A block diagram of a complete system is shown in Fig. 1.2. We shall assume that the contiguous filters are ideal with zero attenuation in the pass-band and infinite attenuation in the stop band and that they cover the full frequency range of the input (Fig. 1.2). The assumption of ideal filters is not a consequential limitation, as filters called quadrature mirror filters (QMF) exist which when used in a contiguous filter bank, approximate well the characteristics of ideal ones. (cf. Chapters 3.1.1, 4.4, and 6.1.1)

When the input waveform to the bank of M ideal filters is a 1-D sequence, the output of any one filter whose lower cutoff frequency is an integer multiple of its bandwidth W_m, for any $m = 1, 2, \ldots, M$, is subsampled by a factor equal to $V_m = \pi/W_m$ and is now a full-band sequence from $-\pi$ to π at the lower sampling frequency (Fig. 1.3). This combination of filtering and subsampling is called decimation. We shall assume that the integer bandwidth to lower frequency relationship holds for all filters in the bank so that all outputs are decimated by the appropriate factor. These outputs are called the subband signals or waveforms and their aggregate number of samples equals that in the original input waveform. The original input can be reconstructed exactly from the subband signals. The sampling rate of each subband signal is increased to that of the original input by filling in the appropriate number of zero samples and the zero-filled waveform is fed into an ideal filter with gain equal to the sub-sampling factor covering the original pass-band (called interpolation) (Fig. 1.4). The sum of these interpolated subband signals equals the original input signal.

When the input is a sampled image, subband images are produced by an analogous 2-D filter bank comprised of contiguous, non-overlapping ideal filters covering the two-dimensional frequency range. In two dimensions, there is much more flexibility in the sampling pattern (e.g., rectangular or hexagonal) and the passband shapes (e.g., rectangular, hexagonal, hexagonal sections). We assume that each filter domain is such that there exists a sub-sampling pattern which replicates the filter characteristic so that it appears at the central low-frequency band and becomes full-band when referenced to its sub-sampling pattern. The most common example is a sampling of the original image on a rectangular grid and locating every filter domain so that the co-ordinates of its lower frequency in each dimension are integer multiples of the bandwidths belonging to their respective dimension. Although we shall use the paradigm of rectangular sampling grids for our subsequent analyses, the methods will extend to other kinds of sampling grids as well, such as hexagonal.

A particular type of subband decomposition which has attracted a great deal of attention lately is the pyramid decomposition. We say a pyramid is obtained when the low-frequency band image is recursively split into four equal area subbands which are then sub-sampled by a factor of two in each dimension. So we begin with the original image, filter into four subbands of equal size, sub-sample by a factor of two in each dimension, and then keep repeating the procedure on the low-pass subband until a 2x2 subband image terminates the process. When subband images belonging to different splitting levels are stacked one on top of the another by decreasing size with the largest at the bottom, the figure resembles a pyramid. Actually, we have four pyramids, one to each distinct subband. As we travel from the bottom to the top of any one pyramid, we regress by successive factors of two in resolution for each dimension.

1.2.2 Rate-Distortion Theory for Subband Coding

We now apply the rate-distortion theory of Section 1.1 to subband coding. The 1-D case will be treated in detail first and the generalization to 2-D sampling grids will be indicated (see [9]). Assume 1-D stationary Gaussian sequence with psd $S_U(\omega)$ has been decomposed with ideal filters into subbands with positive lower edge frequencies $\omega_1, \omega_2, \ldots, \omega_M$ with $\omega_1 = 0$ as shown in Fig. 1.2. The frequency range of the mth subband is denoted by $I_m =$

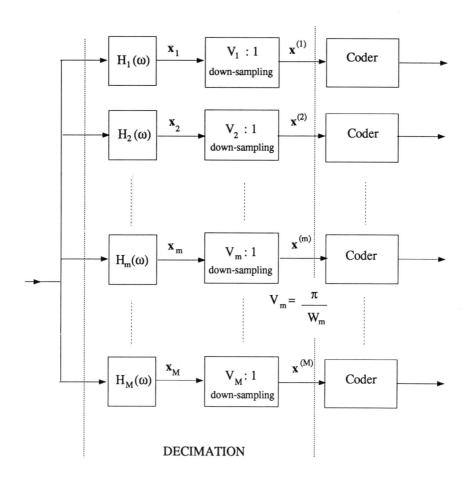

Figure 1.3: Decomposition and coding

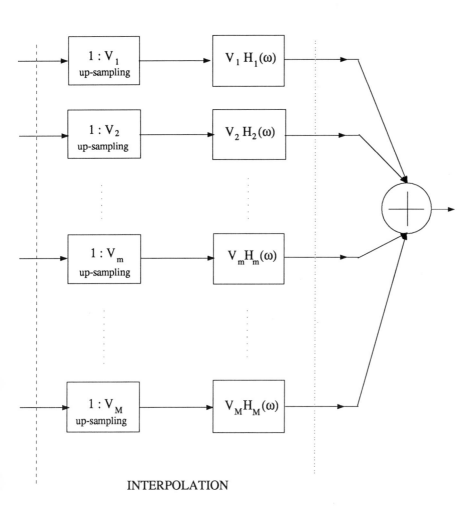

Figure 1.4: Synthesis of reconstructed signal

$[\omega_m, \omega_{m+1}) \bigcup (-\omega_{m+1}, -\omega_m], \omega_{M+1} = \pi, m = 1, 2, \ldots, M$, and its bandwidth by $W_m = \omega_{m+1} - \omega_m = \pi/V_m$. We require that

$$\bigcup_{m=1}^{M} I_m = [-\pi, \pi] \quad \text{and} \quad I_m I_n = \phi \quad \text{for} \quad m \neq n \;, \tag{1.32}$$

and that V_m be a rational number. The ideal filtering of the mth subband filter produces a Gaussian random sequence with psd

$$S_m(\omega) = \begin{cases} S_U(\omega) & , & \omega \; \varepsilon \; I_m \\ 0 & , & \omega \; \not\varepsilon \; I_m \; . \end{cases}$$

Resampling by a factor of V_m results in a random sub-sequence $\mathbf{x}^{(m)}$ with psd

$$S^{(m)}(\omega) = \frac{1}{V_m} S_U\left(\frac{\omega}{V_m} + \omega_m \text{sgn}(\omega)\right) , \tag{1.34}$$

in the frequency interval $-\pi < \omega \leq \pi$ referenced to the new sampling frequency. According to (1.21), the rate-distortion function of each Gaussian subband for the squared-error distortion measure is given by

$$R_{\theta_m}^{(m)} = \frac{1}{2\pi} \int_{-\pi}^{\pi} \max[0, 1/2 \, \log \frac{S^{(m)}(\omega)}{\theta_m}] d\omega \;, \tag{1.35a}$$

$$D_{\theta_m}^{(m)} = \frac{1}{2\pi} \int_{-\pi}^{\pi} \min[\theta_m, S^{(m)}(\omega)] d\omega \;, \quad m = 1, 2, \ldots, M \;. \tag{1.35b}$$

The $1/V_m$ factor in $S^{(m)}(\omega)$ equalizes the subband variances before and after re-sampling so that the mean-squared error per sample of the reconstructed process after re-combining the subbands is

$$D = \sum_{m=1}^{M} D_{\theta_m}^{(m)} \;. \tag{1.36}$$

(The scaling of the subband distortions done in decimation is undone in interpolation.) The contribution of the rate $R_{\theta_m}^{(m)}$ to the overall rate per sample R of the reconstructed process must be reduced by a factor of V_m to reflect the fraction of the samples in the subband. Therefore,

$$R = \sum_{m=1}^{M} \frac{1}{V_m} R_{\theta_m}^{(m)} \;. \tag{1.37}$$

If we substitute $\theta_m = \theta/V_m$, where θ is the parameter solving the rate-distortion function for the original process, we obtain by straightforward manipulations

$$
\begin{aligned}
D &= \sum_{m=1}^{M} \frac{1}{2\pi} \int_{-\pi}^{\pi} \min\left[\theta/V_m, \frac{1}{V_m} S_U\left(\frac{\omega}{V_m} + \omega_m \mathrm{sgn}(\omega)\right)\right] d\omega \\
&= \sum_{m=1}^{M} \frac{1}{2\pi} \int_{\eta \epsilon I_m} \min[\theta, S_U(\eta)] d\eta \\
&= \frac{1}{2\pi} \int_{-\pi}^{\pi} \min[\theta, S_U(\eta)] d\eta = D_\theta,
\end{aligned}
\tag{1.38}
$$

with the substitution $\eta = \omega/V_m + \omega_m \mathrm{sgn}(\omega)$, and similarly,

$$
\begin{aligned}
R &= \sum_{m=1}^{M} \frac{1}{V_m} \cdot \frac{1}{2\pi} \int_{-\pi}^{\pi} \max\left[0, 1/2 \, \log(S_U(\omega/V_m + \omega_m \mathrm{sgn}(\omega))/\theta) d\omega\right] \\
&= \sum_{m=1}^{M} \frac{1}{2\pi} \int_{\eta \epsilon I_m} \max\left[0, 1/2 \, \log \frac{S_U(\eta)}{\theta}\right] d\eta \\
&= \frac{1}{2\pi} \int_{-\pi}^{\pi} \max\left[0, 1/2 \, \log \frac{S_U(\eta)}{\theta}\right] d\eta = R_\theta,
\end{aligned}
\tag{1.39}
$$

where D_θ and R_θ are the distortion and rate of the rate-distortion function of **U**. The meaning of this result is that if ideally filtered subbands of the process **U** are optimally encoded and the rate and distortion are distributed among the subbands according to (1.35) with $\theta_m = \theta/V_m$, the encoding method is optimal.

The result extends readily to two dimensions for sampling on a rectangular grid. The analogous rate and distortion allocation to the mth subband is

$$
R_{\theta_m}^{(m)} = \left(\frac{1}{2\pi}\right)^2 \int_{-\pi}^{\pi} \int_{-\pi}^{\pi} \max\left[0, 1/2 \, \log \frac{S^{(m)}(\omega_1, \omega_2)}{\theta_m}\right] d\omega_1 d\omega_2, \tag{1.40a}
$$

$$
D_{\theta_m}^{(m)} = \left(\frac{1}{2\pi}\right)^2 \int_{-\pi}^{\pi} \int_{-\pi}^{\pi} \min\left[\theta_m, S^{(m)}(\omega_1, \omega_2)\right] d\omega_1 d\omega_2, \tag{1.40b}
$$

for $m = 1, 2, \ldots, M^2$, with $\theta_m = \theta/V_m$ and V_m being the re-sampling factor of the subband. Two common implementations are equal size subbands, $V_m = M^2$, $m = 1, 2, \ldots, M^2$ and recursive splitting and 2x2 sub-sampling of the low-frequency subband to form a pyramid, $V_m \ \varepsilon[2^{2r}, r = 1, 2, \dots, M_R]$, where M_R is the highest level of the splitting corresponding to the lowest resolution. Incidentally for hexagonal grids, the formulas are of the same form except that the integration is over the smallest hexagon which circumscribes a circle of radius π. Since this hexagon has area $2\sqrt{3}\pi^2$, the factor preceeding the integral changes from $(2\pi)^2$ to the reciprocal of this area. Transforms and filters for hexagonal grids are treated in Chapter 4 of this book.

The optimal performance specified by the preceding rate-distortion formulas can be achieved only in the limit as the size of the source process and hence the subband processes grow to infinity. Later in this Chapter, we shall consider the implications of coding the finite size source processes which are encountered in practice.

1.2.3   Frequency Weighting of Noise in Subbands

One of the often stated advantages of subband coding is that the mean squared error in each subband can be weighted according to frequency so that in bands where distortion tends to be more perceptually annoying, it can be attenuated relative to other bands. To express this requirement mathematically, let us assume a non-negative weighting function $P_m(\omega)$ for the mth subband before re-sampling and let $X_m(\omega)$ be the Fourier transform of the output of the subband filter. The weighted distortion for the reproduced subband process Fourier transform $\hat{X}_m(\omega)$ is

$$
\begin{aligned}
D_{P,m} &= E\left[\frac{W_m}{2\pi} \int_{I_m} P_m(\omega) \mid X_m(\omega) - \hat{X}_m(\omega) \mid^2 d\omega\right] \\
&= E\left[\frac{W_m}{2\pi} \int_{I_m} \mid \sqrt{P_m(\omega)}(X_m(\omega) - \hat{X}_m(\omega)) \mid^2 d\omega\right] .
\end{aligned}
\tag{1.41}
$$

Equivalently, one can consider this weighted mean squared error upon $X_m(\omega)$ as ordinary mean squared error between the $\sqrt{P_m(\omega)}X_m(\omega)$ and the reproduced process $\sqrt{P_m(\omega)}\hat{X}_m(\omega)$. Therefore, if one weights the corresponding power spectral density $S_m(\omega)$ by $P_m(\omega)$ in the rate-distortion formulas of equation (1.35), we obtain the correct rate-distortion formulas for frequency weighted squared error. When a contrast sensitivity curve $C_m(\omega)$ is given for human perceptual weighting of frequency, then that is equivalent to weighting the squared-error spectrum by $(C_m(\omega))^2$. Therefore, we equate

$$
P_m(\omega) = (C_m(\omega))^2 ,
\tag{1.42}
$$

for *human contrast sensitivity* weighting.

1.3   Coding of Subbands

We shall now treat actual coding of subbands of a Gaussian source to ascertain the benefit in performance over coding the original full-band (image) source directly. The specific types of coding treated are PCM (pulse code modulation) and DPCM (differential pulse code modulation). The intent is to generalize previously reported formulations [7,11] to unequal size subbands and low rates.

1.3.1   Rate Allocation to Subbands

The first step in coding is allocating a given number of bits among the subbands to achieve the minimum distortion. Distortion is again defined to be mean squared error. Let there be N samples from the image source and a given code rate of R bits/sample. If b_m bits are given to the mth subband, then

$$NR = \sum_{m=1}^{M} b_m \; , \qquad (1.43)$$

because the subbands are statistically independent.

Since the full-band source is stationary, any given subband has stationary samples with variance $\sigma_m^2, m = 1, 2, \ldots, M$. Define $d_m(b)$ to be the mean-squared error obtained for coding a subband of unit variance per sample with b bits. Then the total mean squared error after combining the subbands is

$$ND = \sum_{m=1}^{M} \sigma_m^2 d_m(b_m) \; , \qquad (1.44)$$

where D is the average distortion per sample.

To proceed with an analytical solution, we assume that $d_m(b)$ for each m is a convex monotone, non-increasing function of b. Moreover, $d_m(0) = NW_m/\pi$, the number of samples in the re-sampled subband, since the mean squared error must equal the variance when the subband samples are mapped to zero. The prior constraint equations are re-cast so that the Kuhn-Tucker Theorem can be applied. Let

$$\alpha_m = \frac{b_m}{NR} \; , \qquad (1.45)$$

so that the first constraint (1.43) becomes

$$\sum_{m=1}^{M} \alpha_m = 1 \quad \text{for} \quad \alpha_m \geq 0 \; . \qquad (1.46)$$

Note that the range of $\alpha = (\alpha_1, \alpha_2, \ldots, \alpha_M)$ is a closed convex set. The second equation (1.44) upon substitution of (1.45) becomes

$$D(\alpha) = \frac{1}{N} \sum_{m=1}^{M} \sigma_m^2 d_m(\alpha_m N R) , \qquad (1.47)$$

and $D(\alpha)$ is a convex (upward) function of α. The objective is to find α which minimizes D. The following Kuhn-Tucker Theorem can now be applied [3, Chapter 4].

Kuhn-Tucker Theorem. Let $f(\alpha)$ is a convex $\bigcup$ (upward) function over the convex region defined by $\alpha = (\alpha_1, \ldots, \alpha_M), \alpha_k > 0, \sum_{k=1}^{M} \alpha_k = 1,$. For some constant S, the conditions

$$\frac{\partial f(\alpha)}{\partial \alpha_k} = S \qquad \text{for all } k \text{ such that } \alpha_k > 0 ,$$

$$\frac{\partial f(\alpha)}{\partial \alpha_k} \geq S \qquad \text{for all } k \text{ such that } \alpha_k = 0 , \qquad (1.48)$$

are necessary and sufficient for a minimum point of f.

Applying the conditions of the theorem to $D(\alpha)$, we obtain

$$\sigma_m^2 d'_m(0) \leq \frac{S}{R} \quad , \quad b_m = 0 ,$$

$$\sigma_m^2 d'_m(b_m) = \frac{S}{R} \quad , \quad b_m > 0 , \qquad (1.49)$$

where the prime indicates derivative with respect to the argument of $d_m(\ )$. Note that this derivative is negative.

There are also algorithmic methods for allocating bits among the subbands, which are useful when analytic expressions for $d_m(b)$ are not available. Many of them, however, constrain the number of bits per sample to be an integer. Here we do not impose this restriction, because we wish to consider general subband coders such as entropy-coded quantization, vector quantization, tree coding or trellis coding, all of which do not require integer rates. A recent algorithm [10] does not impose any restrictions on the rate or type of subband coders. It iteratively seeks a point on the lower convex hull in the $D - R$ plane given rate and distortion points of coders in different subbands. An elegant way to implement the same solution, as pointed out in [14], is to use the BFOS algorithm [15] to prune a unitary branching tree with each path attached to the root corresponding to a subband. The rate and distortion of a quantizer are associated with each node such that the lower the rate, the nearer one is to the root node. All interior nodes of a path are examined to find the one with the greatest marginal return from the leaf node and the intervening branches are then pruned. All paths are examined in turn and the process is repeated until the desired rate is reached. For further explanation, see [16]. The rate-distortion point is guaranteed to lie on the convex hull. The analytical solution given here also achieves a point on the convex hull of all (R, D) points satisfying (1.43) and (1.44).

1.3.2 PCM Coding

Thus far, no advantageous properties of subband coding have been revealed, only that optimality can be preserved by coding subbands. Assume now that every sample of every subband is to be quantized independently with the same quantizer and the length of its representative binary codeword either the base two logarithm of the number of levels or the entropy. The distortion (mean squared error) versus rate characteristic of such a scalar quantizer of a unit variance sample can be modeled as

$$\rho(r) = g(r)e^{-ar} \ , \ r \geq 0 \ , \tag{1.50}$$

for $g(r)$ an algebraic function of rate per sample r and a a constant no greater than 2. For larger rates r and/or entropy coded quantizer, $a = 2$ is well justified. Since $\rho(0) = 1, g(0) = 1$. We make the further approximation that since $g(r)$ is a much more slowly varying function than the exponential e^{-ar}, we shall regard it as a constant in the range of interest. Since the quantization of the coefficients in any sub- band is independent and the subbands processes are stationary, the quantization noise at the input of the upsampling and interpolation filters is stationary and white. The variance per sample in a subband noise processs is therefore $\sigma_m^2 \rho(b_m/n_m)$ and the total quantization MSE is $n_m \sigma_m^2 \rho(b_m/n_m)$. After the (ideal) interpolation filter prior to the final recombining of the subbands, the reconstruction MSE of the subband becomes $V_m n_m \sigma_m^2 \rho(b_m/n_m) = N \sigma_m^2 \rho(b_m/n_m)$. Therefore, we take $d_m(b_m) = N \rho(b_m/n_m)$ in (1.49) to obtain

$$(\sigma_m^2/n_m)e^{-ab_m/n_m} = \theta \qquad \text{for all} \quad b_m > 0$$
$$\sigma_m^2/n_m \leq \theta \qquad \text{for all} \quad b_m = 0 \ , \tag{1.51}$$

with $\theta \equiv -S/gaNR$. Solving for b_m in terms of θ, we obtain for $r_m = b_m/n_m$, the rate per sample,

$$r_m = \begin{cases} 1/a \ \log \ [(\sigma_m^2/n_m)/\theta] & , & \sigma_m^2/n_m > \theta \\ 0 & , & \sigma_m^2/n_m \leq \theta \ , \end{cases} \tag{1.52}$$

where a is usually taken to be 2. Solving for θ using the rate constraint of eqn. (1.43), and letting $J_c = \{m : \sigma_m^2/n_m > \theta\}$ be the index set of non-zero rate subbands, we obtain the subband rate allocations

$$
r_m = \begin{cases} NR/N_c & + 1/a \, \log\left[(\sigma_m^2/n_m)/(\pi_{k\varepsilon J_c}(\sigma_k^2/n_k)^{n_k})^{1/N_c}\right] , & m\varepsilon J_c \\ 0 & , \quad m \not\varepsilon J_c , \end{cases} \tag{1.53}
$$

where $N_c = \sum_{m\varepsilon J_c} n_m$ is the total number of samples in the non-zero rate subbands and the number of samples in the mth subband is $n_m = NW_m/\pi$. For this rate allocation the MSE after reconstruction is

$$
D_{SB/PCM} = \sum_{m=1}^{M} \sigma_m^2 g e^{-a r_m} = \sum_{m\varepsilon J_c} g n_m \theta + \sum_{m\not\varepsilon J_c} \sigma_m^2 , \tag{1.54}
$$

where θ is given through the rate constraint and the factor g in the second sum is equated to 1.

Consider now the special case of rate R sufficiently large that $\sigma_m^2/n_m > \theta$ for all m. The rate per sample for the mth subband r_m, after solving for θ,

$$
\theta = \sigma_{WGM}^2 e^{-aR} \quad , \quad \sigma_{WGM}^2 = \pi_{m=1}^{M}(\sigma_m^2/n_m)^{n_m/N} ,
$$

becomes

$$
r_m = R + \frac{1}{a} \, \log \frac{\sigma_m^2/n_m}{\sigma_{WGM}^2} \quad , \tag{1.55a}
$$

and the MSE after reconstruction

$$
D_{SB/PCM} = gN\sigma_{WGM}^2 e^{-aR} . \tag{1.55b}
$$

In addition, when we specialize to equal width subbands,

$$
r_m = R + \frac{1}{a} \, \log \frac{\sigma_m^2}{\sigma_{GM}^2} \quad \text{and} \quad D_{SB/PCM} = Mg\sigma_{GM}^2 e^{-aR} .
$$

When the samples of the full-band source are directly and independently coded using the same technique, the average distortion is

$$
D_{FB/PCM} = g\sigma_U^2 e^{-aR} = g\left(\sum_{m=1}^{M} \sigma_m^2\right) e^{-aR} . \tag{1.56}
$$

The subband coding gain for PCM is defined to be

$$G_{SB/PCM} = \frac{D_{FB/PCM}}{D_{SB/PCM}} \quad ,$$

the ratio of the fullband to subband MSE's. When the rate R is sufficiently large that $r_m > 0$ for all $m = 1, 2, \ldots, M$, this gain, by substitution of (1.56) and (1.55b) is expressed as

$$G_{SB/PCM} = \frac{\sum\limits_{m=1}^{M} \sigma_m^2}{N \sigma_{WGM}^2} = \frac{\sigma_U^2}{\pi_{m=1}^{M} (\sigma_m^2 V_m)^{1/V_m}} \cdot \tag{1.57}$$

Noting that $\sum\limits_{m=1}^{M} (1/V_m) = 1$, the convexity of the logarithm yields

$$\sum_{m=1}^{M} \frac{1}{V_m} \log(\sigma_m^2 V_m) \le \log\left(\sum_{m=1}^{M} \frac{1}{V_m} \cdot \sigma_m^2 V_m \right) = \log \sigma_U^2 , \tag{1.58}$$

with equality if and only if $\sigma_m^2 V_m$ is the same for all subbands. Therefore

$$G_{SB/PCM} \ge 1 , \tag{1.59}$$

with equality if and if $\sigma_m^2 V_m$ is a constant independent of m. Since V_m is inversely proportional to bandwidth W_m, the equality condition is satisfied if the input process is white, even for subbands of different bandwidths.

Therefore, PCM coding of subbands is advantageous over direct PCM coding of the full-band signal even in the case of unequal size subbands. This result is a generalization of previous formulations [11] which treated only the case of equal size subbands.

1.3.3 DPCM Coding

In general, the subbands do not have flat-topped power spectra and therefore are not memoryless. One way to take advantage of this memory to reduce the encoding rate is to encode the difference between a source sample and linear prediction of this sample from past reconstructed sample values. The sequence of independently encoded differences is called differential pulse code modulation (DPCM). The prediction coefficients are optimized, however, for actual past values of the samples, which are not available to the decoder. This type of prediction is therefore suboptimal and approaches optimality only in the limit of large encoding rate, when the mean squared error between the actual and reconstructed samples approaches zero. This large rate approximation for

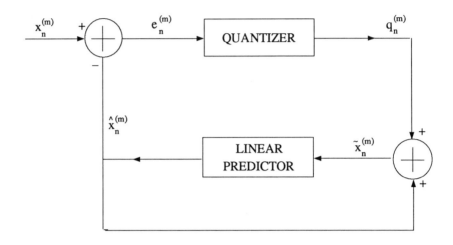

Figure 1.5: DPCM coder for the mth subband.

optimality of the predictor is much more tenable for subband coding than for full-band coding, because the subbands are coded at rates which are higher than their contributions to the overall rate by a factor equal to the resampling factor V_m. Moreover, the flatter spectra of the subbands enable more accurate predictors of lower order than that for the full-band image.

Ignoring the inherent suboptimality of predictors from past signal values and that predictors of a given order in subbands are likely to more accurately matched to the input process than the full-band predictor of the same order, we calculate the mean squared error resulting from DPCM coding of subbands and then compare it with that of DPCM coding of the full-band signal under the condition of optimal prediction from past values throughout.

In a DPCM coder, as shown in Fig. 1.5, the mean squared reconstruction error per sample is equal to the mean squared quantization error per sample. In the mth subband, the difference between the input sample $x_n^{(m)}$ and its prediction $\hat{x}_n^{(m)}$ is scalar quantized to a value $q_n^{(m)}$ with rate $r_m > 0$ for all m to yield a mean squared error (MSE) by (1.50) of

$$E[(e_n^{(m)} - q_n^{(m)})^2] = \sigma_{e,m}^2 g(r_m)e^{-ar_m} , \qquad (1.60)$$

and this error is equal to the reconstruction error, i.e.,

$$E[(e_n^{(m)} - q_n^{(m)})^2] = E[(x_n^{(m)} - \tilde{x}_n^{(m)})^2] . \tag{1.61}$$

The variance of the quantizer input, $\sigma_{e,m}^2$, is the variance of the prediction error and $g(r_m)$ is again considered to be approximately a constant g for all subbands. The total reconstruction MSE per sample $D_{SB/DPCM}$ is therefore

$$D_{SB/DPCM} = \sum_{m=1}^{M} E[(x_n^{(m)} - \tilde{x}_n^{(m)})^2] = \sum_{m=1}^{M} g\sigma_{e,m}^2 e^{-ar_m} . \tag{1.62}$$

The optimal rate allocation $\{r_m\}$ for a given rate R depends on the variances of the elements being quantized and (1.55b) is therefore applicable with $\sigma_{e,m}^2$ replacing σ_m^2 and $\sigma_{e,WGM}^2 = \pi_{m=1}^{M}(\sigma_{e,m}^2/n_m)^{n_m/N}$ replacing σ_{WGM}^2. Substituting these correspondences into the above expression (1.62) for $D_{SB/DPCM}$ results in

$$D_{SB/DPCM} = gN\sigma_{e,WGM}^2 \ e^{-aR} , \tag{1.63}$$

where the MSE is equally distributed across the M subbands.

When the rate R is sufficiently low, there may be some subbands assigned zero rate, making the above formula no longer applicable. In this case, we use the rate allocation eqn (1.52) and distortion eqn (1.54) with $\sigma_m^2 = \sigma_{e,m}^2$ to obtain

$$D_{SB/DPCM} = [g\theta(\sum_{m \varepsilon J_c} n_m) + \sum_{m \notin J_c} \sigma_{e,m}^2] \tag{1.64}$$

with θ determined by the overall rate constraint,

$$R = \frac{1}{N} \sum_{m \varepsilon J_c} n_m r_m \tag{1.65a}$$

and

$$J_c = \{m : \sigma_{e,m}^2/n_m > \theta\} . \tag{1.65b}$$

The corresponding full-band DPCM coding for a given rate R results in the reconstruction MSE of

$$D_{FB/DPCM} = g\sigma_e^2 e^{-aR} , \qquad (1.66)$$

where σ_e^2 is the variance of e, the quantizer input, which is the prediction error. To assess whether there is any benefit of DPCM coding of subbands over DPCM coding of the full-band signal, consider the ratio of their distortions as the gain

$$G_{SB/DPCM} = \frac{D_{FB/DPCM}}{D_{SB/DPCM}} . \qquad (1.67)$$

In the high rate case this gain becomes

$$G_{SB/DPCM} = \frac{\sigma_e^2}{N\sigma_{e,WGM}^2} . \qquad (1.68)$$

In order to evaluate this subband coding gain. We must introduce the fact that the MSE prediction error in the limit as N goes to infinity is

$$\sigma_e^2 = \exp\left\{ \frac{1}{2\pi} \int_{-\pi}^{\pi} \log_e S_U(\omega)d\omega \right\} \qquad (1.69)$$

for the full-band and

$$\sigma_{e,m}^2 = \exp\left\{ \frac{1}{2\pi} \int_{-\pi}^{\pi} \log_e S^{(m)}(\omega)d\omega \right\}, \quad m = 1,2,\ldots,M \qquad (1.70)$$

for the subbands, where we recall that

$$S^{(m)}(\omega) = \frac{1}{V_m} S_U\left(\frac{\omega}{V_m} + \omega_m \mathrm{sgn}(\omega) \right) . \qquad (1.34)$$

We now proceed to show that

$$\lim_{N \to \infty} N\sigma_{e,WGM}^2 = \sigma_e^2 \qquad (1.71)$$

so that

$$G_{SB/DPCM} = 1 \qquad (1.72)$$

in the limit.

First, by definition of $\sigma^2_{e,WGM}$,

$$N\sigma^2_{e,WGM} = \Pi^M_{m=1}\left(\sigma^2_{e,m}/(n_m/N)\right)^{n_m/N} = \Pi^M_{m=1}(V_m\sigma^2_{e,m})^{1/V_m}.$$

Taking the limit and substituting (1.70) and (1.34), this becomes

$$\lim_{N\to\infty} N\sigma^2_{e,WGM} = \Pi^M_{m=1}\left(V_m \exp\left\{\frac{1}{2\pi}\int_{-\pi}^{\pi}\log_e S^{(m)}(\omega)d\omega\right\}\right)^{1/V_m}$$

$$= \Pi^M_{m=1}\left(V_m \exp\left\{\frac{V_m}{2\pi}\int_{I_m}\log_e(S_u(\eta)/V_m)d\eta\right\}\right)^{1/V_m}.$$

Then by expansion of the logarithm in the integrand,

$$\lim_{N\to\infty} N\sigma^2_{e,WGM} = \Pi^M_{m=1}(V_m)^{1/V_m}\exp\left\{-\frac{W_m}{\pi}\log_e V_m\right\}\exp\left\{\frac{1}{2\pi}\int_{I_m}\log_e S_U(\eta)d\eta\right\}.$$

Denoting

$$A_m = (V_m)^{1/V_m}\exp\left\{-\frac{W_m}{\pi}\log_e V_m\right\}$$

and using the fact that $V_m = \pi/W_m$,

$$A_m = (V_m)^{1/V_m}\left(\frac{1}{V_m}\right)^{W_m/\pi} = 1.$$

Therefore,

$$\lim_{N\to\infty} N\sigma^2_{e,WGM} = \Pi^M_{m=1} A_m \exp\left\{\frac{1}{2\pi}\int_{I_m}\log_e S_U(\eta)d\eta\right\}$$

$$= \exp\left\{\frac{1}{2\pi}\sum_{m=1}^M\int_{I_m}\log_e S_U(\eta)d\eta\right\}$$

$$= \exp\left\{\frac{1}{2\pi}\int_{-\pi}^{\pi}\log_e S_U(\eta)d\eta\right\} = \sigma^2_e,$$

which proves (1.72) in the case of unequal size subbands and generalizes the same result in [11] obtained for equal size subbands. The unequal size subband case is of considerable interest, because of the current flurry of activity in coding of image pyramids. The general conclusion is that subband DPCM coding offers no advantage over full-band DPCM coding. One must bear in mind that this derivation assumes optimum predictors both in the subbands and in the full-band source. Such predictors are theoretically able to remove the memory in the sample to be coded, so that in subbands and in the full-band, memoryless error samples are quantized. This is the reason for the pessimistic conclusion of zero performance gain for subband coding. If, however, one would pursue the analysis for finite p-th order predictors in the subbands and the full-band source, one would find that subband error samples to be quantized are more nearly memoryless than the full-band ones. There would, therefore, be a gain for subband DPCM coding. This subject of memory reduction in subbands is pursued in the next section.

1.4   Information Theory for Performance Gains in Subband Coding

In this Section we shall explore the advantages of subband decomposition and coding from an information theory perspective. The gain found for PCM coding may be viewed in this perspective. When one codes the samples of a source independently, the lowest rate achievable is either the first order entropy for distortionless coding of a discrete-amplitude source or the first-order rate-distortion function. Lower rates can be achieved, however, if one encodes p-tuples from the source. Then the p-th order entropy or p-th order rate-distortion function becomes the minimum achievable rate. With this viewpoint in mind, we shall examine first the difference between first order differential entropy and entropy rate for a stationary Gaussian source. For convenience, the rates that follow in this Section are in units of nats per source letter.

1.4.1   Scalar Entropy Reduction

The first order differential entropy and entropy rate for the Gaussian source sequence $\mathbf{U} = (U_0, U_1, U_2, \ldots,)$ are, respectively

$$h_1(U) = 1/2 \ \log_e \ 2\pi e\sigma_U^2, \quad \sigma_U^2 = \frac{1}{2\pi} \int_{-\pi}^{\pi} S_U(\omega)d\omega \ , \qquad (1.73)$$

and

$$h_\infty(\mathbf{U}) = 1/2 \ \log_e \ 2\pi e Q_U, \quad Q_U = \exp\left\{\frac{1}{2\pi} \int_{-\pi}^{\pi} \log_e S_U(\omega)d\omega\right\} \ , \qquad (1.74)$$

where σ_U^2 is the source variance and Q_U is the source entropy power. The difference between these quantities is really the average mutual information between one member of the sequence and all the rest of the sequence [12],

$$
\begin{aligned}
\Delta_\infty^U &= h_1(U) - h_\infty(\mathbf{U}) \\
&= \lim_{N\to\infty} \Delta_N^U = \lim_{N\to\infty} [h(U_0) - \frac{1}{N} h(U_1, \ldots, U_N] \\
&= \lim_{N\to\infty} I(U_0, U_1, U_2, \ldots, U_N) \ .
\end{aligned} \tag{1.75}
$$

As such, Δ_∞^U is a measure of the memory of the sequence and for a Gaussian source, it becomes

$$
\Delta_\infty^U = 1/2 \ \log_e \ \frac{\sigma_U^2}{Q_U} = 1/2 \ \log_e \ \gamma_U^{-2} \ , \tag{1.76}
$$

where γ_U^2 is the spectral flatness measure of (1.25). The quality Δ_∞^U is in a sense a measure of spectral peakedness. A "peakier" power spectrum means a greater difference between the first order entropy and entropy rate.

If one wishes to reproduce source values to a given average distortion $D \geq 0$, it can be shown in general, moreover, that

$$
\Delta_\infty^U \geq R_1(D) - R(D) \ , \quad D \geq 0 \ , \tag{1.77}
$$

with equality if $D \leq D_c$, where $D_c = \min_\omega S_U(\omega)$ for a Gaussian source [12,9]. $R_1(D)$ is the rate distortion function of a memoryless source with the same first order probability distribution. Therefore, Δ_∞^U, the difference between first order entropy and entropy rate, is the difference in minimum achievable rate between encoding source values independently and encoding large blocks where the memory of the source is exploited.

Consider now the decomposition of the Gaussian source into subbands and encoding the decimated subbands. Let us evaluate the same entropy differences as above for the subbands. In the mth decimated process $\mathbf{x}^{(m)} = (\ldots, x_{-1}^{(m)}, x_0^{(m)}, x_1^{(m)}, \ldots)$

$$
\Delta_\infty^{(m)} = 1/2 \ \log_e \ \frac{\sigma_m^2}{Q_U^{(m)}} \quad , \tag{1.78a}
$$

where

$$
\sigma_m^2 = \frac{1}{2\pi} \int\limits_{-\pi}^{\pi} S^{(m)}(\omega) d\omega = \frac{1}{2\pi} \int_{I_m} S_U(\omega) d\omega \tag{1.78b}
$$

and

$$
Q_U^{(m)} = \exp\left\{ \frac{1}{2\pi} \int\limits_{-\pi}^{\pi} \log_e \ S^{(m)}(\omega) d\omega \right\}
$$

$$= \exp\left\{\frac{V_m}{2\pi} \int_{I_m} \log_e (S_U(\omega)/V_m)d\omega\right\} . \tag{1.78c}$$

The same expressions appear in (1.73). Since rates combine as in (1.37), the composite rate difference for subband coding is

$$
\begin{aligned}
\Delta_\infty^{SB} &= \sum_{m=1}^{M} \Delta_\infty^{(m)}/V_m \\
&= \sum_{m=1}^{M} \frac{1}{2V_m} \log_e \frac{\sigma_m^2}{Q_U^{(m)}} , \\
&= \frac{1}{2}\left(\sum_{m=1}^{M} \frac{1}{V_m} \log_e V_m\sigma_m^2 - \frac{1}{2\pi} \int_{I_m} \log_e S_U(\omega)d\omega\right) , \\
&= \frac{1}{2} \sum_{m=1}^{M} \frac{1}{V_m} \log_e V_m\sigma_m^2 - \frac{1}{4\pi} \int_{-\pi}^{\pi} \log_e S_U(\omega)d\omega ,
\end{aligned}
$$

where the next-to-last step follows by substitution of (1.78b) and (1.78c) and $V_m W_m = \pi$. A more compact expression for this rate difference for subbband coding is

$$\Delta_\infty^{SB} = \frac{1}{2} \log_e \frac{\pi_{m=1}^{M}(V_m\sigma_m^2)^{1/V_m}}{Q_U} . \tag{1.79}$$

Now taking the difference between Δ_∞^U and Δ_∞^{SB} in (1.76) and (1.79) reveals

$$\Delta_\infty^U - \Delta_\infty^{SB} = \frac{1}{2} \log_e \frac{\sigma_U^2}{\pi_{m=1}^{M}(V_m\sigma_m^2)^{1/V_m}} . \tag{1.80}$$

Just as in (1.58),

$$\sum_{m=1}^{M} \frac{1}{V_m} \log_e V_m\sigma_m^2 \le \log_e \sum_{m=1}^{M} \frac{1}{V_m} * V_m\sigma_m^2 = \log_e \sigma_U^2 \tag{1.81}$$

with equality iff $V_m\sigma_m^2$ is constant with m.

Substituting (1.81) into (1.80), we obtain the result that

$$\Delta_\infty^U - \Delta_\infty^{SB} \ge 0 . \tag{1.82}$$

This result means that independent encoding of sample values of a subband decomposition requires less rate than independent encoding of values from the full-band source, because the difference between its composite first order entropy and its composite entropy rate is smaller than the corresponding difference for the full band source. In fact, if one compares (1.57) and (1.80), one concludes that

$$\Delta_\infty^U - \Delta_\infty^{SB} = 1/2 \, \log_e \, G_{SB/PCM} \, , \qquad (1.83)$$

which shows that the distortion gain ratio in PCM coding of subband results solely from the reduction of the difference between first order entropy and entropy rate from that of the full-band source. We have shown here that this result for PCM is part of a more general framework of entropy reduction in subband or pyramid decompositions given in (1.80).

1.4.2 Entropy Reduction for Block Coding

Block coding, such as vector quantization, tree coding, or trellis coding, is more rate efficient than scalar coding in meeting a given distortion and is therefore a candidate for encoding subbands. One may ask the same question for block coding as for scalar coding: is block coding of p-tuples in subbands more efficient than block coding of p-tuples in the full-band for any p? Intuitively, one would presume this to be the case, because memory support is smaller in the subbands and a size p block would embrace more of this support in subbands than in the full-band signal. One can follow the natural extension of the development used for scalar encoding. Consider the difference between the per sample p^{th} order entropy and the entropy rate:

$$^p\Delta_\infty^U = h_p(\mathbf{U}) - h_\infty(\mathbf{U}) \, . \qquad (1.84)$$

The generalization of (1.75), which is also proved in [12], is that

$$^p\Delta_\infty^U = \lim_{N \to \infty} \left[\frac{1}{p} I\Big(U_1, U_2, \dots, U_p; U_{p+1}, \dots, U_{p+N}\Big) \right] , \qquad (1.85)$$

which is the amount of information per letter that the rest of the sequence conveys about the first p samples. Therefore $^p\Delta_\infty^U$ is a measure of the memory about the first p samples contained in the samples beyond. One can also show that ([12] and [9])

$$^p\Delta_\infty^U \geq R_p(D) - R(D) \, , \qquad (1.86)$$

with equality for $D \leq D_c$. The quantity $^p\Delta_\infty^U$ is the difference in the minimum achievable rates between encoding p-tuples and infinitely long blocks. For Gaussian sources, the pth order per letter differential entropy (see (1.28)) is

$$h_p(\mathbf{U}) = \frac{1}{2} \log_e 2\pi e \mid \mathbf{\Phi}_p \mid^{1/p} , \qquad (1.87)$$

where $\mathbf{\Phi}_p$ is the pth order covariance matrix of the source. By substituting (1.87) and (1.76) into (1.84), the minimum rate difference for a Gaussian source is

$$^p\Delta_\infty^U = 1/2 \log_e \frac{\mid \mathbf{\Phi}_p \mid^{1/p}}{Q_U} . \qquad (1.88)$$

Turning now to subbands, the corresponding entropy difference for the mth subband of the Gaussian source is

$$^p\Delta_\infty^{(m)} = h_p(\mathbf{x}^{(m)}) - h_\infty(\mathbf{x}^{(m)} ,$$

$$= 1/2 \log_e \frac{\mid \mathbf{\Phi}_p^{(m)} \mid^{1/p}}{Q_U^{(m)}} \qquad m = 1, 2, \ldots, M , \qquad (1.89)$$

where $Q_U^{(m)}$ is given in (1.78c) and $\mathbf{\Phi}_p^{(m)}$ is the pth order covariance matrix of the mth decimated subband process $\mathbf{x}^{(m)}$. The rate differences combine as in (1.37) to yield the composite subband rate difference

$$^p\Delta_\infty^{SB} = \sum_{m=1}^M {}^p\Delta_\infty^{(m)}/V_m ,$$

$$= \sum_{m=1}^M \frac{1}{2V_m} \log_e \frac{\mid \mathbf{\Phi}_p^{(m)} \mid^{1/p}}{Q_U^{(m)}} . \qquad (1.90)$$

The comparative rate advantage for encoding p-tuples of subbands versus full-band, using (1.89) and (1.87) and some algebraic manipulations, is

$$^p\Delta_\infty^U - {}^p\Delta_\infty^{SB} = 1/2 \log_e \frac{\mid \mathbf{\Phi}_p \mid^{1/p}}{\pi_{m=1}^M (V_m \mid \mathbf{\Phi}_p^{(m)} \mid^{1/p})^{1/V_m}} . \qquad (1.91)$$

The above equation is a formula for the minimal rate advantage for encoding p-tuples of subbands versus the same encoding of the full-band source. It is indeed a rate advantage if the argument of the logarithm is greater than one. One can prove it for certain cases of $p = 1$ (as we have done) and p large and for other values of p only for certain spectral classes [9]. The general proof remains elusive until the generic relationship of the pth order eigenvalues of subbands to those of the full-band source is established ($\mid \mathbf{\Phi}_p \mid = \lambda_1 \lambda_2 \ldots \lambda_p$ and $\mid \mathbf{\Phi}_p^{(m)} \mid = \lambda_1^{(m)} \lambda_2^{(m)} \ldots \lambda_p^{(m)}$).

In view of (1.86), one can also interpret ${}^p\Delta^U_\infty - {}^p\Delta^{SB}_\infty \geq 0$ in (1.91) as a statement that p-tuple encoding of subbands brings one closer to the rate-distortion bound than p-tuple encoding of the full-band source. Such an interpretation may help explain the superior results achieved in practice with subband coding.

1.5   Conclusion

The superior results in coding of subbands compared to the full-band source have been attributed to a reduction of the difference between finite order entropy and entropy rate. This means that the decimated subband processes on the average have less "memory" or, for Gaussian processes, have flatter spectra, than the full-band source. The coding gain has been related to this entropy reduction. These formulas and others given are generally applicable to pyramids and other multi-resolution (wavelet) decompositions, because there has been no assumption of equal size subbands. It must be borne in mind, however, that there has been an assumption throughout of ideally flat filters with infinitely sharp cutoff ("brick wall" filters). The reasons for this assumption are the requirements for independent, Gaussian subband processes and non-intersecting spectral occupancy. Orthogonal QMF filters will satisfy the former requirement, but not the latter one. The combination of aliasing factors for these filters cancels out the overlapping signal components, but not the noise ones. So the formulas derived here can only be deemed approximate to a degree determined by the application, but they can convey guidelines as to what performance to expect and the methods of analysis can help to lay the foundation for future work.

Acknowledgements   This work was supported in part by the National Science Foundation under Grant Numbers MIP-8610029 and MIP-8519948. The author has benefited greatly from his collaborations with his former and current graduate students, Dr. Sanjiv Nanda, Balakrishnan Mahesh, and R. Padmanabha Rao and wishes to thank them. Thanks are also due to Charmaine Darmetko for typing the manusript and to Sudhakar Rao for drawing the figures. The author is likewise grateful for the fruitful interactions with his colleague, Dr. John W. Woods, who inspired his interest in subband coding and invited him to write this Chapter.

Bibliography

[1] D.A. Huffman, "A method for construction of minimum redundancy codes", *Proc. IRE*, vol. 40, pp. 1098-1101, Sept. 1952.

[2] R.E. Blahut, "Computation of channel capacity and rate-distortion function", *IEEE Trans. Inform. Theory*, vol. IT-18, pp. 460-473, July 1972.

[3] R.G. Gallager, *Information Theory and Reliable Communication*, New York: J. Wiley and Sons, Inc., 1968.

[4] W.A. Pearlman and P. Jakatdar, "A transform tree code for stationary Gaussian sources", *IEEE Trans. Inform. Theory*, vol. IT-31, pp. 761-768.

[5] B. Mazor and W.A. Pearlman, "A trellis code construction and coding theorem for stationary Gaussian sources", *IEEE Trans. Inform. Theory*, vol. IT-29, pp. 924-930, Nov. 1983.

[6] B. Mazor and W.A. Pearlman, "A tree coding theorem for stationary Gaussian sources and the squared-error distortion measure", *IEEE Trans. Inform. Theory*, vol. IT-32, pp. 156-165, March 1986.

[7] N.S. Jayant and P. Noll, *Digital Coding of Waveforms*, Englewood Cliffs, NJ: Prentice-Hall, 1984.

[8] R. Bellman, *Introduction to Matrix Analysis*, 2nd Ed., New York: McGraw-Hill, 1970.

[9] S. Nanda, *A Tree Coding Theorem and Tree Coding of Subbands with Application to Images*, Ph.D. dissertation, Rensselaer Polytechnic Institute, Troy, NY, October 1988.

[10] P.H. Westerink, "An optimal bit allocation for subband coding", *Proc. 1988 IEEE Int. Conf. Acoust., Speech, Signal Process.*, New York, April 1988, pp. 757-760.

[11] J.W. Woods and S.D. O'Neil, "Subband coding of images", *IEEE Trans. Acoustics, Speech, and Signal Processing*, vol. ASSP-34, pp. 1278- 1286, October 1986.

[12] A.D. Wyner and J. Ziv, "Bounds on the rate-distortion function for stationary sources with memory", *IEEE Trans. Inform. Theory*, vol. IT- 17, pp. 508-513, September 1971.

[13] R.P. Rao and W.A. Pearlman, "On entropy rate for source encoding on a pyramid structure", *Abstracts of 1990 IEEE International Symposium on Information Theory*, January 1990, San Diego, CA, pp. 57-58; also R.P. Rao and W.A. Pearlman, "On entropy rate of pyramid structures", submitted to *IEEE Trans. Inform. Theory*.

[14] P. Chou, T. Lookabaugh, and R.M. Gray, "Optimal pruning with applications to tree-structured source coding and modeling", *IEEE Trans. Inform. Theory*, vol. IT-35, pp. 299-315, March 1989.

[15] L. Breiman, J.H. Friedman, R.A. Olshen, and C.J. Stone, *Classification and Regression Trees*, Monterey, CA: Wadsworth, 1984.

[16] B. Mahesh and W.A. Pearlman, "Multiple rate code book design for vector quantization of image pyramids", *Applications of Digital Image Processing XIII*, A. Tescher, Ed., Proc. SPIE 1349, July 1990.

[17] J.W. Modestino, V. Bhaskaran, and J.B. Anderson, "Tree encoding of images in the presence of channel errors", *IEEE Trans. Inform. Theory*, vol. IT-27, pp. 677-697, Nov. 1981.

[18] J.B. Anderson and J.B. Brodie, "Tree encoding of speech", *IEEE Trans. Inform. Theory*, vol. IT-21, pp. 379-387, July 1975.

[19] H.G. Fehn and P. Noll, "Multipath search coding of stationary signals with applications to speech", *IEEE Trans. Commun.*, vol. COM-30, pp. 687-701, April 1982.

[20] P. Kroon and E.F. Deprettere, "A class of analysis-by-synthesis predictive coders for high quality speech coding at rates between 4.8 and 16 kbits/s", *IEEE J. Select. Area Commun.*, vol. 6, pp. 353-363, Feb. 1988.

Chapter 2

Multirate Filter Banks for Subband Coding

by: Martin Vetterli
Department of Electrical Engineering
and Center for Telecommunications Research
Columbia University, New York, NY 10027

This chapter develops the theory of multirate signal processing as used in subband coding systems. Multirate operations are reviewed, multirate filter banks are analyzed and structures and design methods for perfect reconstruction filter banks are indicated. Special attention is given to the multidimensional case. We do not consider how to actually code the subband signals, and it will be assumed that the subband signals are perfectly transmitted.

First, multirate signal processing is reviewed [8]. Sampling rate conversion is described, with upsampling and downsampling by integer factors as well as by rational factors. A multirate interpretation of some signal analysis schemes is also given. Then, we proceed with the analysis of multirate signal processing systems. These are linear periodically time-varying (LPTV) systems which can be represented in polyphase or modulation domains as will be shown. The relation between the two domains (through the Fourier transform) as well as the polyphase transform will be explained.

Multirate filter banks are then studied per se, first in the polyphase domain and then in the modulation domain. The classical two band QMF scheme will be taken as a case study of how aliasing appears in the subbands but is then cancelled at the reconstruction. Some key results on perfect and aliasing free reconstruction, especially in the case of FIR analysis and synthesis, will be given.

The problem of designing filters for filter banks is addressed next. In the polyphase domain, this is the problem of designing an N-port satisfying certain constraints. Unfortunately, the constraints are not given in the polyphase domain (but on the filters themselves and on the determinant of the N-port). One elegant solution is given by paraunitary or lossless systems, which can be seen as a generalization of unitary matrices or of all-pass functions. Some results which go back to classical network synthesis are given, for example on factorization of paraunitary matrices. However, some important cases have no paraunitary solutions (for example, the two band, linear phase filter bank case). Thus, we examine general systems as well, and give some useful structures for non-paraunitary filter banks.

Finally, we examine the two-dimensional (2-D) case in some detail. We consider the non-separable case since the separable one follows easily from one-dimensional (1-D) solutions. We generalize some relevant 1-D results to two dimensions. The polyphase decomposition, which now depends on the sampling lattice, is derived. Lossless systems in two dimensions are given, and their factorization is discussed. Finally, examples of both paraunitary and general perfect reconstruction systems are shown. Applications to both image and video processing are indicated. Also in this volume, Chapter 3, authored by M.J.T.Smith, covers similar material, but with an emphasis on recursive filters.

Before developing the theory, let us briefly recall the history of the subject. Multirate structures have been used in telecommunications [5] and signal processing [8] for quite some time. However, the discovery of an aliasing free scheme for subband coding of speech in 1976 [9] launched a whole new activity. This lead to the discovery of perfect reconstruction filter banks [47, 49, 64, 53] as well as to mathematical framework for the analysis of multirate filter banks [44, 48, 63, 50, 65, 55, 69]. The development of multidimensional filter banks [62] and its application to image coding [75, 1, 26, 74, 27, 16] soon followed, as well as investigations of perfect reconstruction schemes [56, 2, 21, 72, 23, 73]. Initial work on video is also available [19, 20, 28, 70]. The application of perfect reconstruction filter banks to non-stationary signal analysis seems also promising, especially in light of the recent developments in wavelets [10]. The connection with multiresolution signal analysis [6, 32], used for example in pattern analysis and computer vision, indicates also the important role played by multirate signal processing and filter banks outside of coding.

2.1 Multirate Digital Signal Processing

In multirate digital signal processing, several different sampling rates are used simultaneously, unlike in conventional digital signal processing where all sig-

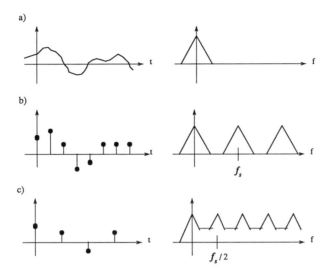

Figure 2.1: Subsampling seen as resampling at lower sampling frequency: (a) original signal, (b) signal sampled with frequency f_s, (c) signal subsampled by 2, or resampled at $f_s/2$.

nals are sampled at a single rate [8]. Such multirate systems appear for example when connecting systems with different sampling rates (e.g 50 Hz and 60 Hz television), in transform and subband coding (of speech, image or video information), in block processing schemes (like overlap-add FFT based convolution), and in efficient narrow band filtering (where the output, being very narrow band, can be subsampled). The two key operations in multirate digital processing are the subsampling and upsampling process.

2.1.1 Subsampling by an integer factor N

When subsampling by N one disregards $N-1$ out of N samples, that is, the input-output relation is:

$$y(n) = x(Nn) \qquad (2.1)$$

from where it is clear that subsampling is a linear periodically time-varying operation. In the frequency domain, subsampling can be viewed as a resampling at a lower sampling frequency. Since sampling corresponds to time-domain multiplication with a train of Dirac pulses (at multiples of f_s, the sampling frequency), subsampling by N corresponds to a convolution of the original spectrum with a train of Dirac pulses spaced f_s/N apart, as depicted in Figure 2.1 (subsampling is indicated as a sampling rate of $1/N$). In z-transform

a)

b)

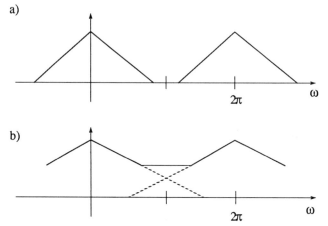

Figure 2.2: Decimation by 2 in normalized frequency: (a) original spectrum, (b) spectrum after subsampling by 2.

domain, one can verify that (2.1) becomes:

$$Y(z) = \frac{1}{N} \sum_{k=0}^{N-1} X(W^k z^{1/N}), \quad W = e^{-j\frac{2\pi}{N}} \tag{2.2}$$

which indicates that the ouput of a decimator by N is the superposition of the input as well as $N-1$ aliased versions of it. The power $1/N$ of z means that, in normalized frequency, the original spectrum gets stretched by a factor N, as shown in Figure 2.2. From the above, it is clear that a decimator should be preceded by an appropriate lowpass filter if one wants to retain an unaltered part of the original spectrum. This is shown schematically in Figure 2.3, where $LP - 1/N$ indicates a lowpass filter with a passband of width $\frac{2\pi}{N}$. Note also that subsampling a highpass signal brings it down to baseband where it might look just like a subsampled lowpass signal. From the above discussion, it is also clear that subsampling is in general a non-reversible operation.

2.1.2 Upsampling by an integer factor N

Conversely to subsampling, upsampling by N consists in placing $N-1$ zeros between every sample of the input signal that is:

$$y(n) = x(\frac{n}{N}), \quad n = Nn' \qquad 0 \text{ else} \tag{2.3a}$$

or

$$Y(z) = X(z^N) \quad = x_0 + x_1 z^N + x_2 z^{2N} + ... \tag{2.3b}$$

Obviously, this is a reversible operation, since no information is lost and the original signal can be simply recovered by subsampling by N. In spectral

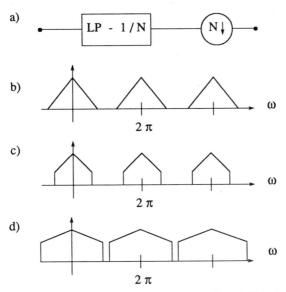

Figure 2.3: Lowpass filtering followed by subsampling: (a) block diagram, (b) original spectrum, (c) spectrum after filtering, (d) spectrum after subsampling.

domain, the N-th power of z will create $N - 1$ images as shown in Figure 2.4. These spectral images can be filtered out by an appropriate lowpass (or interpolation) filter (see Figure 2.5). Note that putting zeros between the input samples in the upsampling process is the "least information" filling procedure [8], since other upsampling schemes can be obtained from it. For example, repeating samples (called pixel replication in image processing) is obtained by a trivial mean lowpass filter of length N $(H(z) = 1 + z^{-1} + ... + z^{-N+1})$.

2.1.3 Sampling rate changes by a rational factor

By cascading an upsampling stage by L with a downsampling stage by M, one obtains a rational sampling rate change with rate L/M. The two lowpass filters now follow each other (see Figure 2.6) and can be merged, which means essentially that the narrower of the two is kept.

2.1.4 Some basic identities in multirate digital signal processing

In the rational sampling rate case, one would generally assume that L and M are relatively prime (a common factor could be cancelled). In this case, it is easy to verify that down and upsampling can be interchanged without

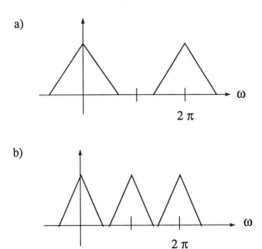

Figure 2.4: Upsampling process, shown in the case $N = 2$: (a) original spectrum, (b) spectrum after upsampling.

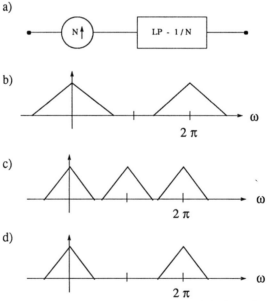

Figure 2.5: Upsampling followed by interpolation: (a) block diagram, (b) original spectrum, (c) spectrum after upsampling, containing spectral images, (d) spectrum after lowpass filtering, with spectral images removed.

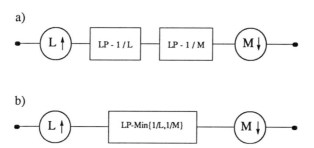

Figure 2.6: Rational sampling rate change: (a) upsampling by L and interpolation followed by lowpass filtering and subsampling by M, (b) same as above, with the 2 filters merged.

affecting the input-output relationship. To see this, assume the subsampling by M comes first. Then using (2.2) and (2.3b):

$$Y'(z) = \frac{1}{M} \sum_{k=0}^{M-1} X(W^{kL} z^{L/M}) \qquad W = e^{-j\frac{2\pi}{M}} \quad . \qquad (2.4a)$$

Conversely, using (2.3b) and then (2.2):

$$Y''(z) = \frac{1}{M} \sum_{k=0}^{M-1} X(W^k z^{L/M}) \quad . \qquad (2.4b)$$

Now, (2.4a) and (2.4b) are equivalent if and only if L and M are relatively prime, because then W^{kL} and W^k cover the same set of powers of the $M-th$ root of unity [61] ($kL \pmod{M}$ is a permutation of the roots of unity if and only if L and M are coprime [4]).

Another pair of useful identities gives conditions when time-invariant filters can be passed across sampling rate changes. For the subsampling case, a filter with a z-transform having only $N - th$ powers of z, $H(z^N)$, can be moved to the subsampled domain since (2.2) indicates that subsampling will leave its impulse response unchanged. Therefore, any filter in the subsampled domain can be represented in the upsampled domain by simply upsampling its impulse response. Very similarly, a filter with z-transform $H(z)$ placed in front of an upsampling by N can be moved past the sampling rate change and represented as $H(z^N)$. Figure 2.7 shows the above identities pictorially.

2.1.5 Multirate filter banks for coding and transmultiplexing

We have seen that in general, there is no way to recover an original signal from a subsampled version of it. The best we could do is to preserve intact

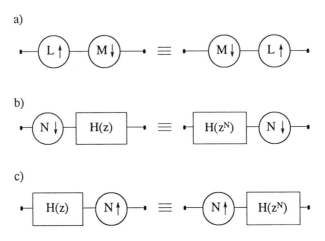

Figure 2.7: Some useful identities in multirate signal processing ($H(z)$ is rational): (a) upsampling by L and subsampling by M can be interchanged if and only if L and M are relatively prime, (b) a filter after a subsampling by N can be represented in its upsampled version in front of the subsampling, (c) a filter in front of an upsampling by N can be represented in its upsampled version after the upsampling.

the part of the spectrum corresponding to the new sampling rate, that is, avoid aliasing. Now, if we consider several subsampled versions of the input, it is intuitively clear that one should be able to recover the original signal if globally, the sampling rate is preserved. To make this statement precise will require some effort, but first, we introduce the two basic devices that will be used, namely the analysis and the synthesis filter bank.

A size M analysis filter bank with subsampling by N is depicted in Figure 2.8. The input is divided into M filtered versions (typically bandpass) which are then subsampled by N. When $M = N$, we call it a critically sampled filter bank because the total number of samples in the subbands is the same as in the input. Conversely, a size M synthesis filter bank with upsampling by N starts by upsampling the M input signals, interpolating them before summing them to create the output signal. This is shown in Figure 2.9.

Main applications of such multirate filter banks are found in subband coding schemes and transmultiplexers. In the former case, an analysis bank divides the input into frequency bands which are usually critically sampled. These bands are then quantized and coded individually. After decoding, the reconstructed signal is obtained with a synthesis filter bank. In the latter case, several individual signals are multiplexed onto a single channel using a synthesis filter bank, and the signals are recovered using an analysis filter bank.

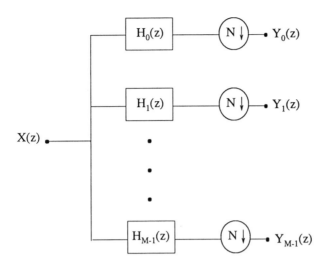

Figure 2.8: Size M analysis filter bank with subsampling by N.

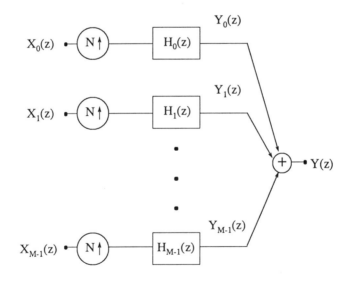

Figure 2.9: Size M synthesis filter bank with upsampling by N.

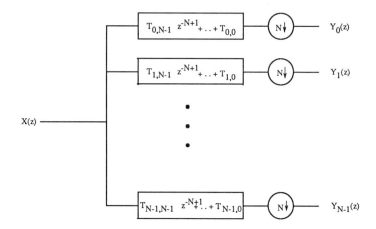

Figure 2.10: Interpretation of size-N block transform as a filter bank with N filters of length-N subsampled by N.

It is intuitively clear that these two systems are dual, and indeed, solutions for one case are applicable for the other [65].

2.1.6   Multirate interpretation of some classical schemes

It turns out that a number of classical schemes of digital signal processing can be cast into the multirate framework, thus giving a unified treatment of them.

(a) *Block transforms*: Such transforms, which are used frequently in coding (e.g. the discrete cosine transform) can be represented as a bank of N filters, each of them having length N and being subsampled by N. Each filter corresponds to a row of the transform matrix (in reverse order). Figure 2.10 gives such a pictorial representation of a block transform, showing that block transforms are a particular case of filter banks. Note that the property that the transform matrix is unitary, and therefore its inverse is simply its transpose, can be extended to general filter banks as well.

(b) *Block processing schemes*: Such schemes are often used for computational or architectural reasons. A classic example is overlap-add/save computation of running convolution [40]. Essentially, a block of input is processed at a time, typically with frequency domain circular convolutions, and the output is merged so as to achieve true linear running convolution. Since the processing advances by steps (which corresponds to subsampling the input by the step-size), block processing schemes are multirate in nature and thus side effects like aliasing can appear [67]. Actually, alias-free block processing schemes are

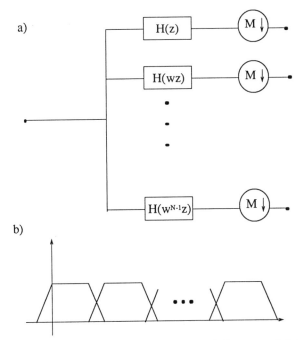

Figure 2.11: Short-time Fourier transform analysis as a modulated analysis filter bank: (a) filter bank with modulated prototype, (b) amplitude of the frequency response.

characterized by a pseudo-circulant transfer matrix [58], which is also a condition for alias-free subband coding schemes, thus indicating the close connection between the two.

(c) *Short-time Fourier transform (STFT)* [43, 36]: This popular tool for non-stationary signal analysis has an immediate filter bank interpretation. Assume a window function with corresponding z-transform $H(z)$. This window function is a prototype lowpass filter with a bandwidth of $2\pi/N$. This prototype lowpass filter is then modulated evenly over the frequency spectrum using consecutive powers of the $N - th$ root of unity:

$$H_i(z) = H(W^i z) \qquad . \qquad (2.5)$$

The resulting filter bank is usually called a modulated filter bank. In the STFT, the window is advanced by M samples at time, which corresponds to a subsampling by M of the corresponding filter bank. This analysis filter bank interpretation of the STFT analysis is depicted in Figure 2.11.

The STFT synthesis is achieved similarly with a modulated synthesis filter bank. Note that because all filters are obtained from a single prototype by modulation with a root of unity, an efficient computational algorithm consists in windowing the signal first and then applying a fast Fourier transform [5, 8].

(d) *Signal analysis with Gabor functions* [14]: if the windowing function in the STFT is chosen as a Gaussian, then the output of the STFT or the corresponding filter bank represents a decomposition of the signal into Gabor functions. The subsampling factor M correspond to the shift between two consecutive locations of the basis functions.

(d) *Wavelet transforms* [17, 10, 32]: An improved frequency resolution in low frequencies as well as an improved time resolution in high frequencies is achieved by taking a single prototype wavelet (which is typically a windowed sinusoid) and stretching or compressing it as well as translating it:

$$h_{a,b}(t) = H(\frac{t-b}{a}) \qquad . \tag{2.6}$$

Thus, b indicates the shift in time and a is the dilatation factor of the wavelet. To get a good time-frequency resolution trade-off, the short, high frequency, wavelet outputs are sampled more often than the output corresponding to long, low frequency one. Typically, a and b are powers of 2, which means that one obtains a logarithmic resolution in both time and frequency, thus achieving a better trade-off than what the STFT or the Gabor functions can achieve. Now, discrete wavelet transforms can be implemented with multirate banks as depicted in Figure 2.12.

The high and lowpass filters are half-band filters, and thus, the iteration on the lower half of the spectrum leads to a logarithmic frequency resolution. The length of the filters increases as one goes to lower branches, since the filter producing the i-th output has a transfer function equal to (using (2.3b)):

$$H_i(z) = G(z^{2^i}) \cdot H(z^{2^{i-1}}) \cdot H(z^{2^{i-2}}) \cdot \ldots \cdot H(z^2) \cdot H(z), \qquad i \geq 1 \tag{2.7a}$$

$$H_0(z) = G(z), \qquad i = 1 \tag{2.7b}$$

where $H(z)$ and $G(z)$ are the z-transforms of the low and highpass filters respectively. From a complexity point of view, note that such a structure has at most twice the complexity of the first stage, since subsequent stage are subsampled by 2. This leads to a complexity of the form $C = (1+1/2+1/4+...)$ where C stands for the complexity of the first stage. Note that if the scheme is iterated to infinity, it is of interest to see what the resulting lowpass filter becomes:

$$H_\infty(z) = \prod_{i=0}^{\infty} H(z^{2^i}) \qquad . \tag{2.7c}$$

If the filter $H(z)$ satisfies a certain regularity condition [10], $H_\infty(z)$ corresponds to a continuous function, otherwise it becomes a fractal function.

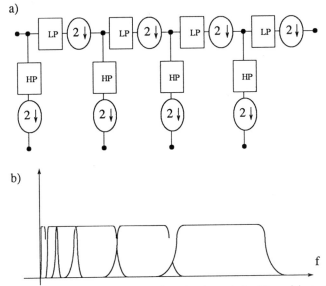

Figure 2.12: Discrete wavelet transform implemented with multirate filter banks: (a) two channel filter bank iterated infinitely on the lowpass output, (b) typical frequency response corresponding to an octave band (or logarithmic) decomposition.

(f) *Pyramid and multiresolution schemes* [6, 32]: These schemes perform an iterative approximation of a signal and derive a hierarchy of better and better approximations. They are thus closely related to both wavelets and filters banks. The main difference is that there is an increase in the number of samples in these schemes (100% in one dimension, and 33% in two dimensions). However, the filter design problem is almost trivial, since nearly any lowpass filter can be used to derive a low resolution approximation.

2.2 Analysis of Multirate Digital Signal Processing Systems

2.2.1 Linear periodically time-varying systems

The presence of subsampling devices in multirate systems makes them periodically time-variant, since the transmission of the subsampler is 1 at multiples of N and 0 otherwise. Thus, a more general analysis techniques is required than for time-invariant systems. Since the subsampler is a linear device, and that we assume all filters to be linear and time-invariant, the overall multirate system is linear periodically time-variant (LPTV), with a period equal to

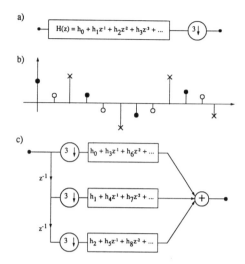

Figure 2.13: Polyphase decomposition of a filter followed by decimation: (a) original filter followed by decimation by 3, (b) impulse response of the filter divided into 3 polyphase components, (c) alternate representation of the filter with decimation by 3. The input is divided into components which trigger the various polyphase components of the filter.

the least common multiple of the various sub-periods present in the system. Various analysis techniques are possible for LPTV systems, and they all rely on the fact that such systems can be represented by N linear time-invariant subsystems (N being the period).

2.2.2 Polyphase representations

Consider a lowpass filter followed by a decimation by N (see Figure 2.3). For the sake of example, assume $N = 3$ and a filter impulse response given by $(h_0, h_1, h_2, ...)$. Then, impulses at time $0, -1$ and -2 will produce the following outputs:

$$
\begin{aligned}
\delta(n) &: \quad h_0, h_3, h_6, h_9, \ldots \\
\delta(n+1) &: \quad h_1, h_4, h_7, h_{10}, \ldots \\
\delta(n+2) &: \quad h_2, h_5, h_8, h_{11}, \ldots
\end{aligned}
\tag{2.8}
$$

An impulse at time -3 will produce the same impulse response as the one at time 0, thus this system is characterized completely by 3 impulse responses. This is shown in Figure 2.13, where the filter impulse response is subdivided into 3 sub-responses, so-called polyphase components, each of which is triggered by impulses which differ modulo 3.

The input signal is thus divided into 3 components:

$$\{x_{-5}, x_{-4}, x_{-3}, x_{-2}, x_{-1}, x_0, x_1, x_2, x_3 \dots\}$$

$$= \{x_{-3}, x_0, x_3, \dots\} \cup \{x_{-4}, x_{-1}, x_2, \dots\} \cup \{x_{-5}, x_{-2}, x_1, \dots\} \qquad (2.9)$$

Note that the polyphase components of the input signal are defined in reverse order of the ones of the filter, since the phases have to cancel so as to be synchronous with the subsampler. From the above example (see (2.8)) and generalizing to a general decimation by N, we see that the z-transform of a filter $H_0(z)$ can be decomposed into polyphase components:

$$H(z) = \sum_{j=0}^{N=1} z^{-j} H_j(z^N) \qquad (2.10a)$$

$$H_j(z) = \sum_{n=-\infty}^{\infty} h_{Nn+j} z^{-n} \qquad (2.10b)$$

where $\{h_{j+nN}\}$ is the j-th polyphase component of the i-th filter. Note that the j-th polyphase component corresponds to the impulse response advanced by j and subsampled by N. Therefore, the polyphase decomposition is a function of the subsampling factor. Similarly, the polyphase decomposition of the input signal (see (2.9)) is given by (note the ordering):

$$X(z) = \sum_{j=o}^{N-1} z^j X_j(z^N) \qquad (2.11a)$$

$$X_j(z) = \sum_{n=-\infty}^{\infty} x_{nN-j} z^{-n} \qquad (2.11b)$$

This is the polyphase transform of the input signal, shown in Figure 2.14 together with its (non-causal) inverse. To get a causal inverse, an additional delay of $N - 1$ samples is required. To verify the inversion property, it is sufficient to put forward and inverse polyphase transforms back to back and verify that N impulses at time $0, 1, ..., -N+1$ are transmitted, each with zero delay, as seen in Figure 2.14(c). Note that the (non-causal) inverse polyphase transform is really the dual of the forward polyphase transform: subsampling becomes upsampling, branching becomes summation and delays (z^{-1}) become forward shifts (z). Note that polyphase transforms can be represented with commutators [8]. For example, the forward polyphase transform in Figure 2.14(a) corresponds to a counter-clockwise commutator of size N. With the polyphase definitions of (2.10b) and (2.11b), the output of a subsampled filter with z-transform $H(z)$ equals:

$$Y(z) = \sum_{j=0}^{N-1} H_j(z) \cdot X_j(z) \qquad (2.12a)$$

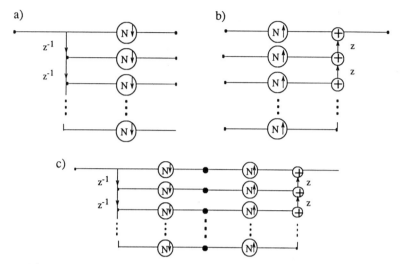

Figure 2.14: Polyphase transforms: (a) forward polyphase transform, (b) inverse polyphase transform (non-causal version), (c) back to back configuration showing the inversion property. Note the zero delay for input to output, which is achieved because of the non-causal inverse. A causal inverse would produce a delay of $N - 1$ samples.

or in matrix notation:

$$Y(z) = (H_0(z) \quad H_1(z) \quad \ldots \quad H_{N-1}(z)) \cdot \begin{pmatrix} X_0(z) \\ X_1(z) \\ \vdots \\ X_{N-1}(z) \end{pmatrix} \qquad (2.12b)$$

The polyphase decomposition can also be applied to the upsampling followed by interpolation (see Figure 2.15(a)). By deriving the dual circuit to the one in Figure 2.13(c), one obtains Figure 2.15(b). Note that the polyphase components are defined in reverse order, that is:

$$G(z) = \sum_{j=0}^{N-1} z^{-(N-1-j)} G_j(z^N) \qquad (2.13a)$$

$$G_j(z) = \sum_{n=-\infty}^{\infty} g_{nN+N-1-j} z^{-n} \qquad (2.13b)$$

Now, the interpolator in Figure 2.15(b) is non-causal because of the forward shifts, and a causal version, obtained by applying a delay of $N - 1$ samples at the output, is shown in Figure 2.15(c). It is easily verified that the circuit of

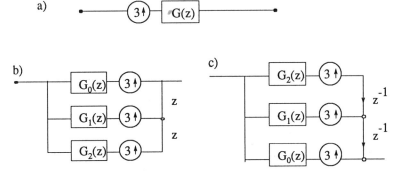

Figure 2.15: Polyphase representation of interpolator: (a) upsampling followed by interpolation, (b) polyphase components followed by upsampling and summation. (This is the non-causal dual version of Figure 2.13(c).), (c) causal version of previous circuit.

Figure 2.15(c) is equivalent to that of Figure 2.15(a) when using the polyphase filters as defined in (2.13b), that is again with $N = 3$ for the sake of example:

$$
\begin{aligned}
G_0(z) &= g_2 + g_5 z^{-1} + g_8 z^{-2} + \cdots \\
G_1(z) &= g_1 + g_4 z^{-1} + g_7 z^{-2} + \cdots \\
G_2(z) &= g_0 + g_3 z^{-1} + g_6 z^{-2} + \cdots
\end{aligned} \tag{2.14}
$$

If the z-transform of the input to the interpolator is given by $X(z)$, then the output equals:

$$
Y(z) = \sum_{j=0}^{N-1} z^{-(N-1-j)} G_j(z^N) \cdot X(z^N) \tag{2.15a}
$$

where the N-th power of z is due to the upsampling. In matrix notation, (2.15a) becomes:

$$
Y(z) = \begin{pmatrix} z^{-N+1} & z^{-N+2} & \cdots & z^{-1} & 1 \end{pmatrix} \cdot \begin{pmatrix} G_0(z^N) \\ G_1(z^N) \\ \vdots \\ G_{N-1}(z^N) \end{pmatrix} \cdot X(z^N) \,. \tag{2.15b}
$$

The various polyphase representations discussed above will prove especially useful when dealing with multirate filter banks later in this chapter.

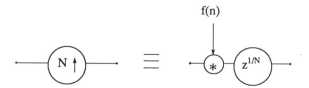

Figure 2.16: Subsampling by N seen as a modulation by $f(n)$ (see equation 2.16) followed by a time contraction by N.

2.2.3   Modulation representation

Let us go back to (2.2) and give an interpretation of it. Essentially, subsampling by N can be seen as a modulation by a function $f(n)$ given by:

$$f(n) = \frac{1}{N} \sum_{k=0}^{N-1} W^{nk} \tag{2.16}$$

(which is a train of dirac pulses, that is, 1 at multiples of N and zero everywhere else) followed by a time "contraction" by N ($z^{1/N}$). This is shown in Figure 2.16. From (2.16) and the modulation theorem of the z-transform [40],equation (2.2) follows readily. Obviously, the output of a subsampling device will contain components corresponding to the input signal as well as $N-1$ modulated versions of it. Therefore, it is natural to expand the input signal and the filter into a vector containing these N components. For example, the output of a filter subsampled by N is equal to:

$$Y(z) = \frac{1}{N} \sum_{k=o}^{N-1} H(W^k z^{1/N}) \cdot X(W^k z^{1/N}) \tag{2.17a}$$

or, in vector notation:

$$Y(z) = \frac{1}{N} \left(H(z^{\frac{1}{N}}) \quad H(W z^{\frac{1}{N}}) \quad \cdots \quad H(W^{N-1} z^{\frac{1}{N}}) \right) \cdot \begin{pmatrix} X(z^{\frac{1}{N}}) \\ X(W z^{\frac{1}{N}}) \\ \vdots \\ X(W^{N-1} z^{\frac{1}{N}}) \end{pmatrix}$$
$$\tag{2.17b}$$

where the above vectors are called modulation expansions (of order N) of $H(z)$ and $X(z)$ respectively (the power $1/N$ attached to z is only a scaling of the z-plane). Thus, a LPTV system can be represented by N linear filters, each driven by one modulated component of the input signal.

2.2.4 Relation between polyphase and modulation representation through the Fourier transform

In some sense, one can see the polyphase decomposition as a "time domain" view of LPTV systems, while the modulation decomposition indicates a "frequency" domain view. Thus, it comes as no surprise that the 2 representations are related by the Fourier transform. If we denote the polyphase decomposition of the signal as:

$$\mathbf{x}_p(z) = (X_0(z) \quad X_1(z) \quad \ldots \quad X_{N-1}(z))^T \tag{2.18a}$$

where $X_i(z)$ is defined as in (2.11), and the modulation representation as:

$$\mathbf{x}_m(z) = (X(z) \quad X(Wz) \quad \ldots \quad X(W^{N-1}z))^T \tag{2.18b}$$

then one can verify that:

$$\mathbf{x}_m(z) = \mathbf{F} \cdot \mathbf{D}(z) \cdot \mathbf{x}_p(z^N) \tag{2.19a}$$

where:

$$\mathbf{F}_{ij} = W^{ij} \tag{2.19b}$$

is the usual Fourier transform matrix, and

$$\mathbf{D}(z) = diag[1 \quad z \quad z^2 \ldots z^{N-1}] \tag{2.19c}$$

introduces the appropriate delays (see (2.11a)). Because the polyphase decomposition of the filter (see (2.10)) is in inverse order, we get:

$$\mathbf{h}_p(z^N) = \frac{1}{N}\mathbf{D}(z) \cdot \mathbf{F} \cdot \mathbf{h}_m(z) \tag{2.20}$$

where $\mathbf{h}_p(z)$ and $\mathbf{h}_m(z)$ are the polyphase and modulation vectors of the filter $H(z)$ following (2.10) and (2.17). There is a similar relation for the interpolation filter $G(z)$ (see (2.13)), but with a reversal of the order which corresponds to a multiplication by an anti-diagonal matrix. To illustrate the above relations, let us briefly verify that the modulation (2.17b) and polyphase representation of the decimated filter are indeed equivalent. Starting with (2.17) and using (2.19, 2.20) we have:

$$
\begin{aligned}
Y(z) &= \frac{1}{N}\mathbf{h}_m^T(z^{1/N}) \cdot \mathbf{x}_m(z^{1/N}) \\
&= \frac{1}{N}[N \cdot \mathbf{h}_p^T(z) \cdot \mathbf{D}(z^{-1})\mathbf{F}^{-1} \cdot \mathbf{F} \cdot \mathbf{D}(z) \cdot \mathbf{x}_p(z)] \\
&= \mathbf{h}_p^T(z) \cdot \mathbf{x}_p(z)
\end{aligned}
\tag{2.21}
$$

which is equal to (2.12).

In conclusion to this section, let us recall that a single input, single output LPTV system with period N can be mapped into an N input, single output linear time-invariant system, and this by decomposing the input signal into either polyphase or modulation components. This is schematically indicated in Figure 2.17.

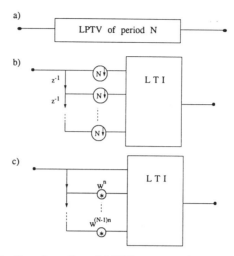

Figure 2.17: Transformation of LPTV systems of period N into N input LTI system: (a) original LPTV system, (b) expansion of input into polyphase components that feed into an N input, single output LTI system, (c) expansion of input into N modulation components that feed into a N input, single output LTI system.

2.4 Multirate Filter Banks

As mentioned in the introduction, one of the main application of multirate digital signal processing is found in subband coding and transmultiplexing schemes. Both use multirate filter banks, which will be analyzed next.

2.4.1 Polyphase domain analysis

Consider a size M analysis filter banks with outputs subsampled by N (see Figure 2.8). Calling $H_{ij}(z)$ the z-transform of the j-th polyphase component of the i-th filter, then following (2.12b), the i-th channel signal equals:

$$Y_i(z) = (\, H_{i,0}(z) H_{i,1}(z) \ldots H_{i,N-1}(z) \,) \cdot \begin{pmatrix} X_0(z) \\ X_1(z) \\ \vdots \\ X_{N-1}(z) \end{pmatrix} \qquad (2.22a)$$

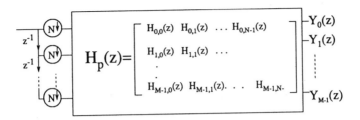

Figure 2.18: Size M analysis filter bank with outputs subsampled by N represented by a size M by N polyphase matrix $H_p(z)$.

and the vector of the M channel signals equals:

$$
\begin{pmatrix} Y_0(z) \\ Y_1(z) \\ \vdots \\ Y_{M-1}(z) \end{pmatrix} = \begin{pmatrix} H_{0,0}(z) & H_{0,1}(z) & \cdots & H_{0,N-1}(z) \\ H_{1,0}(z) & H_{1,1}(z) & \cdots & H_{1,N-1}(z) \\ \vdots & \vdots & \cdot & \vdots \\ H_{M-1,0}(z) & H_{M-1,1}(z) & \cdots & H_{M-1,N-1}(z) \end{pmatrix} \cdot \begin{pmatrix} X_0(z) \\ X_1(z) \\ \vdots \\ X_{N-1}(z) \end{pmatrix}
$$

(2.22b)

or, in matrix notation:

$$ \mathbf{y}_p(z) = \mathbf{H}_p(z) \cdot \mathbf{x}_p(z) \tag{2.22c} $$

where $\mathbf{H}_p(z)$ is an M by N matrix called the polyphase matrix of the analysis filter bank. This representation of the filter bank by a polyphase matrix is shown in Figure 2.18.

Therefore, the filter bank is represented by a forward polyphase transform followed by an N-input, M-output LTI system. Similarly, a size M synthesis filter bank upsampled by N can be represented by a polyphase matrix. Assume the i-th input is upsampled by N and interpolated by a filter with z-transform $G_i(z)$. Calling $G_{ij}(z)$ the j-th polyphase component of that interpolation filter, and following (2.15b), it is clear that the i-th input contribute the following to the output signal $Y(z)$:

$$
\left(z^{-N+1} z^{-N+2} \ldots z^{-1} 1 \right) \cdot \begin{pmatrix} G_{i,0}(z^N) \\ G_{i,1}(z^N) \\ \vdots \\ G_{i,N-1}(z^N) \end{pmatrix} \cdot X_i(z^N) \quad . \tag{2.23a}
$$

The total output being the superposition of the contributions from $X_0(z)$ ... $X_{M-1}(z)$, it can be written as:

$$ Y(z) = \left(z^{-N+1} z^{-N+2} \ldots z^{-1} 1 \right) \cdot $$

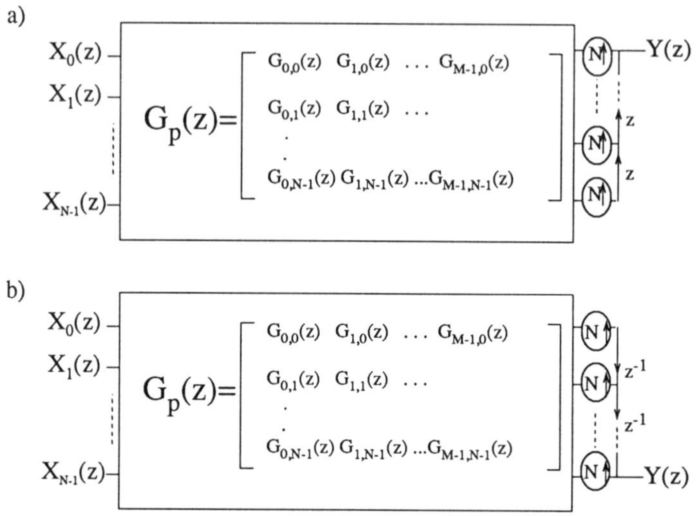

a)

b)

Figure 2.19: Size M synthesis filter bank with inputs upsampled by N represented by a size N by M polyphase matrix $G_p(z)$: (a) non-causal version, (b) causal version.

$$
\begin{pmatrix}
G_{0,0}(z^N) & G_{1,0}(z^N) & \cdots & G_{M-1,0}(z^N) \\
G_{0,1}(z^N) & G_{1,1}(z^N) & \cdots & G_{M-1,1}(z^N) \\
\vdots & \vdots & \cdot & \vdots \\
G_{0,N-1}(z^N) & G_{1,N-1}(z^N) & \cdots & G_{M-1,N-1}(z^N)
\end{pmatrix}
\cdot
\begin{pmatrix}
X_0(z^N) \\
X_1(z^N) \\
\vdots \\
X_{M-1}(z^N)
\end{pmatrix}
\tag{2.23b}
$$

or, in matrix notation:

$$
Y(z) = \mathbf{d}(z) \cdot \mathbf{G}_p(z^N) \cdot \mathbf{x}(z^N)
\tag{2.23c}
$$

where $\mathbf{G}_p(z)$ is a N by M matrix called the polyphase matrix of the synthesis filter bank. Recall that it differs from the one of the analysis filter bank both in its size (which is transposed) and in the definition of the polyphase components (which are reversed). The synthesis filter bank can thus be represented by a M input N output LTI system, followed by an inverse polyphase transform, as shown in Figure 2.19.

Now, a subband coding system consists of an analysis bank, a coding-transmission-decoding part, and a synthesis filter bank. Assuming for now lossless coding and transmission, the system is the cascade of an analysis and a synthesis filter bank, as shown in Figure 2.20(a).

Equivalently, in polyphase representation, it is a forward polyphase transform, followed by 2 polyphase matrices in cascade as well as an inverse polyphase

a)

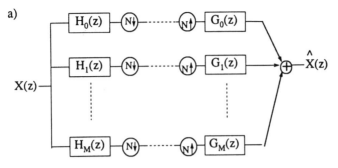

b)

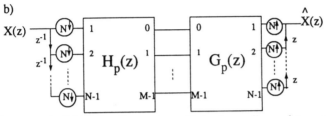

Figure 2.20: Subband coding system with lossless coding and transmission: (a) initial system, (b) polyphase representation.

transform, as depicted in Figure 2.20(b). Thus, the transmission matrix of size N by N $\mathbf{T}(z)$ defined as:

$$\mathbf{T}(z) = \mathbf{G}_p(z) \cdot \mathbf{H}_p(z) \tag{2.24}$$

characterizes the subband coding system. Note that we dropped the subscript p which is implicit in this definition of $\mathbf{T}(z)$. Obviously, if $\mathbf{T}(z) = \mathbf{I}$, then the forward polyphase transform is followed by its inverse (as in Figure 2.14(c)) and perfect reconstruction is achieved. Let us explore what other matrices $\mathbf{T}(z)$ lead to perfect reconstruction. The input-output relation is obtained by cascading (2.22b) with a non-causal version of (2.23b) (which is just (2.23b) multiplied by z^{N-1}):

$$Y(z) = (\, 1 \ z \ z^2 \ \ldots \ z^{N-1}\,) \cdot \mathbf{T}(z^N) \cdot \begin{pmatrix} X_0(z^N) \\ X_1(z^N) \\ \vdots \\ X_{N-1}(z^N) \end{pmatrix}. \tag{2.25}$$

Obviously, $\mathbf{T}(z) = z^{-i}\mathbf{I}$ will lead to perfect reconstruction with an input-output delay of z^{-iN}. But what about delays which are not multiples of N? Consider an input-output delay of z^{-1}. It is easy to see that the matrix $\mathbf{T}_1(z)$

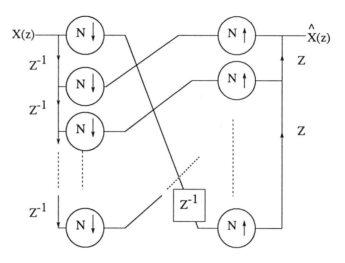

Figure 2.21: Delay by 1 sample implemented by a pseudo-circulant shift polyphase transfer matrix

equal to:

$$
\mathbf{T}_1(z) = \begin{pmatrix}
0 & 1 & 0 & 0 & \ldots & 0 & 0 \\
0 & 0 & 1 & 0 & \ldots & 0 & 0 \\
0 & 0 & 0 & 1 & \ldots & 0 & 0 \\
\vdots & \vdots & \vdots & \vdots & \ddots & \vdots & \vdots \\
0 & 0 & 0 & 0 & \ldots & 0 & 1 \\
z^{-1} & 0 & 0 & 0 & \ldots & 0 & 0
\end{pmatrix}
\tag{2.26}
$$

will produce a delay of 1, since an impulse at time -1 comes out at time 0, and so on, up to an impulse at time $-N+1$ which comes out at time $-N+2$. Finally, an impulse at time 0 is delayed by N samples (due to the z^{-1} in the last line of $\mathbf{T}_1(z)$) and comes out at time 1. Such a transfer matrix $\mathbf{T}_1(z)$ which produces a delay of 1 sample is shown in Figure 2.21. Note that it is a circulant matrix with an extra-twist: the wrapped-around term is multiplied by z^{-1}. This is an example of a pseudo-circulant matrix [58].

Now it is clear how to obtain an arbitrary delay by k samples: write k as

$$
k = \lfloor k/N \rfloor + k \quad (\text{mod } N) = k_1 + k_2
\tag{2.27}
$$

where $k_1 = \lfloor k/N \rfloor$ and $k_2 = k \pmod{N}$, then $\mathbf{T}_k(z)$ equals:

$$
\mathbf{T}_k(z) = z^{-k_1}[\mathbf{T}_1(z)]^{k_2}
\tag{2.28a}
$$

One can verify that ($k = 0..N - 1$):

$$
[\mathbf{T}_1(z)]^k = \begin{pmatrix}
0 & \mathbf{I}_{N-k} \\
z^{-1}\mathbf{I}_k & 0
\end{pmatrix}
\tag{2.28b}
$$

where $\mathbf{I}_l$ is the l by l identity matrix. The above construction indicates how to implement an arbitrary time-invariant filter as a transfer function matrix in polyphase domain. First decompose $F(z)$ into its polyphase components $F_i(z)$. These polyphase components correspond to filters with an initial delay of i followed by non-zero taps at integral multiples of N. Thus, the i-th polyphase component can be implemented with a transfer matrix equal to:

$$[\mathbf{T}_1(z)]^i \cdot F_i(z) = \begin{pmatrix} 0 & F_i(z) \cdot \mathbf{I}_{N-i} \\ z^{-1} \cdot F_i(z) \cdot \mathbf{I}_i & 0 \end{pmatrix} . \qquad (2.29)$$

The transfer matrix corresponding to the total filter is just the superposition of the polyphase components, and is thus of the form:

$$\mathbf{T}_F(z) = \begin{pmatrix} F_0(z) & F_1(z) & \cdots & F_{N-1}(z) \\ z^{-1}F_{N-1}(z) & F_0(z) & \cdots & F_{N-2}(z) \\ \cdot & \cdot & \cdot & \cdot \\ z^{-1}F_1(z) & z^{-1}F_2(z) & \cdots & F_0(z) \end{pmatrix} \qquad (2.30)$$

which is a general pseudo-circulant matrix [58]. We are now ready to prove an important result for subband coding systems, namely, in which cases they will be free of aliasing. Aliasing appears if the system is not time-invariant but instead periodically time-variant. Conditions for aliasing cancellation are given by the following theorem.

Theorem 4.1 [58]: Aliasing in a one-dimensional subband coding system will be cancelled if and only if the transfer function matrix is pseudo-circulant.

Proof: aliasing will be cancelled if and only if the subband coding system is equivalent to a time invariant filter $F(z)$. If the system is a time-invariant filter, then the transfer function matrix is pseudo-circulant as shown in (2.29-2.30). It is thus sufficient to show that any pseudo-circulant matrix will produce a time-invariant filter. To verify this, evaluate (2.25) with a matrix as in (2.30):

$$Y(z) = (F(z) \quad zF(z)z^2F(z) \quad \cdots \quad z^{N-1}F(z)) \cdot \begin{pmatrix} X_0(z^N) \\ X_1(z^N) \\ \vdots \\ X_{N-1}(z^N) \end{pmatrix} \qquad (2.31)$$

where:

$$F(z) = F_0(z^N) + z^{-1}F_1(z) + \ldots + z^{-N+1}F_{N-1}$$

and thus, (2.31) is indeed a time-invariant system.

Corollary 4.1: perfect reconstruction in a subband coding system is achieved if and only if the transfer function matrix is a pseudo-circulant delay (i.e. $F(z) = z^{-l}$).

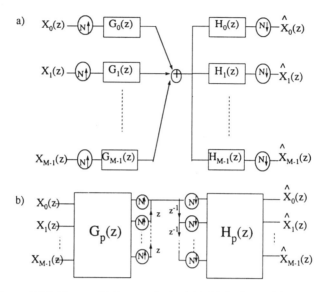

Figure 2.22: Transmultiplexer analysis: (a) initial scheme, (b) polyphase representation.

The proof follows either from the above theorem (a single polyphase component is different from zero and equal to a delay) or from (2.27-2.28).

The other main application of multirate filter banks is transmultiplexing, where M signals are multiplexed onto a channel with N times higher sampling rate. This is achieved with a synthesis filter bank of size M, upsampled by N. To recover the signals, a converse analysis filter bank is used, and the complete system is shown in Figure 2.22(a).

Note that the transmultiplexer is a time-invariant device, because subsampling is preceded by upsampling. Obviously, in polyphase representation, the inverse polyphase transform followed by a forward polyphase transform cancel each other, and an equivalent representation is given in Figure 2.22(b). The transfer matrix is now:

$$\mathbf{T}(z) = \mathbf{H}_p(z) \cdot \mathbf{G}_p(z) \tag{2.32}$$

and of size M by M. The equivalent of aliasing cancellation is absence of crosstalk, and this is achieved if $\mathbf{T}(z)$ is diagonal (with arbitrary diagonal elements). Perfect reconstruction is obtained if $\mathbf{T}(z)$ is a diagonal matrix of delays. Note that a permutation of the above matrices will simply cause a permutation of the signals from input to output.

2.4.2 Modulation Domain Analysis

Instead of expanding filters and signals into their respective polyphase representations, one can use the modulation representation instead (see (2.17)). An analysis filter bank of size M produces M outputs given by (following (2.17b)):

$$
\begin{pmatrix} Y_0(z) \\ Y_1(z) \\ \vdots \\ Y_{M-1}(z) \end{pmatrix} = \frac{1}{N} \cdot
$$

$$
\begin{pmatrix} H_0(z^{1/N}) & H_0(Wz^{1/N}) & \cdots & H_0(W^{N-1}z^{1/N}) \\ H_1(z^{1/N}) & H_1(Wz^{1/N}) & \cdots & H_1(W^{N-1}z^{1/N}) \\ \vdots & \vdots & \vdots & \vdots \\ H_{M-1}(z^{1/N}) & H_{M-1}(Wz^{1/N}) & \cdots & H_{M-1}(W^{N-1}z^{1/N}) \end{pmatrix} \begin{pmatrix} X(z^{1/N}) \\ X(Wz^{1/N}) \\ \vdots \\ X(W^{n-1}z^{1/N}) \end{pmatrix}
$$

$$
\tag{2.33a}
$$

$$
= \frac{1}{N} \mathbf{H}_m(z^{1/N}) \cdot \mathbf{x}_m(z^{1/N}) \quad . \tag{2.33b}
$$

The synthesis filter bank leads then to the following output:

$$
\hat{X}(z) = \frac{1}{N} \begin{pmatrix} G_0(z) & G_1(z) & \cdots & G_{n-1}(z) \end{pmatrix} \cdot \mathbf{H}_m(z) \cdot \mathbf{x}_m(z) \quad . \tag{2.34}
$$

Thus, a necessary and sufficient condition for aliasing cancellation is that the vector of synthesis filters times $\mathbf{H}_m(z)$ is equal to a vector with only the first component different from zero. If we now expand the synthesis filter vector into a matrix by modulation:

$$
\mathbf{G}_m(z) = \begin{pmatrix} G_0(z) & G_1(z) & \cdots & G_{M-1}(z) \\ G_0(Wz) & G_1(Wz) & \cdots & G_{M-1}(Wz) \\ G_0(W^{N-1}z) & G_1(W^{N-1}z) & \cdots & G_{M-1}(W^{N-1}z) \end{pmatrix} \tag{2.35}
$$

then it is easy to see that the aliasing cancellation condition becomes:

$$
\mathbf{T}_m(z) = \mathbf{G}_m(z) \cdot \mathbf{H}_m(z) = diag[F(z)F(Wz)..F(W^{N-1}z)] \quad . \tag{2.36}
$$

The matrix $\mathbf{T}_m(z)$ is also sometimes called the aliasing cancellation matrix [50]. Now, by using (2.20) for transforming the decimation filters into their polyphase components, as well as its equivalent for the interpolation filter (which equal to (2.20) left-multiplied by $\mathbf{J}$), it follows that:

$$
\mathbf{G}_p(z^N) \cdot \mathbf{H}_p(z^N) = \frac{1}{N^2} \mathbf{J} \cdot \mathbf{D}(z) \cdot \mathbf{F} \cdot diag[F(z)F(Wz)..F(W^{N-1}z)] \cdot \mathbf{F} \cdot \mathbf{D}(z).
$$

$$
\tag{2.37}
$$

One can verify that the right side of (2.37) is a pseudo-circulant matrix, and thus (2.36) is equivalent to the result of Theorem 4.1.

For the sake of illustration, consider a two filter analysis-synthesis system. The two channel signals equal:

$$\begin{pmatrix} Y_0(z) \\ Y_1(z) \end{pmatrix} = 1/2 \begin{pmatrix} H_0(z^{1/2}) & H_0(-z^{1/2}) \\ H_1(z^{1/2}) & H_1(-z^{1/2}) \end{pmatrix} \cdot \begin{pmatrix} X(z^{1/2}) \\ X(-z^{1/2}) \end{pmatrix} \qquad (2.38a)$$

and the reconstructed output is:

$$\hat{X}(z) = \begin{pmatrix} G_0(z) & G_1(z) \end{pmatrix} \cdot \begin{pmatrix} Y_0(z^2) \\ Y_1(z^2) \end{pmatrix} . \qquad (2.38b)$$

The classic quadrature mirror filter (QMF) solution [9] poses $H_0(z) = H_p(z)$, $H_1(z) = H_p(-z)$, $G_0(z) = H_p(z)$ and $G_1(z) = -H_p(-z)$, where $H_p(z)$ is a linear phase half-band FIR prototype filter. Then, (2.38a-b) becomes:

$$\hat{X}(z) = \frac{1}{2} \begin{pmatrix} H_p(z) & -H_p(-z) \end{pmatrix} \begin{pmatrix} H_p(z) & H_p(-z) \\ H_p(-z) & H_p(z) \end{pmatrix} \cdot \begin{pmatrix} X(z) \\ X(-z) \end{pmatrix}$$

$$= \frac{1}{2} \begin{pmatrix} H_p^2(z) - H_p^2(-z) & 0 \end{pmatrix} \cdot \begin{pmatrix} X(z) \\ X(-z) \end{pmatrix} \qquad (2.38c)$$

which indeed cancels aliasing. This can also be verified by multiplying out the corresponding polyphase matrices. Note that $H(z) = H_{p,0}(z^2) + z^{-1}H_{p,1}(z^2)$ and thus $H(-z) = H_{p,0}(z^2) - z^{-1}H_{p,1}(z^2)$. Then:

$$\begin{aligned} \mathbf{G}_p(z) \cdot \mathbf{H}_p(z) &= \begin{pmatrix} H_{p,1}(z) & H_{p,1}(z) \\ H_{p,0}(z) & -H_{p,0}(z) \end{pmatrix} \cdot \begin{pmatrix} H_{p,0}(z) & H_{p,1}(z) \\ H_{p,0}(z) & -H_{p,1}(z) \end{pmatrix} \\ &= \begin{pmatrix} 2H_{p,0}(z)H_{p,1}(z) & 0 \\ 0 & 2H_{p,0}(z)H_{p,1}(z) \end{pmatrix} \end{aligned} \qquad (2.38d)$$

which is obviously pseudo-circulant.

2.4.3 Time Domain Analysis

Rather than going to z-transforms and then into polyphase or modulation domain, it is also possible to consider the operations performed by multirate filter banks directly in the time-domain. [35, 69]. The matrix representing the operation of a size M filter bank subsampled by N is block Toeplitz or block circulant (depending on the extension at the boundaries):

$$\begin{pmatrix} \vdots \\ y_0(0) \\ y_1(0) \\ \vdots \\ y_{M-1}(0) \\ y_0(1) \\ y_1(1) \\ \vdots \end{pmatrix} = \begin{pmatrix} \ddots & \ddots & \ddots & \ddots & & \ddots & & \ddots \\ 0 & \mathbf{A}_0 & \mathbf{A}_1 & \dots & \mathbf{A}_{K-1} & 0 & & \dots \\ \dots & 0 & \mathbf{A}_0 & \mathbf{A}_1 & \dots & \mathbf{A}_{K-1} & 0 & \\ \dots & \dots & 0 & \mathbf{A}_0 & \mathbf{A}_1 & \dots & \mathbf{A}_{K-1} & \\ \ddots & \ddots & \ddots & \ddots & \ddots & & \ddots & \ddots \end{pmatrix} \begin{pmatrix} \vdots \\ x(0) \\ x(1) \\ x(2) \\ \vdots \\ \\ \\ \vdots \end{pmatrix}$$

$$(2.39)$$

where $\mathbf{A}_i$ is of size M by N and we assumed FIR filters of length NK for simplicity. Define:

$$\mathbf{A}(z) = \sum_{i=0}^{K-1} \mathbf{A}_i z^{-i} \tag{2.40}$$

and rewrite the polyphase matrix $\mathbf{H}_p(z)$ as a matrix polynomial (instead of the matrix of polynomials as done up to now):

$$\mathbf{H}_p(z) = \sum_{i=0}^{K-1} \mathbf{H}_{pi} \cdot z^{-i} \quad . \tag{2.41}$$

Because convolution involves time-reversal, it is easy to verify the following relation between the matrix coefficients of $\mathbf{A}(z)$ and $\mathbf{H}_p(z)$ [69]:

$$\mathbf{A}_{K-k-1} = \mathbf{H}_{pk} \cdot \mathbf{J} \tag{2.42a}$$

$$\mathbf{A}(z) = z^{-K+1} \cdot \mathbf{H}_p(z^{-1}) \cdot \mathbf{J} \quad . \tag{2.42b}$$

Similarly, the synthesis filter bank of size M with upsampling by N can be written in matrix form as (the upsampled input sequences are interleaved):

$$
\begin{pmatrix} \vdots \\ \hat{x}(0) \\ \hat{x}(1) \\ \vdots \\ \vdots \end{pmatrix}
=
\begin{pmatrix}
\ddots & \ddots & \ddots & \ddots & \ddots & & \ddots & \ddots \\
0 & \mathbf{B}_0 & \mathbf{B}_1 & \ldots & \mathbf{B}_{K-1} & 0 & & \ldots \\
\ldots & 0 & \mathbf{B}_0 & \mathbf{B}_1 & \ldots & \mathbf{B}_{K-1} & 0 & \\
\ldots & \ldots & 0 & \mathbf{B}_0 & \mathbf{B}_1 & \ldots & \mathbf{B}_{K-1} & \\
\ddots & \ddots & \ddots & \ddots & \ddots & & \ddots & \ddots
\end{pmatrix}
\begin{pmatrix} \vdots \\ y_0(0) \\ y_1(0) \\ \vdots \\ y_0(1) \\ y_1(1) \\ \vdots \end{pmatrix}
\tag{2.43}
$$

where $\mathbf{B}_i$ is of size N by M. Similarly to (2.40), define:

$$\mathbf{B}(z) = \sum_{i=0}^{K-1} \mathbf{B}_i z^{-i} \quad . \tag{2.44}$$

With $\mathbf{G}_p(z)$ written as a matrix polynomial:

$$\mathbf{G}_p(z) = \sum_{i=0}^{K-1} \mathbf{G}_{pi} \cdot z^{-i} \quad . \tag{2.45}$$

A comparison between (2.44) and (2.45) leads to:

$$\mathbf{B}_{K-k-1} = \mathbf{J} \cdot \mathbf{G}_{pk} \tag{2.46a}$$

$$\mathbf{B}(z) = z^{-K+1} \cdot \mathbf{J} \cdot \mathbf{G}_p(z^{-1}) \tag{2.46b}$$

where the reverse ordering of polyphase components in the synthesis bank was taken into account.

2.4.4   Relation to Unitary Transforms

As a simple example, take the case of a unitary size N by N transform with subsampling by N. Obviously, the inverse of the analysis transform is just the transpose, $\mathbf{B}_0 = \mathbf{A}_0^T$, and $\mathbf{B}_0 \cdot \mathbf{A}_0 = \mathbf{I}$. Thus, from (2.42) and (2.46):

$$\mathbf{B}_0 \cdot \mathbf{A}_0 = \mathbf{J} \cdot \mathbf{G}_0 \cdot \mathbf{H}_0 \cdot \mathbf{J} = \mathbf{I} \tag{2.47a}$$

where J is the antidiagonal matrix. Thus:

$$\mathbf{G}_0 = \mathbf{H}_0^T \tag{2.47b}$$

since $\mathbf{H}_0$ is unitary. In a more general case, assume the matrix in (39) is such that when premultiplied by its transpose it is equal to the identity (if $M = N$, it means that the matrix is unitary):

$$\begin{pmatrix} \mathbf{A}_0^T & \mathbf{A}_1^T & \mathbf{A}_2^T & \cdots & \cdots \\ \cdots & \mathbf{A}_0^T & \mathbf{A}_1^T & \mathbf{A}_2^T & \cdots \\ \cdots & \cdots & \mathbf{A}_0^T & \mathbf{A}_1^T & \mathbf{A}_2^T \end{pmatrix} \cdot \begin{pmatrix} \mathbf{A}_0 & \mathbf{A}_1 & \mathbf{A}_2 & \cdots & \cdots \\ \cdots & \mathbf{A}_0 & \mathbf{A}_1 & \mathbf{A}_2 & \cdots \\ \cdots & \cdots & \mathbf{A}_0 & \mathbf{A}_1 & \mathbf{A}_2 \end{pmatrix} = \begin{pmatrix} \mathbf{I} & & \\ & \mathbf{I} & \\ & & \mathbf{I} \end{pmatrix}. \tag{2.48}$$

This means that:

$$\sum_{i=0}^{K-1} \mathbf{A}_i^T \cdot \mathbf{A}_i = \mathbf{I} \tag{2.49a}$$

$$\sum_{i=0}^{K-1} \mathbf{A}_{i+j}^T \cdot \mathbf{A}_i = \mathbf{0}, \qquad j = 1..K - 1 \tag{2.49b}$$

which can also be written as:

$$\mathbf{A}^T(z^{-1}) \cdot \mathbf{A}(z) = \mathbf{I} \quad . \tag{2.50}$$

Equations (2.49) and (2.50) can also be interpreted as matrix autocorrelations, and the fact that the left inverse of the matrix in (39) is its transpose means that the matrix autocorrelation is a "matrix" impulse. Because of (2.42), (2.50) can be written as:

$$\mathbf{H}_p^T(z^{-1})\mathbf{H}_p(z) = \mathbf{I} \quad . \tag{2.51}$$

This property is an extension of the unitary property to matrices of rational functions. Matrices satisfying (2.51) are called paraunitary [55] (or lossless if all elements are stable). This property will be discussed in detail later, but obviously, a perfect reconstruction subband system is obtained by choosing:

$$\mathbf{G}_p(z) = \mathbf{H}_p^T(z^{-1}) \tag{2.52}$$

since $\mathbf{T}(z) = \mathbf{I}$. Note that the synthesis filters are not causal. If the analysis filters are FIR, a finite delay will make the synthesis filters causal.

2.4.5 Some Basic Results

With the above developments, we are in a position to prove some basic results on analysis/synthesis systems. Define the normal rank of a matrix $\mathbf{M}(z)$ as the rank that $\mathbf{M}(z)$ has almost everywhere on the complex plane (except at singularities).

(a) *Given an analysis filter bank subsampled by N with polyphase matrix* $\mathbf{H}_p(z)$, *then aliasing free reconstruction is possible if and only if the normal rank of* $\mathbf{H}_p(z)$ *is equal to N.*

Proof: it is sufficient since one can take N channels corresponding to N independent rows of $\mathbf{H}_p(z)$. Call the corresponding matrix $\mathbf{H}'_p(z)$. Then choose:

$$\mathbf{G}'_p(z) = cofactor[\mathbf{H}'_p(z)] \qquad (2.53a)$$

and thus:

$$\mathbf{T}(z) = \mathbf{G}'_p(z) \cdot \mathbf{H}'_p(z) = det[\mathbf{H}'_p(z)] \cdot \mathbf{I} \qquad (2.53b)$$

which is pseudo-circulant. It is necessary because, from Theorem 4.1, $\mathbf{T}(z)$ has to be pseudo-circulant and has thus rank N, and since the rank of a matrix product is upperbounded by the ranks of the terms, $\mathbf{H}_p(z)$ cannot have rank lower than N. In practice, this means that aliasing cancellation is easily achievable.

(b) *Given a critically sampled FIR analysis filter bank, perfect reconstruction with FIR filters is possible if and only if* $det[\mathbf{H}_p(z)]$ *is a pure delay.*

Proof: it is sufficient, since choosing:

$$\mathbf{G}_p(z) = cofactor[\mathbf{H}_p(z)] \qquad (2.54a)$$

leads to perfect reconstruction with FIR filters. It is necessary, because $\mathbf{T}(z)$ has to be a pseudo-circulant shift, that is:

$$det(\mathbf{T}(z)) = det[\mathbf{G}_p(z)] \cdot det[\mathbf{H}_P(z)] = z^{-i} \qquad (2.54b)$$

and therefore has i poles at $z = 0$. Since the synthesis has to be FIR, $det[\mathbf{G}_p(z)]$ has only zeroes (or poles at the origin). Therefore, $det[\mathbf{H}_p(z)]$ cannot have any zeroes (except possibly at the origin if it is non-causal).

(c) *Perfect reconstruction with identical (within time reversal) analysis and synthesis filters is possible if and only if the polyphase matrix* $\mathbf{H}_p(z)$ *is paraunitary.* This follows from (2.51).

An issue of importance is the complexity of the reconstruction, given an analysis filter bank. Note that the cofactor matrix contains in general higher degree polynomials than the original matrix, thus corresponding to longer synthesis filters. There are ways to enforce that the synthesis filters are of the same length than the analysis filter, but the conditions are not necessary [69]. For example, if (2.49b) is satisfied, but:

$$\sum_{i=0}^{K-1} \mathbf{A}_i^T \cdot \mathbf{A}_i = \mathbf{C} \qquad (2.55a)$$

and that $\mathbf{C}$ has rank N, then:

$$\mathbf{B}(z) = \mathbf{C}^{-1} \cdot \mathbf{A}^T(z^{-1}) \qquad (2.55b)$$

will achieve perfect reconstruction with same length filters. Other methods are based on factorizations of the polyphase matrix and will be pursued in the next section.

In conclusion to this section, we have seen how to analyze multirate filter banks. The polyphase decomposition of the filters plays a central role, and multirate filter banks can be represented by a time-invariant matrix transfer function. A traditional time-domain view is also possible, and the concept of unitary matrices was generalized to the matrix transfer function case.

2.5   Synthesis of Filters for Filter Banks

As we have seen, designing filter banks for subband coding systems means finding analysis and synthesis filters such that the product of their polyphase matrices, $\mathbf{T}_p(z) = \mathbf{G}_p(z) \cdot \mathbf{H}_p(z)$ satisfies certain properties. At the same time, the filters themselves have to satisfy some properties as well (frequency selectivity, complexity, delay characteristic). All this together makes it a complex design problem. In the following, we will mostly concentrate on the case where both analysis and synthesis filter are FIR, as well as on the critically sampled case ($M = N$, which means $\mathbf{H}_p(z)$ and $\mathbf{G}_p(z)$ are square matrices). Also, we concentrate on the perfect reconstruction case, since aliasing cancellation alone is easely achieved as demonstrated in Subsection 2.4.5(a).

2.5.1   Paraunitary and Lossless Systems

When we first encountered paraunitary matrices in (2.51), we had tacitly assumed that the coefficients were real. The general definition of a paraunitary matrix is [60]:

$$\mathbf{H}_*^T(z^{-1}) \cdot \mathbf{H}(z) = c\,\mathbf{I} \qquad c \neq 0 \qquad (2.56)$$

where the subscript $*$ means conjugation of the coefficients but not of z. If all entries are stable, such a matrix is called lossless. The interpretation of losslessness, a concept very familiar in classical circuit theory [3], is that the "energy" of the signals is conserved through the system given by $\mathbf{H}(z)$. Note that the losslessness of $\mathbf{H}(z)$ implies that $\mathbf{H}(e^{jw})$ is unitary:

$$\mathbf{H}^*(e^{jw}) \cdot \mathbf{H}(e^{jw}) = c\,\mathbf{I} \tag{2.57}$$

where the superscript $*$ stands for Hermitian conjugation (note that $\mathbf{H}^*(e^{jw}) = \mathbf{H}_*(e^{-jw})$). For the scalar case (single-input/single output), lossless transfer functions are all-pass filters given by [40]:

$$F(z) = \frac{a(z)}{z^{-k}a_*(z^{-1})} \tag{2.58}$$

where $k = deg[a(z)]$ (possibly, there is a multiplicative delay and scaling factor equal to $c \cdot z^{-k}$). Thus, to any zero at $z = a$ corresponds a pole at $z = 1/a^*$, that is, at a mirror location with respect to the unit circle. This guarantees a perfect transmission at all frequencies (in amplitude) and only phase distortion. It is easy to verify that (2.58) is lossless (assuming all poles inside the unit circle) since:

$$F_*(z^{-1}) \cdot F(z) = \frac{a_*(z^{-1})}{z^k a(z)} \cdot \frac{a(z)}{z^{-k}a_*(z^{-1})} = 1 \ . \tag{2.59}$$

Obviously, non-trivial scalar all-pass functions are IIR, and are thus not linear-phase. Interestingly, matrix all-pass functions exist that are FIR, and linear-phase behavior is possible. Trivial examples of matrix all-pass functions are unitary matrices, as well as diagonal matrices of delays. Then, cascades of unitary and diagonal matrices of delays will be lossless as well:

$$\mathbf{H}(z) = \mathbf{U}_{K-1} \prod_{i=K-2}^{0} \mathbf{D}_i(z)\mathbf{U}_i \tag{2.60a}$$

since:

$$\mathbf{H}_*^T(z^{-1}) \cdot \mathbf{H}(z) = [\prod_{i=0}^{K-2} \mathbf{U}_i^*\mathbf{D}_i(z^{-1})]\mathbf{U}_{K-1}^*\mathbf{U}_{K-1} \prod_{i=K-2}^{0} \mathbf{D}_i(z)\mathbf{U}_i \ . \tag{2.60b}$$

It turns out that the above factorization is general [11]. More precisely, a real lossless FIR matrix $\mathbf{H}(z)$ of size N by N can be written as in (2.60a) with $\mathbf{U}_0 \ldots \mathbf{U}_{K-2}$ special orthogonal matrices, $\mathbf{U}_{K-1}$ a general orthogonal matrix, and delay matrices of the form:

$$\mathbf{D}_i(z) = diag[z^{-1}\ 1\ 1\ .\ .\ 1] \ . \tag{2.61}$$

Such a general real lossless FIR N input N output system is shown in Figure 2.23, and Figure 2.24 indicates the form of the matrices $\mathbf{U}_0..\mathbf{U}_{K-1}$ and $\mathbf{U}_{K-1}$.

Note that $\mathbf{U}_{K-1}$ is characterized by $(\frac{N}{2})$ rotations [34] while the other orthogonal matrices are characterized by $N-1$ rotations. Thus, a real FIR

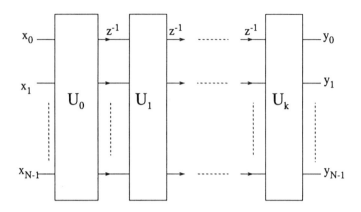

Figure 2.23: General lossless transfer matrix $H(z)$ of size N by N. If $H(z)$ is real, the unitary matrices become rotation matrices.

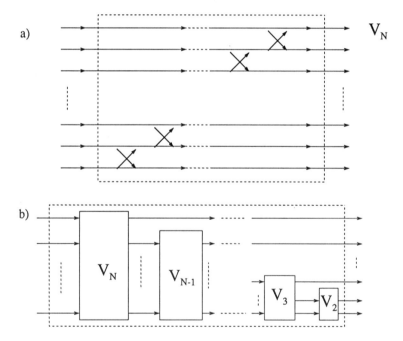

Figure 2.24: Parametrization of the orthogonal matrices appearing in Figure 2.23: (a) constrained orthogonal matrix for $U_0, ..., U_{K-2}$, (b) general orthogonal matrix U_{K-1}.

lossless system of degree $k - 1$ has the following number of free parameters [11, 59]:

$$p = (K - 1)(N - 1) + (\frac{N}{2}) \ . \qquad (2.62)$$

From (2.60b) it is clear that these structures are lossless, and the completeness is demonstrated in [11]. Now, in order to obtain good filters, one can optimize the various angles in the rotation matrices, derive the filters corresponding to the resulting polyphase matrix, and evaluate an objective cost function measuring the quality of the filters (like the stopband energy). Note that because we design a lossless polyphase filter matrix $H_p(z)$, the synthesis filters are simply time-reversed versions of the analysis filters, and it is sufficient to design the analysis bank alone (this will not be the case any more with non-paraunitary filter matrices).

An alternative representation of FIR lossless systems, which turns out to be more convenient for optimization, was presented in [59]. There it is shown that a N by N causal FIR system of degree $K - 1$ is lossless if and only if it can be written in the form:

$$\mathbf{H}_{N-1}(z) = \mathbf{V}_{K-1}(z) \cdot \mathbf{V}_{K-z}(z) \cdots \mathbf{V}_1(z)\mathbf{H}_0 \qquad (2.63a)$$

where $\mathbf{H}_0$ is a general N by N unitary (orthogonal in the real case) matrix and:

$$\mathbf{V}_k(z) = [\mathbf{I} - \mathbf{v}_k\mathbf{v}_k^*(1 - z^{-1})] \qquad (2.63b)$$

with $\mathbf{v}_k$ a size N vector of unit norm (superscript $*$ denotes hermitian conjugation). It is easy to verify that $\mathbf{V}_k(z)$ is lossless:

$$\mathbf{V}_{k*}^T(z^{-1})\mathbf{V}_k(z) = [\mathbf{I} - (1 - z)\mathbf{v}_k\mathbf{v}_k^*] \cdot [\ I - (1 - z^{-1})\mathbf{v}_k\mathbf{v}_k^*]$$

$$= \mathbf{I} + \mathbf{v}_k\mathbf{v}_k^*((z - 1) + (z^{-1} - 1) + (1 - z)(1 - z^{-1})) = \mathbf{I} \qquad (2.63c)$$

where we used $\mathbf{v}_k\mathbf{v}_k^*\mathbf{v}_k\mathbf{v}_k^* = \mathbf{v}_k\mathbf{v}_k^*$, and for the completeness issues, we refer to [59]. Note that these structures can be extended to the IIR case as well, simply by replacing the delay element z^{-1} with a first order scalar all-pass section $(1 - az^{-1})/(z^{-1} - a^*)$. Again, it is easy to verify that such structures are lossless (assuming $|a| > 1$) and for the completeness can be shown similarly to the FIR case.

It is interesting to consider the lossless property in state space description as well. Call $\mathbf{v}(n)$ the state vector, then a state space description is given by [18]:

$$\mathbf{v}(n + 1) = \mathbf{A}\mathbf{v}(n) + \mathbf{B}\mathbf{x}(n) \qquad (2.64a)$$

$$\mathbf{y}(n) = \mathbf{C}\mathbf{v}(n) + \mathbf{D}\mathbf{x}(n) \qquad (2.64b)$$

where $\mathbf{A}$ is of size d by d $(d \geq K - 1$, the degree of the system), $\mathbf{D}$ of size M by N, $\mathbf{C}$ of size M by d and $\mathbf{B}$ of size d by N. A minimal realization satisfies $d = K - 1$. The transfer function matrix is equal to:

$$\mathbf{H}(z) = \mathbf{D} + \mathbf{C}(z\mathbf{I} - \mathbf{A})^{-1}\mathbf{B} \qquad (2.64c)$$

and the impulse response is given by:

$$(\mathbf{D}, \mathbf{CB}, \mathbf{CAB}, \mathbf{CA^2B}, ...) \quad . \qquad (2.64d)$$

The fundamental nature of the losslessness property appears in the following result [52, 60]. A stable transfer matrix $\mathbf{H}(z)$ is lossless if and only if there exists a minimal realization such that:

$$\mathbf{R} = \begin{pmatrix} \mathbf{A} & \mathbf{B} \\ \mathbf{C} & \mathbf{D} \end{pmatrix} \qquad (2.64e)$$

is unitary. This gives another way to parametrize lossless transfer function matrices. In particular, $\mathbf{H}(z)$ will be FIR if $\mathbf{A}$ is lower triangular with a zero diagonal, and thus, it is sufficient to find orthogonal matrices with an upper right triangular corner of size $K - 1$ with only zeros to find all lossless transfer matrices of a given size and degree [11].

After this general look at lossless systems, let us consider some particular cases of interest:

(a) *Two channel FIR lossless solution*

From (2.60-2.61), a general 2 by 2 real lossless FIR transfer matrix can be written as [54, 57]:

$$\mathbf{H}_p(z) = \begin{pmatrix} cos(\alpha_{K-1}) & -sin(\alpha_{K-1}) \\ sin(\alpha_{K-1}) & cos(\alpha_{K-1}) \end{pmatrix} \prod_{k=K-2}^{0} \begin{pmatrix} 1 & 0 \\ 0 & z^{-1} \end{pmatrix} \begin{pmatrix} cos(\alpha_k) & -sin(\alpha_k) \\ sin(\alpha_k) & cos(\alpha_k) \end{pmatrix}.$$
$$(2.65)$$

The corresponding filters are of length $L = 2K$. As an example, take $L = 4$ and use the shorthand $cos(\alpha_i) = a_i$ and $sin(\alpha_k) = b_i$. Then:

$$H_p(z) = \begin{pmatrix} a_0 a_1 z^{-1} - b_0 b_1 & -a_0 b_1 z^{-1} - b_0 a_1 \\ b_0 a_1 z^{-1} + a_0 b_1 & -b_0 b_1 z - 1 + a_0 a_1 \end{pmatrix} \qquad (2.65b)$$

and the corresponding filters have z-transforms equal to:

$$H_0(z) = -b_0 b_1 - b_0 a_1 z^{-1} + a_0 a_1 z^{-2} - a_0 b_1 z^{-3} \qquad (2.65c)$$

$$H_1(z) = a_0 b_1 + a_0 a_1 z^{-1} + b_0 a_1 z^{-2} - b_0 b_1 z^{-3} . \qquad (2.65d)$$

As one can verify:

$$H_1(z) = z^{-L+1} \cdot H_0(-z^{-1}) \qquad (2.66)$$

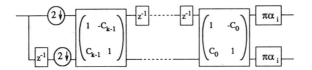

Figure 2.25: Lossless two channel filter bank using denormalized rotation matrices ($C_i = sin(x_i)/cos(x_i)$).

Causal synthesis filters are obtained from:

$$\mathbf{G}_p(z) = z^{-K+1}\mathbf{H}_p^T(z^{-1}) \tag{2.67a}$$

and are equal to:

$$G_i(z) = z^{-L+1} \cdot H_i(z^{-1}) . \tag{2.67b}$$

One powerful feature of the structure given by (2.64) is that it is structurally lossless, that is, it will be lossless even if coefficients are quantized [54, 57]. For computational complexity reason, one can denormalize the rotation matrices by dividing with $cos(\alpha_i)$ (assuming it different from zero). All these factors can be gathered as a single factor which can be multiplied at the end. Such a structure is indicated in Figure 2.25.

As it turns out, the same filters were obtained earlier [47, 33, 49] with the factorization of a special autocorrelation sequence into $H_0(z)$ and $H_1(-z)$. For optimization purposes, the cascade structure is more convenient than a method relying on factorization of polynomials. Note that the $H_i(z)$'s will never have linear phase behavior (except in the trivial case $L = 2$) and that $G_i(z)$ has the converse phase response of $H_i(z)$. Therefore, the input-output delay is equal to that obtained with linear-phase filters, namely $L-1$ samples.

(b) *Case $L = 2N$*

An interesting case appears if one restricts the length of the filters to be equal to twice the subsampling factor. In time domain matrix notation, this leads to an analysis matrix which has a double block diagonal:

$$\mathbf{A} = \begin{pmatrix} \ddots & \ddots & \ddots & \ddots & \ddots & \ddots & \ddots & \ddots \\ \cdots & 0 & \mathbf{A}_0 & \mathbf{A}_1 & 0 & \cdots & \cdots & \cdots \\ \cdots & \cdots & 0 & \mathbf{A}_0 & \mathbf{A}_1 & 0 & \cdots & \cdots \\ \cdots & \cdots & \cdots & 0 & \mathbf{A}_0 & \mathbf{A}_1 & 0 & \cdots \\ \ddots & \ddots & \ddots & \ddots & \ddots & \ddots & \ddots & \ddots \end{pmatrix} . \tag{2.68a}$$

The orthogonality condition on $\mathbf{A}$ (or equivalently, the lossless condition on $\mathbf{H}_p(z)$) leads to:

$$\mathbf{A}_0^T \cdot \mathbf{A}_0 + \mathbf{A}_1^T \cdot \mathbf{A}_1 = \mathbf{I} \tag{2.68b}$$

$$\mathbf{A}_0^T \cdot \mathbf{A}_1 = 0 \ . \tag{2.68c}$$

Note that (2.68b) means that the filters form a set of N length-$2N$ orthonormal vectors. Then, because $\mathbf{A}$ is orthonormal, $\mathbf{A} \cdot \mathbf{A}^T = \mathbf{I}$ and (68c) can also be written as:

$$\mathbf{A}_1 \cdot \mathbf{A}_0^T = \mathbf{A}_0 \cdot \mathbf{A}_1^T = 0 \ . \tag{2.68d}$$

This means that the "tails" of the filters are orthogonal. This was already the underlying property in (2.50), simply here, this fact is easy to visualize.

Because of the reduced length of the filters, the design problem is simpler, and a number of solutions have been proposed, like the lapped orthogonal transform (LOT) [7, 30, 31] and modulated filters [41, 42, 68, 69]. As an example, the frequency response of a perfect reconstruction filter bank of 8 filters of length 16 is shown in [24]. This filter bank resembles to the operation of the discrete cosine transform (DCT) because the prototype filter is modulated by cosines evenly spaced over the frequency spectrum. The main difference is that $2N$ data points are considered through a smooth window, thus avoiding the blocking effects of the DCT and reducing the spectral leakage as well. One potential problem is that linear phase filters are not possible. Such filter banks, as well as LOT's, have been used in image compression [7, 30, 24].

(c) IIR case

As mentioned earlier, using all-pass sections instead of delays in the diagonal delay matrix (2.61) would lead to paraunitary (or lossless if stable) IIR systems. Obviously, any diagonal matrix with all-pass elements is paraunitary, since following (2.59):

$$\mathbf{D}_*^T(z^{-1}) \cdot \mathbf{D}(z) = diag[\ldots F_{i*}(z^{-1}) \cdot F_i(z) \ldots] = \mathbf{I} \tag{2.69}$$

where $F_i(z)$ is an all-pass function as in (2.58). A simple such example for the case $N = 2$ was given in [45]:

$$\mathbf{H}_p(z) = \begin{pmatrix} 1 & 1 \\ 1 & -1 \end{pmatrix} \cdot \begin{pmatrix} F_0(z) & \\ & F_1(z) \end{pmatrix} \tag{2.70a}$$

and even with such simple all-pass functions as:

$$F_0(z) = \frac{\alpha + z^{-1}}{1 + \alpha z^{-1}}, \qquad F_1(z) = 1 \tag{2.70b}$$

one obtains good half-band filters at very low complexity ($F_0(z)$ can be implemented with only 1 multiplier). The z-transforms of the resulting filters is:

$$
\begin{aligned}
H_0(z) &= F_0(z^2) + z^{-1} F_1(z^2) = \frac{\alpha + z^{-2}}{1 + xz^{-2}} + z^{-1} \\
&= \frac{\alpha + z^{-1} + z^{-2} + \alpha z^{-3}}{1 + \alpha z^{-2}} \tag{2.71}
\end{aligned}
$$

and $H_1(z) = H_0(-z)$ is the mirror filter. The main advantage is a good frequency selectivity at low computational complexity, and the price paid is non-causality of the synthesis filters. However, if the signals involved are finite length (like in image coding), the non-causality can be taken care of by time-reversal of the signals between analysis and synthesis. Nevertheless, the output of the analysis bank is infinite length even if the input is finite length. This last problem can be taken care of in two ways. Either one stores the state of the filters after the end of the input signal and uses this as an initial state for the synthesis filters, or one takes advantage of the fact that the outputs of the analysis filter bank decay rapidly after the input is zero, and stores only a finite extension of these signals. While the former technique is exact, the latter is usually a good enough approximation. Note that when such filters are used for image coding, they are usually designed such as to obtain quasi linear phase behavior in the passband.

2.5.2 General systems

While the lossless solutions we have seen above possess an unquestionable elegance, they are sometimes too restricted. A classical example is that in the two channel case, there is no (non-trivial) linear phase lossless solution. One has therefore to depart from the well-defined lossless framework, where issues like completeness and factorizations are settled, and search for more general solutions. We will concentrate on the critically sampled FIR case, so from Subsection 2.4.5(b), we will search for polyphase matrices with determinants equal to a delay. However, this does not guarantee (except in the case $N = 2$) that the synthesis filters will be of equal complexity. In general, given polyphase components of degree $K - 1$ (corresponding to filters of size $N \cdot K$) in the analysis filter matrix, the cofactor matrix will contain polynomials of degree up to $(N - 1) \cdot (K - 1)$, and thus, the synthesis filters can be of size as large as $N(N - 1)(K - 1) + N$, that is, roughly $(N - 1)$ times longer. A way out of this complexity problem is to impose additional structure so that the synthesis filters are guaranteed to be of the same length as the analysis filters. Typically, if a factorization is imposed like in (2.60a) but with the relaxed constraint that the matrices U_i are simply non-singular [71], then it is clear that $G(z)$ is easily obtained as:

$$G(z) = [\prod_{i=0}^{K-2} U_i^{-1} D_i(z^{-1})] U_{K-1}^{-1} \tag{2.72}$$

and therefore, the synthesis filters are automatically not longer than the analysis filters.

Another problem is that the frequency behavior of the synthesis filters can be quite exotic [64, 50] since there is no direct relation between the analysis and synthesis filters like in the lossless case. In general, this means that both analysis and synthesis filters have to be designed concurrently. Again, imposed structures like the one in (2.72) tend to ease this problem [71], at least if the non-singular matrices $\mathbf{U}_i$ are well-conditioned.

In general, a polyphase matrix $\mathbf{H}_p(z)$ has a determinant equal to a delay if its Smith form [15, 18]:

$$\mathbf{H}_p(z) = \mathbf{L}(z) \cdot \mathbf{D}(z) \cdot \mathbf{R}(z) \tag{2.73}$$

has the diagonal matrix $\mathbf{D}(z)$ with increasing delays (rather than general polynomials). In (2.73), $\mathbf{L}(z)$ and $\mathbf{R}(z)$ are unimodular, that is, their determinant is a constant rather than a function of z. The problem with the Smith form is that it is more an analysis than a synthesis tool for polynomial matrices, a fact related among other things to the non-uniqueness of the unimodular matrices $\mathbf{L}(z)$ and $\mathbf{R}(z)$ [18].

(a) *Two channel case*

In the two channel case, the two problems associated with non-paraunitary solutions are avoided. The synthesis filters are related to the analysis filters by:

$$G_0(z) = H_1(-z) \tag{2.74a}$$

$$G_1(z) = -H_0(-z) \tag{2.74b}$$

which follows from inverting $\mathbf{H}_p(z)$ and making $\mathbf{G}_p(z)$ causal. Thus, the synthesis filters are of the same length, and if the $H_i(z)$'s are good half-band filter, so will the $G_i(z)'s$ be. The condition of $det[\mathbf{H}_p(z)]$ equal a delay means equivalently that $det[\mathbf{H}_m(z)]$ has to be a delay. Writing this out:

$$det[\mathbf{H}_m(z)] = H_0(z)H_1(-z) - H_0(-z)H_1(z) = P(z) - P(-z) = z^{-2k-1} \tag{2.74c}$$

where $P(z) = H_0(z) \cdot H_1(-z)$. Thus, the most general synthesis procedure consists, after choosing the desired delay $2k + 1$, to choose polynomials $P(z)$ with arbitrary even coefficients and all odd coefficients except $2k + 1$ equal to zero and then to factor them into parts $H_0(z)$ and $H_1(-z)$ [64]. Note that $H_0(z)$ and $H_1(-z)$ are both half-band lowpass filters, and thus $P(z)$ should be a half-band lowpass filter as well. One problem consists in assigning zeroes of $P(z)$ to one or the other filter, since the number of possibilities is combinatorial. One possibility is to choose $P(z)$ such that all zeroes appear in pairs $(a, 1/a^*)$ which means that $P(z)$ is an autocorrelation sequence. Splitting the pairs between $H_0(z)$ and $H_1(-z)$ then just leads to the lossless solutions found earlier [47, 33, 48]. Another general procedure is to choose $H_0(z)$ and then to solve the linear system of equations for $H_1(z)$ in order to meet (2.74) [64]. Both methods, while complete in the sense that they can reach all possible perfect reconstruction solutions, are not practical for large filter lengths because of the indirect way the filters are obtained. For small filter sizes and under more constraints (linear phase and integer coefficients), the factorization method was used to derive filters which have been quite successful in image and video coding applications [26, 27].

One case of interest where no lossless solution is possible is the two channel bank with linear phase filters. For this case, we will develop a cascade structure which will automatically generate linear phase filters and guarantee perfect reconstruction. First, note that when the two filters are of the same length, the length has to be even [64]. Then, $\mathbf{H}_p(z)$ has to satisfy the following linear phase test:

$$\begin{pmatrix} 1 & 0 \\ 0 & -1 \end{pmatrix} \cdot [z^{-k} \cdot \mathbf{H}_p(z^{-1})] \cdot \begin{pmatrix} 0 & 1 \\ 1 & 0 \end{pmatrix} = \mathbf{H}_p(z) \tag{2.75}$$

where k is the highest degree in $\mathbf{H}_p(z)$. We will now develop a cascade form to obtain linear phase filters of any even length and which guarantee perfect reconstruction:

(i) assume $\mathbf{H}_p(z)$ satisfies (2.75)

(ii) then $\mathbf{H}'_p(z)$ given by:

$$\mathbf{H}'_p(z) = \mathbf{H}_p(z) \cdot \begin{pmatrix} 1 & 0 \\ 0 & z^{-1} \end{pmatrix} \cdot \begin{pmatrix} 1 & \alpha \\ \alpha & 1 \end{pmatrix} \tag{2.76}$$

satisfies (2.75) as well.

The proof is straightforward by replacing (2.76) into (2.75) and verifying that it holds indeed. Note that for $L = 2$, the two only possible filters are given by $H_0(z) = 1 + z^{-1}$ and $H_1(z) = 1 - z^{-1}$ (or scaled versions thereof) and therefore, a possible way to obtain length $L = 2K$ linear phase perfect reconstruction filters is by writing $\mathbf{H}_p(z)$ as [66, 38, 39, 69]:

$$\mathbf{H}_p(z) = \begin{pmatrix} 1 & 1 \\ 1 & -1 \end{pmatrix} \cdot \prod_{k=1}^{K-1} \begin{pmatrix} 1 & 0 \\ 0 & z^{-1} \end{pmatrix} \begin{pmatrix} 1 & \alpha_k \\ \alpha_k & 1 \end{pmatrix} \tag{2.77a}$$

and $\mathbf{G}_p(z)$ as:

$$\mathbf{G}_p(z) = \frac{1}{2} \cdot \begin{pmatrix} 1 & 1 \\ 1 & -1 \end{pmatrix} \cdot \prod_{k=1}^{K-1} \begin{pmatrix} z^{-1} & 0 \\ 0 & 1 \end{pmatrix} \begin{pmatrix} 1 & -\alpha_k \\ -\alpha_k & 1 \end{pmatrix} \cdot \frac{1}{1 - \alpha_k^2} . \tag{2.77b}$$

Therefore, (2.54b) is satisfied with $i = (K - 1)$. Note that while this is not a paraunitary solution, the synthesis filters are simply related to the analysis filters by a modulation with $(-1)^n$:

$$G_0(z) = H_1(-z) \qquad G_1(z) = -H_0(-z) . \tag{2.78}$$

L	Scaling factor	Analysis	Filters
2	2	$H_0(z) = 1 + z^{-1}$ $H_1(z) = 1 - z^{-1}$	
4	$2(1-\alpha_1^2)$	$H_0(z) = 1 + \alpha z^{-1} + \alpha z^{-2} + z^{-3}$ $H_1(z) = 1 + \alpha z^{-1} - \alpha z^{-2} - z^{-3}$	
6	$2(1-\alpha_1^2)(1-\alpha_2^2)$	$H_0(z) = 1 + \alpha_2 z^{-1} + (\alpha_1 + \alpha_1\alpha_2)z^{-2} + (\alpha_1 + \alpha_1\alpha_2)z^{-3} + \alpha_2 z^{-4} + z^{-5}$ $H_1(z) = 1 + \alpha_2 z^{-1} + (-\alpha_1 + \alpha_1\alpha_2)z^{-2} + (\alpha_1 - \alpha_1\alpha_2)z^{-3} - \alpha_2 z^{-4} - z^{-5}$	

Table 2.1: First few linear phase perfect reconstruction filters of even length obtained with the cascade structure.

The first few filters obtained from (2.77a) are given in Table 2.1. Note that length $2K$ filters are defined by $K - 1$ free parameters.

A similar structure is possible for odd length filters as well, leading to filters of length $2k + 1$ and $2k + 3$ respectively [39, 69]. In this case, the cascade structure is given by:

$$\mathbf{H}_p(z) = \prod_{k=1}^{K} \left(\begin{matrix} 1 + z^{-1} & a_k \\ 1 + b_k z^{-1} + z^{-2} & a_k(1 + z^{-1}) \end{matrix} \right), \quad a_k \neq 0, \quad b_k \neq 2 . \quad (2.79)$$

The computational complexity of these structures is very low. An elementary block in (2.77) is a size 2 circular convolution which requires 2 multiplications. However, it can be denormalized and written as:

$$\frac{2}{1+a} \left(\begin{matrix} 1 & a \\ a & 1 \end{matrix} \right) = \left(\begin{matrix} 1 & 1 \\ 1 & -1 \end{matrix} \right) \cdot \left(\begin{matrix} 1 & \\ & \frac{1-a}{1+a} \end{matrix} \right) \cdot \left(\begin{matrix} 1 & 1 \\ 1 & -1 \end{matrix} \right), \quad a \neq \{1, -1\}. \quad (2.80)$$

By merging all scaling factors, this leads to 1 multiplication per block, plus one scaling at the beginning or the end. This gives a total of $(k+1)/2$ multiplications per input sample for length $2K$ filters. This is half as many as required by paraunitary solutions or classical QMF solutions as shown in Table 2.2.

A complete picture is obtained when contrasting computational complexity with achievable filter quality, and this was done in [39]. A hardware structure for the cascade structures in (2.77a-b) is given in Figure 2.26. Another question of interest is the completeness of the structure. Obviously, in (2.77), $\alpha = 1$ or -1 are forbidden values since they create singular blocks. However, one can construct linear phase perfect reconstruction filter banks which, when one attempts to write them in cascade form as in (2.77), would lead to singular blocks [69]. Thus, such solutions can only be approximated with the cascade form, and then only in an ill-conditioned manner. This shows that the cascade

Type of Filter bank	Normalized block	Denormalized block
paraunitary	3K/2	K+1
Linear Phase	K-1	(K+1)/2
Classical QMF	K	-

Table 2.2: Computational complexity of various two channel filter banks.

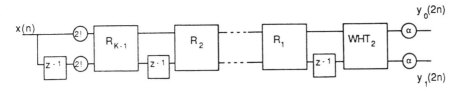

Figure 2.26: Linear phase, perfect reconstruction two channel filter bank with even length filters (analysis filter bank). The computational blocks can be implemented with one multiplier.

form is note complete, however, the "singular" cases seem of little practical interest.

(b) *General Case with Symmetries*

The linear phase test in (2.75) can easily be extended to polyphase matrices of size larger than 2. This leads to cascade structures generating automatically linear phase filters [69]. Note that for $N > 2$, paraunitary solutions are possible. Another symmetry of interest is the mirror property of filters in the bank, that is :

$$H_{N-i-1}(z) = H_i(-z) \quad . \tag{2.81}$$

This restriction is often desirable, and reduces the number of free parameters in half thus leading to improved and faster design procedures [37]. In a similar fashion to the linear phase case, one can also impose a cascade structure which automatically produces mirror filters [69]. The point of the above brief discussion is to show that desirable features can be imposed on the filter bank, and the price paid is a corresponding reduction of the degree of freedoms.

2.6   Multidimensional Subband Systems

When dealing with subband coding of images, one needs to extend the results obtained so far to two dimensional signals and filters. The presentation below follows [21, 22, 23], to where we refer for details. If one assumes separable subsampling and separable filters, the extension is trivial since one can consider the two dimensions independently. This has been done in a number of image subband coding schemes. The more interesting case appears when the filters are non-separable and/or the sub-sampling is non-separable. We will consider mostly two-dimensional systems, and also touch upon the three dimensional case.

2.6.1   Sampling of Two-Dimensional Signals

Sampling theory in multiple dimensions is more involved than in one dimension, and is related to the theory of lattices. For a thorough discussion, we refer to [13, 12]. We are mostly concerned with subsampling of multidimensional sampled signals, as well as subsequent upsampling. Note that we will restrict the discussion to sampling patterns which can be represented as lattices (as opposed to union of lattices). To describe the subsampling process, call $\mathbf{n} = (n_1, n_2)$ and $\mathbf{u} = (u_1, u_2)$ the locations on the input and subsampling lattice respectively. Then, a location on the subsampling lattice corresponds to the following location on the input lattice:

$$\begin{pmatrix} n_1 \\ n_2 \end{pmatrix} = \begin{pmatrix} d_{00} & d_{01} \\ d_{10} & d_{11} \end{pmatrix} \begin{pmatrix} u_1 \\ u_2 \end{pmatrix} = \mathbf{D} \cdot \mathbf{u} \qquad (2.82)$$

where d_{ij} are integers. Without loss of generality, one can choose $d_{10} = 0$ (making thus u_1 and n_1 collinear) and $d_{00} > 0$, $d_{11} > 0$ and $0 \leq d_{01} < d_{00}$. Other forms of the subsampling matrix are related to this one by multiplication with a matrix of integers with determinant equal to 1. Familiar examples of subsapling matrices are hexagonal subsampling:

$$\mathbf{D}_h = \begin{pmatrix} 2 & 1 \\ 0 & 2 \end{pmatrix} \qquad (2.83a)$$

and quincunx subsampling:

$$\mathbf{D}_q = \begin{pmatrix} 2 & 1 \\ 0 & 1 \end{pmatrix} \qquad (2.83b)$$

which are shown in Figure 2.27 (a) and (b) respectively. The subsampling rate is given by $Det[\mathbf{D}]$:

$$Det[\mathbf{D}] = d_{00} \cdot d_{11} \quad . \qquad (2.84)$$

a)   b)

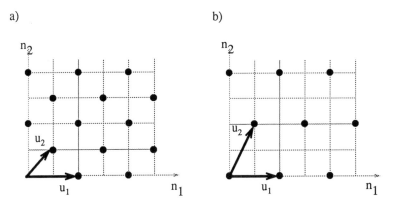

Figure 2.27: Subsampling patterns: (a) hexagonal subsampling, (b) quincunx subsampling.

In the z-transform domain, downsampling by N according to the subsampling matrix $\mathbf{D}$ results in:

$$Y(z_1, z_2) = \frac{1}{N} \sum_{k=0}^{d_{00}-1} \sum_{l=0}^{d_{11}-1} X(W_N^{-d_{11}k} z_1^{1/d_{00}}, W_N^{-(d_{00}l-d_{01}k)} z_1^{-d_{01}/N} z_2^{1/d_{11}}) \quad (2.85)$$

where $N = Det[\mathbf{D}]$ and $W_N = e^{-j\frac{2\pi}{N}}$. Upsampling by N according to $\mathbf{D}$ leads to:

$$Y(z_1, z_2) = X(z_1^{d_{00}}, z_1^{d_{01}} z_2^{d_{11}}) . \quad (2.86)$$

One way to verify the above is to replace (2.86) into (2.85) and verify that the result is simply $X(z_1, z_2)$, since upsampling followed by subsampling by the same pattern is an identity operation.

Now, in a subband coding system, the subsampled signal is upsampled to the original sampling grid for reconstruction purposes. This down/upsampling process is equivalent to modulate the input signal with a function that is zero everywhere except at the points belonging to the subsampling lattice where it is one. It can be checked that such a function, which is the two-dimensional equivalent of (2.16), is given by [23]:

$$
\begin{aligned}
f(n_1, n_2) &= \frac{1}{N} \sum_{k=0}^{d_{00}-1} \sum_{l=0}^{d_{11}-1} exp(-j2\pi \begin{pmatrix} n_1 & n_2 \end{pmatrix} (\mathbf{D}^{-1})^T \begin{pmatrix} k \\ l \end{pmatrix}) \\
&= \frac{1}{N} \sum_{k=0}^{d_{00}-1} \sum_{l=0}^{d_{11}-1} W_N^{d_{11} n_1 k + d_{00} n_2 l - d_{01} n_2 k} \quad (2.87)
\end{aligned}
$$

From the modulation theorem, it follows that an input signal with z-transform

$X(z_1, z_2)$ results, after down/upsampling, in:

$$Y(z_1, z_2) = \frac{1}{N} \sum_{k=0}^{d_{00}-1} \sum_{l=0}^{d_{11}-1} X(W_N^{-d_{11}k} z_1, W_N^{-(d_{00}l - d_{01}k)} z_2). \qquad (2.88)$$

The above can also be derived by using (2.85) in (2.86) and it can be verified that indeed $Y(z_1, z_2)$ is a polynomial with powers of $z_1^{d_{00}}$ and $z_1^{d_{01}} z_2^{d_{11}}$ only, which means, it is non-zero only on locations corresponding to (n_1, n_2) given by (2.82).

From (2.86), it follows that the output of a multidimensional subband coding system contains the original signal, as well as $N - 1$ aliased versions thereof.

2.6.2   Polyphase Decomposition of 2-D Filter Banks

Similarly to the periodically time-variant property of 1-D subband coders, 2-D systems will be periodically space variant. The periodicity is given by the subsampling lattice, and a sufficient characterization is given by the N impulse responses corresponding to locations in a basis cell, such as the unit cell [12]. Therefore, it is natural to decompose a filter impulse response into a sum of polyphase components. That is, the z-transform of a filter with coefficients $h(n_1, n_2)$ can be written as [22, 23]:

$$H(z_1, z_2) = \sum_{k=0}^{d_{00}-1} \sum_{l=0}^{d_{11}-1} z_1^{-(d_{00}+d_{01})} z_2^{-d_{11}} H_{k,l}(z_1^{d_{00}}, z_1^{d_{01}} z_2^{d_{11}}) \qquad (2.89a)$$

where

$$H_{k,l}(z_1, z_2) = \sum_{u_1=-\infty}^{\infty} \sum_{u_2=-\infty}^{\infty} h(d_{00}u_1 + d_{01}u_2 + k, d_{11}u_2 + l) \cdot z_1^{-u_1} \cdot z_2^{-u_2}. \,(2.89b)$$

This is a straightforward generalization of the 1-D polyphase decomposition given in (2.10). Note that an impulse at location $(-k_0, -l_0)$ will trigger the polyphase component (k_0, l_0) as an output of the subsampling. Therefore, a filter followed by subsampling can be represented as a 2-D polyphase transform of the filter, as shown in Figure 2.28.

As an example, we indicate a filter and its polyphase decomposition with respect to a hexagonal subsampling grid in Figure 2.29.

Because each polyphase component was chosen so as to have 4 coefficients, the filter impulse is not rectangular.

One can define similarly the polyphase components of a synthesis filter preceded by upsampling, simply with reversed order:

$$G_{k,l}(z_1, z_2) = \sum_{n_1=-\infty}^{\infty} \sum_{n_2=-\infty}^{\infty} g(d_{00}n_1 + d_{01}n_2 + d_{00} - 1 - k, d_{11}n_2 + d_{11} - 1 - l). \,(2.90)$$

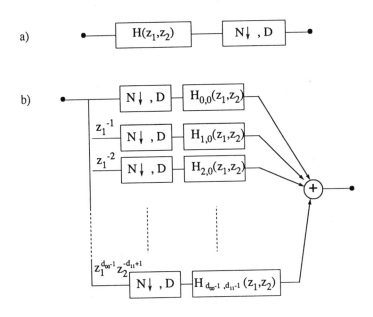

Figure 2.28: Subsampled 2-D filter: (a) original representation, (b) two-dimensional polyphase transform followed by polyphase components of the filter.

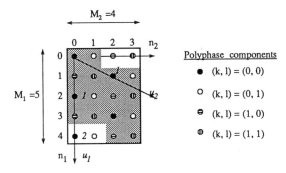

Figure 2.29: Polyphase decomposition with respect to hexagonal subsampling.

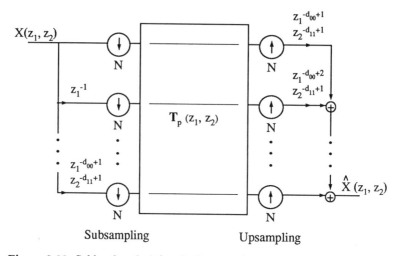

Figure 2.30: Subband analysis/synthesis system in polyphase representation. The transmission matrix $T_p(z_1, z_2)$ is equal to the product of the analysis and synthesis filter bank matrix.

Now, when dealing with analysis and synthesis filter banks, one can define the polyphase matrices $\mathbf{H}_p(z_1, z_2)$ and $\mathbf{G}_p(z_1, z_2)$ exactly as in the 1-D case (see (2.22b) and (2.23b)). Therefore, a complete analysis/synthesis system can be represented as in Figure 2.30, that is, a transmission matrix $\mathbf{T}_p(z_1, z_2)$ equal to:

$$\mathbf{T}_p(z_1, z_2) = \mathbf{G}_p(z_1, z_2) \cdot \mathbf{H}_p(z_1, z_2) \tag{2.91}$$

which appears between a forward and an inverse polyphase transform. Note that we choose a non-causal inverse polyphase transform, which could be made causal by multiplying with $z_1^{-d_{00}+1} \cdot z_2^{-d_{11}+1}$.

2.6.3 Aliasing Cancellation and Perfect Reconstruction

From (2.88), it is clear that in general, the output of an analysis/synthesis system can contain aliased components of the input signal. Conditions for aliasing cancellation are easely derived if the filter matrices are written in the modulation domain, since then, it is necessary and sufficient that $\mathbf{T}_m(z_1, z_2)$ is diagonal. In the polyphase domain, the necessary and sufficient condition is that the vector

$$\left(1 \ z_2 \ \cdots \ z_2^{d_{11}-1} \right) \otimes \left(1 \ z_1 \ \cdots \ z_1^{d_{00}-1} \right) \tag{2.92}$$

is a left eigenvector of $\mathbf{T}_p(z_1^{d00}, z_1^{d01} z_2^{d11})$ [22, 23] ($\otimes$ is the Kronecker product [4]). This is the generalization of the pseudo-circulant property to the 2-D case. Except when the subsampling is separable [29], the structure of matrices satisfying aliasing cancellation is more involved than in the one-dimensional case.

When the above property is met, the system become shift-invariant, and the output is a filtered version of the input:

$$Y(z_1, z_2) = T(z_1, z_2) \cdot X(z_1, z_2) \ . \tag{2.93}$$

Obviously, a necessary and sufficient condition for perfect reconstruction is that $T(z_1, z_2)$ is of the form $z_1^m \cdot z_2^n$, that is, a pure shift in two dimensions. For the sake of illustration, let us consider a subband system with four channels and hexagonal subsampling. The following transmission matrix

$$\mathbf{T}_p(z_1, z_2) = \begin{pmatrix} T_0(z_1, z_2) & T_1(z_1, z_2) & T_2(z_1, z_2) & T_3(z_1, z_2) \\ T_1(z_1, z_2)z_1^{-1} & T_0(z_1, z_2) & T_3(z_1, z_2)z_1^{-1} & T_2(z_1, z_2) \\ T_3(z_1, z_2)z_1^{-1} & T_2(z_1, z_2)z_1 z_2^{-1} & T_0(z_1, z_2) & T_1(z_1, z_2) \\ T_2(z_1, z_2)z_2^{-1} & T_3(z_1, z_2)z_2^{-1} & T_1(z_1, z_2)z_1^{-1} & T_0(z_1, z_2) \end{pmatrix} \tag{2.94a}$$

yields aliasing cancellation, since:

$$(1 \ z_1 \ z_2 \ z_1 z_2)\mathbf{T}_p(z_1^2, z_1 z_2^2) = [T_0\left(z_1^2, z_1 z_2^2\right) + z_1^{-1} T_1\left(z_1^2, z_1 z_2^2\right)$$

$$+ z_2^{-1} T_2\left(z_1^2, z_1 z_2^2\right) + z_1^{-1} z_2^{-1} T_3\left(z_1^2, z_1 z_2^2\right)](1 \ z_1 \ z_2 \ z_1 z_2) \ . \tag{2.94b}$$

Perfect reconstruction is achieved if and only if 3 out of the 4 $T_i(z_1, z_2)$'s in (2.94a) are zero and the 4th one is a monomial.

2.6.4 Filter Design Structures for Perfect Reconstruction

We will restrict our discussion to FIR filter banks, both because then stability is of no concern, and linear phase behavior is easely achievable. The definition of paraunitary systems in two dimensions is that the transfer matrix is unitary on the unit bi-circles ($z_1 = e^{-jw_1}$ and $z_2 = e^{-jw_2}$), thus:

$$\mathbf{H}_{p*}^T(z_1^{-1} z_2^{-1}) \cdot \mathbf{H}_p(z_1, z_2) = \mathbf{I} \ . \tag{2.95}$$

Note that this means that $\mathbf{H}_p(z_1^{d00}, z_1^{d01}, z_2^{d11})$ is paraunitary as well, since this is just another location on the unit bi-circles.

The difficulty with 2-D systems is the absence of a factorization theorem for 2-D polynomials. Thus, while it is easy to derive paraunitary cascade structures by combining elementary paraunitary blocks, a complete such factorization is not known so far, unlike in the 1-D case (see (2.63)). An alternative approach is based on a state space description. A 2-D transfer function can be written as [51]:

$$H_p(z_1, z_2) = D + C \left\{ \begin{pmatrix} z_1 I_{m_1} & 0 \\ 0 & z_2 I_{m_2} \end{pmatrix} - A \right\}^{-1} B, \qquad (2.96)$$

where A is m by m, D is N by N, C is N by m, B is m by N and $m = m_1 + m_2$. The system is FIR if A is of the form

$$A = \begin{pmatrix} 0 & 0 & \cdots & 0 \\ * & 0 & \cdots & 0 \\ \vdots & \ddots & \ddots & \vdots \\ * & \cdots & * & 0 \end{pmatrix} \qquad (2.97)$$

(or if it is upper diagonal with 0 diagonal). It can be verified that $H_p(z_1, z_2)$ is lossless if the matrix

$$R = \begin{matrix} & \overset{N}{} \quad \overset{m}{} \\ \begin{matrix} m \\ N \end{matrix} & \begin{pmatrix} B & A \\ D & C \end{pmatrix} \end{matrix} \qquad (2.98)$$

is unitary. Thus, one can use cascades of rotations to generate unitary matrices as in (2.98) and which satisfy (2.97) as well, very similarly to the one dimensional case (see (2.64) and [11]), except that now completeness is not guaranteed.

Non-paraunitary systems can be designed with techniques similar to the 1-D case. Again, synthesis filters can become more complex than the analysis filters, unless special techniques are used to guarantee equal complexity. This can be done by imposing additional constraints, like forcing a cascade structure. Examples of such constrained design structures can be found in [23], and as in the 1-D case, completeness is not achieved.

Because of the importance of phase in images and thus to avoid phase distortion in subband coding schemes, linear phase filters are often desirable. A 2-D filter has linear phase if it has central symmetry (or antisymmetry):

$$h(n_1, n_2) = h(M_1 - n_1 - 1, M_2 - n_2 - 1) \qquad (2.99)$$

(or a minus sign if it is antisymmetric). M_1 by M_2 is the size of the filter impulse response. If one restricts the filter to be of a size equal to a multiple of unit cells of the subsampling pattern, let's say w_1 and w_2 in directions n_1 and n_2, then the linear phase condition for the entire bank can be written, similarly to (2.77) as:

$$H_p(z_1, z_2) = z_1^{-(w_1-1)} z_2^{-(w_2-1)} \, diag[\pm 1, \cdots \pm 1] \, H_p(z_1^{-1}, z_2^{-1}) \, J, \qquad (2.100)$$

where the diagonal entry is -1 if the corresponding filter has odd symmetry. Now, assuming we have a filter bank that satisfies (2.100) and we want to extend it so as to still satisfy (2.100) and conserve the same symmetries among the filters. The extended filter bank can be written as:

$$\mathbf{H}_p(z_1, z_2) = \mathbf{H}_p(z_1, z_2) \cdot \mathbf{\Delta}(z_1, z_2) \cdot \mathbf{E} \qquad (2.101a)$$

and, in order to meet (2.100), $\mathbf{\Delta}(z_1, z_2)$ and $\mathbf{E}$ have to satisfy:

$$\mathbf{\Delta}(z_1, z_2) = z_1^{-(\mu_1 - w_1)} \cdot z_2^{-(\mu_2 - w_2)} \cdot \mathbf{J}\mathbf{\Delta}(z_1^{-1}, z_2^{-1}) \qquad (2.101b)$$

$$\mathbf{E} = \mathbf{JEJ} \qquad (2.101c)$$

where μ_1 by μ_2 is the new size (in number of unit cells) and is typically equal to $(w_1 + 1)$ by $(w_2 + 1)$. While this structure might not be complete, it is quite practical for design purposes. As a simple example, let us design a polyphase matrix of the form: $\mathbf{H}_p(z_1, z_2)$:

$$
\underbrace{\begin{pmatrix} 1 & 1 & 1 & 1 \\ 1 & -1 & 1 & -1 \\ 1 & 1 & -1 & -1 \\ 1 & -1 & -1 & 1 \end{pmatrix}}_{\mathbf{W}}
\underbrace{\begin{pmatrix} 1 & 0 & 0 & 0 \\ 0 & z_1^{-1} & 0 & 0 \\ 0 & 0 & z_2^{-1} & 0 \\ 0 & 0 & 0 & z_1^{-1}z_2^{-1} \end{pmatrix}}_{\mathbf{\Delta}}
\underbrace{\begin{pmatrix} a_0 & a_2 & a_1 & a_3 \\ a_2 & a_0 & a_3 & a_1 \\ a_1 & a_3 & a_0 & a_2 \\ a_3 & a_1 & a_2 & a_0 \end{pmatrix}}_{\mathbf{E}} .
$$

$$(2.102a)$$

In the case of separable subsampling by 2 in each direction, the impulse response of the filters is:

$$
\mathbf{H}_0 = \begin{pmatrix} a_0 & a_1 & a_1 & a_0 \\ a_2 & a_3 & a_3 & a_2 \\ a_2 & a_3 & a_3 & a_2 \\ a_0 & a_1 & a_1 & a_0 \end{pmatrix}
\qquad
\mathbf{H}_1 = \begin{pmatrix} a_0 & a_1 & a_1 & a_0 \\ a_2 & a_3 & a_3 & a_2 \\ -a_2 & -a_3 & -a_3 & -a_2 \\ -a_0 & -a_1 & -a_1 & -a_0 \end{pmatrix}
$$

$$
\mathbf{H}_2 = \begin{pmatrix} a_0 & a_1 & -a_1 & -a_0 \\ a_2 & a_3 & -a_3 & -a_2 \\ a_2 & a_3 & -a_3 & -a_2 \\ a_0 & a_1 & -a_1 & -a_0 \end{pmatrix}
\qquad
\mathbf{H}_3 = \begin{pmatrix} a_0 & a_1 & -a_1 & -a_0 \\ a_2 & a_3 & -a_3 & -a_2 \\ -a_2 & -a_3 & a_3 & a_2 \\ -a_0 & -a_1 & a_1 & a_0 \end{pmatrix}
$$

Note that the filters are in general non-separable, and that the synthesis filters have the same symmetry. The matrix $\mathbf{E}$ in (2.102a) was chosen so as to obtain the double symmetry in the filters. However, $\mathbf{E}$ cannot be orthogonal, and thus, $\mathbf{H}_p(z_1, z_2)$ is not paraunitary. If we do not require the double symmetry but only linear phase, then the following matrix $\mathbf{E}$ is orthogonal:

$$
\mathbf{E} = \begin{pmatrix} a_0 & a_1 & a_2 & a_3 \\ a_1 & -a_0 & -a_3 & a_2 \\ a_2 & -a_3 & -a_0 & a_1 \\ a_3 & a_2 & a_1 & a_0 \end{pmatrix}, \quad \text{with}
$$

$$
a_2 = \sqrt{a_0^2 \frac{1 - a_0^2 - a_1^2}{a_0^2 + a_1^2}}, \qquad a_3 = -\sqrt{a_1^2 \frac{1 - a_0^2 - a_1^2}{a_0^2 + a_1^2}} .
$$

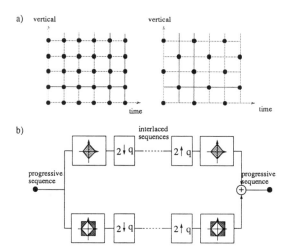

Figure 2.31: Progressive to interlaced video transformation using perfect reconstruction filter bank with quincunx subsampling: (a) progressive and interlaced grids, (b) analysis and synthesis filter banks. If the input is progressive, the channels are interlaced, while if the input is interlaced, the channels are progressive.

This leads to a paraunitary system with non-separable linear phase filters. Note that there is no separable solution, since there is no paraunitary linear phase two-channel filter bank. This shows the greater freedom of non-separable filter banks. This freedom is also demonstrated in the number of free parameters available for the design of filters, which is linked to the fact that separable filters of size L by L have $2L$ free parameters while non-separable ones have L^2. Of course, constraints like paraunitary or linear phase reduce this number. A comparison of the number of degrees of freedom is given in [23].

2.6.5   Application to 3-D Processing

The techniques developed above can be applied to 3-D signal processing as well. As an example, separable subband division is useful in low complexity image and video compression [19, 20]. Experiences with filter design for video coding can be found in a number of contributions and are mostly intraframe based methods [28, 25]. We will briefly describe a technique where quincunx subsampling is applied to go from progressive to interlaced video and back [70] which has applications in compatibility and coding of video signals. Such a system is shown in Figure 2.31.

For this case, diamond shaped lowpass and highpass filters are required, and an interesting set is given by the following elementary filters:

$$
\begin{pmatrix} & 1 & \\ b & a & b \\ & 1 & \end{pmatrix}
\qquad
\begin{pmatrix} & & 1 & & \\ & b+c/a & a & b+c/a & \\ bc/a & c & d & c & bc/a \\ & b+c/a & a & b+c/a & \\ & & 1 & & \end{pmatrix}
\qquad (2.103)
$$

which can be cascaded to obtain arbitrary size diamond shaped filters with linear phase. By choosing $b = 1$ and $c = a$ one obtains additional circular symmetry (the property is also retained when cascading the polyphase matrices). A useful example is obtained with $a = -4$ and $d = -28$ where the impulse responses become:

$$
\begin{pmatrix} & 1 & \\ 1 & -4 & 1 \\ & 1 & \end{pmatrix}
\qquad
\begin{pmatrix} & & 1 & & \\ & 2 & -4 & 2 & \\ 1 & -4 & -28 & -4 & 1 \\ & 2 & -4 & 2 & \\ & & 1 & & \end{pmatrix}.
\qquad (2.104)
$$

Other structures are given in [67] to obtain paraunitary solutions as well. Note that the filter bank thus obtained, which guarantees perfect reconstruction, can also be applied to interlaced sequences in order to get progressive channel sequences.

2.6.6 Remarks

Note that our discussion has been general in the sense that sampling rate change between arbitrary lattices was assumed. A particular case appears when the starting lattice is non-separable, like hexagonal, but the subsampling is separable. Therefore, the output will be hexagonally sampled as well. In that case, standard results for separable systems can be used, for example for aliasing cancellation (where the classic QMF solution holds), and this was shown in [46] together with a number of applications.

2.7 Conclusion

We have shown an overview of multirate signal processing as it is used in the design of subband coding systems. The analysis of such systems has been described, indicating various views (polyphase, z-transform and time domain), as well as some basic results on time-invariance (aliasing cancellation) and FIR perfect reconstruction. The synthesis of filter banks has been discussed, indicating possible structures for various types of filter banks. The multidimensional case has been described as a generalization of the one-dimensional case but with some original and partly open questions.

Acknowledgements

This work was supported in part by the National Science Foundation under grants CDR-84-21402 and MIP-88-08277. The author would like to thank Dr. D. LeGall of Bell Communications Research, Dr. G. Karlsson of IBM Research and J. Kovacevic and K. M. Uz of Columbia University for the collaborative work which led to some of the results described in this paper. The fruitful interactions with Prof. P. P. Vaidyanathan of the California Institute of Technology and Prof. M. J. T. Smith of the Georgia Institute of Technology is acknowledged. The author would also like to thank A. Lee and S. Mazumdar for helping in the preparation of the manuscript. Finally, the author would like to thank Prof. H. J. Nussbaumer for his early guidance which led the author to investigate subband coding systems.

Bibliography

[1] E. H. Adelson and E. Simoncelli, "Orthogonal pyramid transforms for image coding," *Proc. SPIE Conf. Visual Commun. Image Proc.*, vol. 845, Cambridge, MA, Oct. 1987, pp. 50-58.

[2] R. Ansari, "Two dimensional IIR filters for exact reconstruction in tree-structured sub-band decomposition," *Electronics Letters*, vol.23, pp.633-634, June 1987.

[3] V. Belevitch, *Classical Network Synthesis*, Holden Day, San Francisco, Ca., 1968.

[4] R. E. Blahut, *Fast algorithms for Digital Signal Processing*, Addison-Wesley, 1984.

[5] M. G. Bellanger, and J. L. Daguet, "TDM-FDM transmultiplexer: Digital polyphase and FFT," *IEEE Trans. on Communications*, vol. COM-22, pp. 1199-1204, Sept. 1974.

[6] P. J. Burt, E. H. Adelson, "The Laplacian pyramid as a compact image code", *IEEE Transactions on Communications*, vol. COM-31, April 1983.

[7] P. Cassereau, *A New Class of Optimal Unitary Transforms for Image Processing*, S.M. Thesis, Dept. EE and CS, Massachusetts Institute of Technology, May 1985.

[8] R. E. Crochiere, and L. R. Rabiner, *Multirate Digital Signal Processing*, Prentice-Hall, Englewood Cliffs, 1983.

[9] A. Croisier, D. Esteban, and C. Galand, "Perfect channel splitting by use of interpolation, decimation, tree decomposition techniques," *Proc. Int. Conf. on Information Sciences/Systems*, Patras, pp. 443-446, Aug. 1976.

[10] I. Daubechies, "Orthonormal bases of compactly supported wavelets," *Comm. on Pure and Applied Mathematics*, vol.XLI, pp.909-996, 1988.

[11] Z. Doğanata, P. P. Vaidyanathan, and T. Q. Nguyen, "General synthesis procedures for FIR lossless transfer matrices, for perfect reconstruction multirate filter bank applications," *IEEE Trans. on ASSP*, vol. ASSP-36, pp. 1561-1574, Oct. 1988.

[12] E. Dubois, "The sampling and reconstruction of time-varying imagery with application in video systems," *Proc. of IEEE*, vol. 73, pp. 502-522, April 1985.

[13] D. Dudgeon and R. Mersereau, *Multidimensional Signal Processing*, Prentice-Hall, 1984.

[14] D. Gabor, "Theory of communication," *J. of the IEE*, Vol.93, pp.429-457, 1946.

[15] F. R. Gantmacher, *The Theory of Matrices*, Chelsea Pub., New York, 1959.

[16] H. Gharavi and A. Tabatabai, "Sub-band coding of monochrome and color images," *IEEE Trans. on Circ. and Systems*, vol. CAS-35, pp. 207-214, February 1988, .

[17] P. Goupillaud, A. Grossman and J. Morlet, "Cycle-octave and related transforms in seismic signal analysis," *Geoexploration*, vol.23, pp.85-102, 1984.

[18] T. Kailath, *Linear Systems*, Prentice-Hall, Englewood Cliffs, 1980.

[19] G.Karlsson and M.Vetterli, "Three dimensional sub-band coding of video," *Proc. of IEEE Int. Conf. on ASSP*, April 1988, pp.1100-1103.

[20] G. Karlsson and M. Vetterli, "Sub-band coding of video for packet networks," *Optical Engineering*, vol.27, pp.574-586, July 1988.

[21] G. Karlsson, M. Vetterli, and J. Kovačević , "Non-separable two-dimensional perfect reconstruction filter banks," *Proc. SPIE Conf. Visual Commun. Image Proc.*, vol. 1001, Cambridge, MA, November 1988, pp. 187-199.

[22] G. Karlsson, *Subband Coding for Packet Video*, Ph.D. Thesis, Columbia University, New York, NY, March 1989.

[23] G. Karlsson and M. Vetterli, "Theory of two-dimensional multirate filter banks," *IEEE Trans. on ASSP*, vol. ASSP-38, p. 925-937, June 1990.

[24] J. Kovacevic, D. Le Gall and M. Vetterli, "Image coding with windowed modulated filter banks," *Proc. ICASSP-89*, pp.1949-1952.

[25] T. Kronander, *Some Aspects of Perception based Image Coding*, PhD Thesis, Dept. of EE, Linköping University, Sweden, 1989.

[26] D. J. LeGall, "Sub-band coding of images with low computational complexity," presented at *The Picture Coding Symposium*, Stockholm, Sweden, June 1987.

[27] D. LeGall and A. Tabatabai, "Subband coding of digital images using symmetric short kernels and arithmetic coding techniques," *Proc. ICASSP-88*, New York, April 1988, pp.761-764.

[28] D. LeGall, H. Gaggioni and C. T. Chen, "Transmission of HDTV signals under 140 Mbits/s using a sub-band decomposition and discrete cosine transform coding," in *Signal Processing of HDTV*, L. Chiariglione, ed., North-Holland, 1988.

[29] V. C. Liu and P. P. Vaidyanathan, "Alias cancellation and distortion elimination in multi-dimensional QMF banks," *Proc. IEEE ISCAS-88*, Espoo, Finland, June 1988, pp. 1281-1284.

[30] H. S. Malvar, *Optimal Pre- and Post-filtering in Noisy Sampled Data Systems*, Ph.D. Thesis, Dept. EE and CS, Mass. Inst. of Technology, Aug. 1986.

[31] H. S. Malvar, "The LOT: a link between block transform coding and multirate filter banks," *Proc. IEEE ISCAS-88*, Espoo, Finland, June 1988, pp. 835-838.

[32] S. Mallat, "A theory of multiresolution signal decomposition: the wavelet representation," *IEEE Trans. on PAMI*, vol.PAMI-11 , pp.674-693, July 1989.

[33] F. Mintzer, "Filters for distortion-free two-band multirate filter banks," *IEEE Trans. on ASSP*, vol. ASSP-33, pp.626-630, June 1985.

[34] F. D. Murnaghan, *The Unitary and Rotation Groups*, Washington, DC, Spartan, 1962.

[35] K. Nayebi, T. P. Barnwell and M. J. T. Smith, "Time domain conditions for exact reconstruction in analysis/synthesis systems based on maximally decimated filter banks," *Proc. Southeastern Symposium on System Theory*, March 1987, pp.498-503.

[36] H. Nawab, "Short time Fourier transform," chapter in *Advanced Topics in Digital Signal Processing*, J.S.Lim and A.V.Oppenheim Eds, Prentice-Hall, 1988.

[37] T. Q. Nguyen and P. P. Vaidyanathan, "Maximally decimated perfect-reconstruction FIR filter banks with pairwise mirror-image analysis (and synthesis) frequency responses," *IEEE Trans.on ASSP*, vol. ASSP-36, pp.693-706, May 1988.

[38] T. Q. Nguyen and P. P. Vaidyanathan, "Perfect reconstruction QMF structures which yield linear phase FIR analysis filters," *Proc. Int. Sympos. on Circ. and Systems*, Helsinki, June 1988, pp.297-300.

[39] T. Q. Nguyen and P. P. Vaidyanathan, "Two-channel perfect-reconstruction FIR QMF structures which yield linear-phase analysis and synthesis filters," *IEEE Trans.on ASSP*, vol. ASSP-37, pp.676-690, May 1989.

[40] A. V. Oppenheim, and R. W. Schafer, *Discrete Time Signal Processing*, Prentice-Hall, 1989.

[41] J. Princen and A. Bradley, "Analysis/synthesis filter bank design based on time domain aliasing cancellation," *IEEE Trans. on ASSP*, vol. ASSP-34, pp.1153-1161, Oct. 1986.

[42] J. Princen, A. Johnson and A. Bradley, "Sub-band/transform coding using filter bank designs based on time domain aliasing cancellation," *Proc. ICASSP-87*, pp.2161-2164, Dallas, April 1987.

[43] M. R. Portnoff "Time-frequency representation of digital signals and systems based on short-time Fourier analysis," *IEEE Trans. on ASSP*, vol. ASSP-28, pp.55-69, Feb. 1980.

[44] T. A. Ramstad, "Analysis/synthesis filter banks with critical sampling," *Proc. Intl. Conf. on DSP*, Florence, Sept.1984, pp. 130-134.

[45] T. A. Ramstad, "IIR filter bank for subband coding of images," *Proc. ISCAS-87*, Espoo, Finland, June 1988, pp.827-830.

[46] E. Simoncelli and E. H. Adelson, "Non-separable extensions of quadrature mirror filters to multiple dimensions," submitted for publication, 1989.

[47] M. J. T. Smith, T. P. Barnwell, "A procedure for designing exact reconstruction filter banks for tree structured sub-band coders," *Proc. IEEE ICASSP-84*, San Diego, March 1984.

[48] M. J. T. Smith, T. P. Barnwell, "A unifying framework for analysis/synthesis systems based on maximally decimated filter banks," *Proc. IEEE ICASSP-85*, Tampa, March 1985, pp. 521-524.

[49] M. J. T. Smith, T. P. Barnwell, "Exact reconstruction for tree-structured subband coders," *IEEE Trans. on ASSP*, vol. ASSP-34, pp. 434-441, June 1986.

[50] M. J. T. Smith, T. P. Barnwell, "A new filter bank theory for time-frequency representation," *IEEE Trans. on ASSP*, vol. ASSP-35, pp. 314-327, March 1987.

[51] R. P. Roesser, "A discrete state-space model for linear image processing," *IEEE Trans Auto. Control*, vol. AC-20, pp. 1-10, February 1975.

[52] P. P. Vaidyanathan, "The discrete time bounded-real lemma in digital filtering," *IEEE Trans. on Circ. and Systems*, vol. CAS-32, Sept. 1985, pp.918-924.

[53] P. P. Vaidyanathan, "Theory and design of M-channel maximally decimated quadrature mirror filters with arbitrary M, having perfect reconstruction property," *IEEE Trans. on ASSP*, vol. ASSP-35, pp.476-492, April 1987.

[54] P. P. Vaidyanathan and P.-Q. Hoang, "The perfect reconstruction QMF bank: new architectures, solutions, and optimization strategies," *Proc. IEEE ICASSP*, Dallas, pp.2169-2172, April 1987.

[55] P. P. Vaidyanathan, "Quadrature mirror filter banks, M-band extensions and perfect-reconstruction technique," *IEEE ASSP Magazine*, vol. 4, pp.4-20, July 1987.

[56] P. P. Vaidyanathan, "Perfect reconstruction QMF banks for two-dimensional applications," *IEEE Trans.on Circ. and Systems*, vol. CAS-34, pp. 976-978, Aug. 1987.

[57] P. P. Vaidyanathan and P.-Q. Hoang, "Lattice structures for optimal design and robust implementation of two-band perfect reconstruction QMF banks," *IEEE Trans. on ASSP*, vol. ASSP-36, pp.81-94, Jan. 1988.

[58] P. P. Vaidyanathan and S. K. Mitra, "Polyphase networks, block digital filtering, LPTV systems, and alias-free QMF banks: a unified approach based on pseudo-circulants," *IEEE Trans. on ASSP*, vol. ASSP-36, pp. 381-391, March 1988.

[59] P. P. Vaidyanathan, T. Q. Nguyen, Z. Doǐganata and T. Saramäki, "Improved technique for design of perfect reconstruction FIR QMF banks with lossless polyphase matrices," *IEEE Trans. on ASSP*, vol. ASSP-37, pp.1042-1056, July 1989.

[60] P. P. Vaidyanathan and Z. Doǧanata, "The role of lossless systems in modern digital signal processing," *IEEE Trans.on Education, Special issue on Circuits and Systems*, vol. 32, pp.181-197, Aug. 1989.

[61] P. P. Vaidyanathan "Multirate digital filters, filter banks, polyphase networks, and applications: a tutorial,' to appear, *Proc. IEEE*, Dec. 1989.

[62] M. Vetterli, "Multi-dimensional sub-band coding: some theory and algorithms," *Signal Processing*, vol. 6, pp. 97-112, Feb. 1984.

[63] M. Vetterli, "Splitting a signal into subsampled channels allowing perfect reconstruction," *Proc. of the IASTED Conf. on Applied Signal Processing and Digital Filtering*, Paris, June 1985.

[64] M. Vetterli, "Filter banks allowing perfect reconstruction," *Signal Processing*, vol.10, pp.219-244, April 1986.

[65] M. Vetterli, "A theory of multirate filter banks," *IEEE Trans.on ASSP*, vol. ASSP-35, pp. 356-372, March 1987.

[66] M. Vetterli and D. Le Gall, "Analysis and design of perfect reconstruction filter banks satisfying symmetry constraints," *Proc. of the Princeton Conference on Information Sciences and Systems*, March 1988, pp.670-675.

[67] M. Vetterli, "Running FIR and IIR filtering using multirate filter banks," *IEEE Trans. on ASSP*, vol. ASSP-36 pp. 730-738, May 1988.

[68] M. Vetterli and D. Le Gall, "Perfect reconstruction FIR filter banks: lapped transforms, pseudo-QMF's and paraunitary matrices," *Proc. Int. Symp. on Circ. and Systems*, Helsinki, June 1988, pp. 2249-2253.

[69] M. Vetterli and D. J. Le Gall, "Perfect reconstruction FIR filter banks: some properties and factorizations," *IEEE Trans. on ASSP*, vol. ASSP-37, pp.1057-1071, July 1989.

[70] M. Vetterli, J. Kovacevic and D. Le Gall, "Perfect reconstruction filter banks for HDTV representation and coding," *Proc. Third Intern. Workshop on HDTV*, Torino, Italy, Aug. 1989.

[71] E. Viscito and J. Allebach, "The design of tree-structured *M*-channel filter banks using perfect reconstruction filter blocks," *Proc. ICASSP-88*, New York, April 1988, pp.1475-1478.

[72] E. Viscito and J. Allebach, "Design of perfect reconstruction multidimensional filter Banks using cascaded Smith form matrices," *Proc. IEEE ISCAS-88*, Espoo, Finland, June 1988, pp. 831-834.

[73] E. Viscito and J. Allebach, "The analysis and design of multidimensional FIR perfect reconstruction filter banks for arbitrary sampling lattices," submitted for publication, 1989.

[74] P. H. Westerink, D. E. Boekee, J. Biemond, and J. W. Woods, "Subband coding of images using vector quantization," *IEEE Trans. on Commun.*, vol. COM-36, pp. 713-719, June 1988.

[75] J. W. Woods and S. D. O'Neil, "Sub-band coding of images," *IEEE Trans. on ASSP*, vol. ASSP-34, pp. 1278-1288, May 1986.

Chapter 3

IIR Analysis/Synthesis Systems

by: Mark J. T. Smith
    School of Electrical Engineering
    Georgia Institute of Technology
    Atlanta, Georgia 30332

A subband coding system may be conveniently viewed as having two constituent components: the analysis/synthesis section pair and the coding section pair. The term *analysis* describes the process of splitting the input into critically-sampled frequency related subband signals while the term *synthesis* refers to the dual operation of interpolating and merging the signals to reconstruct the input. The coding section pair, which consists of an encoder and decoder, appears between the analysis and synthesis operations and enables the input to be represented at a reduced bit rate.

Due to the dissociate nature of the constituent section pairs, these subsystems have traditionally been treated separately in the literature. In both one-dimensional (1-D) and two-dimensional (2-D) signal applications a plethora of articles may be found on theoretical issues related to band splitting and merging [5], [51], [26], [35], [7], [53], [30]. Similarly, many papers concerned with basic coding can by cited which later have been applied to coding subbands [18], [10], [15]. Unquestionably, both section pairs are critical since distortions introduced by either will contribute to the overall system error. Each of these section pairs contains a set of unique issues and problems that must be addressed effectively in order to assure high quality operation. Discussion of quantization, bit allocation and other related coding issues may be found in various sources, including [28], [27], [8], and Chapters 5 through 8 of this book. However, this chapter is devoted exclusively to the analysis/synthesis sections, with specific emphasis on systems which employ recursive filter banks.

There are many important issues to be considered in the design of practical analysis/synthesis systems for subband image coding. Conventional spectrally-contiguous filter banks for band splitting will distort the input when the channels are decimated. This distortion takes the form of aliasing, spectral amplitude distortion, and spectral phase distortion. Removing these distortions or reducing them to acceptable levels has been an active area of research in the recent past and was the focus of discussion in Chapter 2. It is well known that a broad class of tree structures composed of two-band filter bank components provides effective solutions to the filtering distortion problems. Such two-band systems may include any one of several sets of lowpass-highpass filter pairs, the most popular being *quadrature mirror filters* or QMFs. Non-tree-structured uniform multiband filter banks have also been proposed that address aliasing and spectral distortion [34], [35], [24], [51], [36]. Collectively, these structures provide many options for analysis/synthesis system configuration.

In addition to distortion-free reconstruction, computational efficiency is another important issue for consideration. Frequently the computational limitations of the system hardware impose constraints on the system design. In these situations, it is often important to have arithmetically efficient analysis/synthesis systems, which may motivate the use of recursive filter banks. Finally, the issue of conserving the minimum number of pixels to be coded must be addressed in some way. If conventional filtering is performed in the filter banks, the resulting ensemble of subband images will contain more pixels than the original image. This expansion of data is a direct result of linear convolution. As we will discuss, there are options to consider regarding the conservation of pixels and issues related to performance that are also relevant. In short, there are a host of factors which should be considered in the design of analysis/synthesis systems for practical subband image coders.

An exhaustive treatment of analysis/synthesis systems is, by no means, the intent of this chapter. Rather it is hoped to provide discussion on a very relevant subset of systems that are particularly well-suited to practical subband image coding systems. Several features characterize the subband coding system in our subset. We assume the systems of interest to contain a relatively small number of bands, typically fewer than twenty. This assumption follows the historical perspective of subband coders being viewed as wideband analysis/synthesis methods. Systems with a large number of bands could also be considered but they tend to more closely resemble transform coding approaches due to their narrowband structure and hence are not part of the mainstream discussion of this chapter. We will focus our discussion on computationally efficient systems, (in this case recursive systems), with high quality performance capabilities. In the sections that follow, we will examine some specific IIR filter banks and compare them with FIR QMF filter banks discussed in the previous chapter. In particular, we will explore issues related to filter design, efficiency of implementation and performance in a subband coding system.

3.1 Filter Banks

Two-dimensional filter banks can easily be constructed to split and merge images without distortion. However, to achieve benefits in data compression, it is often important to subsample the filter bank outputs. In subband coding the filtered images are downsampled to their respective Nyquist rates. The downsampling, which is virtually always performed by integer decimation for practical reasons, introduces unwelcome distortions due to aliasing and filtering. Reconstruction theory for 2-D filter banks demonstrates that alias-free and distortion-free solutions exist [26], [50], [7]. It is noteworthy that, of the solutions presently known, solutions based on separable filtering are the most computationally efficient and generally are the most attractive[1]. Thus we restrict our discussion to separable filters. Conveniently this enables us to treat 2-D systems in terms of 1-D filtering concepts. One popular and straightforward implementation of 2-D separable filtering consists of first filtering and decimating the rows of the image and then applying the same procedure to the columns of the resulting images. Although we will present these concepts in terms of row/column separability, the same principles can apply to filter banks which are separable in a less restrictive sense [32], [52]. In addition to the computational benefits of separable filtering, all the desirable 1-D reconstruction properties carry over directly. This is clearly true since the rows and columns are sequentially reconstructed with 1-D filter banks. The theory of 1-D reconstruction is relatively mature by comparison to the 2-D theory and is conceptually simpler. As we will see, this 1-D theory provides particularly useful solutions in this context. During our treatment of filter banks, we will discuss filter design, computationally efficient implementations, and comparisons of properties and structures.

We begin by discussing filter bank structures. Many possible subband size configurations have been considered in the past and have been shown to work well. In cases where uniform subbands are desired, the pseudo-QMF filter banks and lapped orthogonal transforms discussed in the previous chapter may be attractive, particularly if the number of subband images is large. In systems where a small number of uniform bands or non-uniform bands are desired, tree structures composed of two-band systems are generally effective. We will primarily direct our attention to the latter case.

---

[1]Separability is used here in a general sense and refers to any 2-D filter which can be implemented in terms of series of 1-D filtering operations performed along linearly independent rows of the image.

3.1.1   Tree Structured Systems

Tree structured analysis/synthesis systems are composed of cascaded two-band systems that successively split the signal into lowpass and highpass components until the desired degree of spectral resolution, or, equivalently, the targeted number of subbands is achieved.

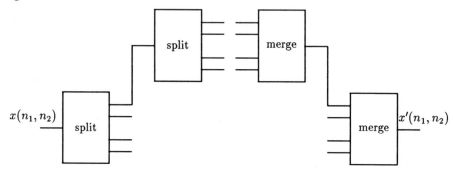

Figure 3.1: Tree-structured analysis/synthesis filter bank.

Figure 3.1 illustrates one of the many tree-structured systems which are possible. The split operator decomposes the input image into four subband images of smaller size. Typically this is done by first filtering each row with a 1-D two-band analysis section and then filtering each column with the same section. Figure 3.2 depicts the internal operations of the basic 1-D two-band analysis building block.

The "merge" operation shown in Figure 3.1 upsamples and interpolates the various lowpass and highpass inputs following the same kind of row/column processing. The outputs are then combined to form a single image. The basic 1-D synthesis block is also shown in Figure 3.2. The salient feature of the two-band block is its convenient reconstruction property. Examining Figure 3.2, we see the relationship between the input, $X(e^{j\omega})$, and the output, $X'(e^{j\omega})$, is

$$X'(e^{j\omega}) = \frac{1}{2}X(e^{j\omega})\overbrace{[H_0(e^{j\omega})G_0(e^{j\omega}) + H_1(e^{j\omega})G_1(e^{j\omega})]}^{\text{frequency response}}$$
$$+ \underbrace{\frac{1}{2}X(e^{j(\omega+\pi)})[H_0(e^{j(\omega+\pi)})G_0(e^{j\omega}) + H_1(e^{j(\omega+\pi)})G_1(e^{j\omega})]}_{\text{aliasing component}} \qquad (3.1)$$

where $H_0(e^{j\omega})$ and $H_1(e^{j\omega})$ are the lowpass and highpass analysis filters and $G_0(e^{j\omega})$ and $G_1(e^{j\omega})$ are the lowpass and highpass synthesis filters respectively.

As shown in (3.1), an aliasing component is present. We can eliminate this term by designing filters so that the aliasing component becomes zero. Equation (3.1) also shows the system frequency response, which could introduce

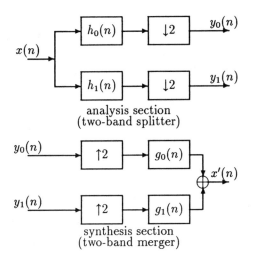

Figure 3.2: Two-band filter bank.

noticeable distortion. Consequently the filters are designed so that the system frequency response is equal to, or approximates, unity. These reconstruction properties are clearly necessary for high quality performance. In addition, we will require the analysis and synthesis filters to have the same order and same degree of arithmetic complexity. Historically, this class of filter banks has been the class of choice for practical subband coding systems. In the next section, the QMF and exactly reconstructing solutions introduced in the previous chapter are reviewed. This is followed by the introduction of IIR analysis/synthesis systems and a discussion of performance related issues.

The Quadrature Mirror Filter Solution

The QMF solution originally proposed by Croisier, Esteban, and Galand, [5] and discussed in the previous chapter is completely specified by a single lowpass filter, $h_0(n)$. The highpass analysis and synthesis filters are formed simply by modulating $h_0(n)$ by $e^{j\pi n}$ or equivalently by $(-1)^n$. In short, the set of analysis/synthesis filters are related as follows:

$$G_0(e^{j\omega}) = H_0(e^{j\omega}) \tag{3.2}$$

$$H_1(e^{j\omega}) = H_0(e^{j(\omega+\pi)}) \tag{3.3}$$

$$G_1(e^{j\omega}) = -G_0(e^{j(\omega+\pi)}). \tag{3.4}$$

The set of QMFs designed by Johnston [12] have been widely used in this context and have good reconstruction properties. These filters are FIR, even in length, and linear phase. When applied in a two-band system, they result in an overall system response with only a negligible degree of distortion.

This QMF solution has the advantage that it can be implemented in an efficient way because the coefficients of the highpass and lowpass filters are identical in magnitude but may have a different sign. This coefficient property may be formally exploited in implementation by using the well-known polyphase structure [48] shown in Figure 3.3.

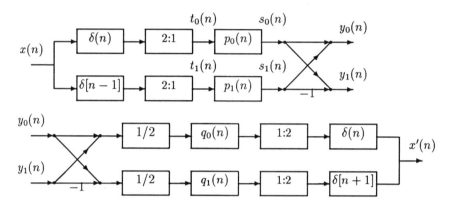

Figure 3.3: Two-band polyphase structure.

The analysis polyphase filters and synthesis polyphase filters are denoted $P_0(z)$, $P_1(z)$, $Q_0(z)$, and $Q_1(z)$ respectively and can be obtained from the impulse response of $H_0(e^{j\omega})$ by the relationships:

$$p_0(n) = h_0(2n) \tag{3.5}$$
$$p_1(n) = h_0(2n + 1) \tag{3.6}$$
$$q_0(n) = g_0(2n) \tag{3.7}$$
$$q_1(n) = g_0(2n + 1). \tag{3.8}$$

In other words, the polyphase filters are decimated lowpass filters defined by (3.5-3.8). It is possible to implement any filter pair $v_0(n)$, $v_1(n)$ in the form shown in Figure 3.3 as long as it satisfies the condition,

$$v_0(n) = (-1)^n v_1(n), \tag{3.9}$$

which is equivalent to saying that the filter pairs must be related by a frequency shift of π. Because QMFs are chosen to satisfy the condition in (3.3) and (3.4), they lend themselves to polyphase implementations and as a result are able to perform with this increased efficiency. If multiplies and adds are considered as a measure of arithmetic complexity, we find that for QMFs of length L, the polyphase implementation requires L/2 multiplies and L/2 adds per output sample.

A number of authors have proposed methods for the design of QMFs [6] [12], [46]. However it is often convenient to use one of the optimized FIR QMFs computed by Johnston. These QMFs may by found in [12] and [3] and include a variety of filters with different attenuations, transition widths, filter lengths and reconstruction errors.

Solutions composed of FIR filters with exact reconstruction are also possible where the lowpass-highpass relationship is given by

$$H_1(z) = H_0(-z^{-1})z^{-K} \tag{3.10}$$

and z^{-K} is a delay whose value is selected to make $H_1(z)$ causal. Such filters, although not amenable to polyphase realizations, can be implemented efficiently. Galand and Nussbaumer showed that an N-order analysis/synthesis filter pair can be implemented with approximately $\frac{3}{4}N$ multiplies and adds [19]. More recently, lattice structures where introduced for this application which reduce the computational demand to approximately N/2 multiplies and adds per output sample.

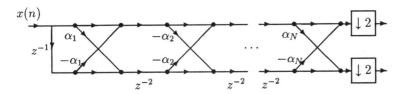

Figure 3.4: Perfect reconstruction two-band lattice structure.

The lattice structure shown in Figure 3.4 has been investigated extensively by Vaidyanathan, Hoang, and Nguyen for implementation as well as for filter design [37], [23], [46]. These FIR lattices have the property that they preserve their exact reconstruction property even when the lattice coefficients are quantized. The effects of the quantization show up as a degradation in the filter's frequency characteristics as opposed to the overall system's reconstruction characteristic. We might also point out that the filters we have discussed are restricted to be even in length. However, QMF solutions and exact reconstruction filters also exist for odd length filters but with slightly different structures [39], [46]. It should also be mentioned that the exactly reconstructing filters we alluded to have non-linear phase. The phase can be made to be approximately linear phase which is often considered preferable for coding at low bit rates. Linear phase can be imposed on the exact reconstruction filters of a given order but at the expense of relative loss in the quality of the magnitude response [49]. If we assume that all of these solutions are implemented in their most computationally efficient form, it can be shown that the complexity (in terms of multiplies and adds) is about the same for each.

The computational cost is often an important issue particularly for coding images and video sequences. As we shall see, we can achieve dramatic computational efficiency gains over these FIR systems by using IIR filters and at the same time achieve comparable reconstruction performance.

3.1.2 IIR Exact Reconstruction Filter Banks

The superior magnitude characteristics associated with IIR filters can be exploited in filter banks. In particular, the same efficient polyphase structure can be used for implementation, but with polyphase filters that are recursive. The requirement is that the filters must satisfy the condition of (3.9). This condition appears in a different form when we examine the transfer function of the recursive filter. Specifically we wish the filters to have the general form

$$H_0(z) = \frac{\sum_{\ell=0}^{J-1} d_\ell z^{-1}}{\prod_{k=1}^{M}(1 - b_k z^{-2})} = \frac{N_0(z)}{D(z^2)} \qquad (3.11)$$

where J is assumed to be even. Based on the condition of equation (3.9), the highpass filter is then given by

$$H_1(z) = H_0(-z) = \frac{N_0(-z)}{D(z^2)}. \qquad (3.12)$$

The important feature here is that the denominators of the filters are functions of z^{-2} and consequently are the same for both the highpass and lowpass filters. Applying the polyphase relationships expressed in (3.5) and (3.6) we obtain

$$P_0(z) = \frac{\sum_{\ell=0}^{J/2-1} d_{2\ell} z^{-1}}{\prod_{k=1}^{M}(1 - b_k z^{-1})} \qquad (3.13)$$

$$P_1(z) = \frac{\sum_{\ell=0}^{J/2-1} d_{2\ell+1} z^{-1}}{\prod_{k=1}^{M}(1 - b_k z^{-1})} \qquad (3.14)$$

This gives us an efficient method for implementing a pair of analysis or synthesis filters. The remaining task is to determine the synthesis filters that will result in an overall system with exact reconstruction. Exactly reconstructing synthesis filters may be obtained directly from the reconstruction theory derived in [24] where it is shown that for any given set of analysis filters, $H_0(e^{j\omega})$ and $H_1(e^{j\omega})$ the synthesis filters $G_0(z)$ and $G_1(z)$ that will reconstruct the input exactly are obtained from the equations

$$G_0(z) = \frac{H_1(-z)}{H_0(z)H_1(-z) - H_1(z)H_0(-z)} \qquad (3.15)$$

$$G_1(z) = \frac{-H_0(-z)}{H_0(z)H_1(-z) - H_1(z)H_0(-z)} \qquad (3.16)$$

and that this solution is unique.

For convenience we will express the analysis filters in terms of the polyphase filters; i.e.,

$$H_0(z) = P_0(z^2) + z^{-1}P_1(z^2) \qquad (3.17)$$
$$H_1(z) = P_0(z^2) - z^{-1}P_1(z^2) \quad . \qquad (3.18)$$

The synthesis filters are obtained by substituting these expressions into (3.15) and (3.16) resulting in

$$G_0(z) = \frac{P_0(z^2) + P_1(z^2)z^{-1}}{2P_0(z^2)P_1(z^2)z^{-1}} = \frac{1}{2}[\frac{z}{P_1(z^2)} + \frac{1}{P_0(z^2)}] \qquad (3.19)$$

and

$$G_1(z) = \frac{P_0(z^2) - P_1(z^2)z^{-1}}{2P_0(z^2)P_1(z^2)z^{-1}} = \frac{1}{2}[\frac{z}{P_1(z^2)} - \frac{1}{P_0(z^2)}] \quad . \qquad (3.20)$$

As before, the synthesis filters $G_0(z)$ and $G_1(z)$ may be implemented in the polyphase form shown in Figure 3.3, where $Q_0(z)$ and $Q_1(z)$ are the synthesis polyphase filters. Since the synthesis polyphase filters are the reciprocals of the analysis polyphase filters it is easy to see, either by examining the equations or the polyphase structure, that exact reconstruction occurs when

$$Q_0(z) = \frac{1}{P_0(z)} \qquad (3.21)$$

and

$$Q_1(z) = \frac{1}{P_1(z)}. \qquad (3.22)$$

The causality requirement which is commonly in force for speech processing and other 1-D applications renders this particular solution impractical because causality implies that the synthesis filters are most likely unstable. Fortunately, such causality constraints are not necessary for image filtering which makes this solution attractive.

With the reconstruction properties in place, the next step is to design our filters. IIR filter bank design has been considered previously by a number of researchers and thus many design methods have been reported. For example, Barnwell designed analysis filters using a modified version of the Martinez and Parks algorithm [2], [41]. Göckler based his design procedure on the Remez exchange algorithm [40]. Smith and Tracy [43] designed filters of this type by minimizing an error function. In all cases, these filters are of the form of (3.11), (i.e., with denominator polynomial a function of z^{-2}), and have magnitude characteristics significantly better than FIR QMFs requiring a comparable number of multiplies and adds. In practice, care should be exercised in implementing these filters as they are often sensitive to finite word length effects and numerical accumulation errors. It is noteworthy that these filters were not designed for implementation using this exact reconstruction solution. The filtering process contributed small amounts of phase and/or frequency distortion to the system. Thus, in general, a direct application of these IIR filter banks is not desirable in our case.

For subband image coding, we might consider another family of IIR filters to be the filters of choice in terms of having excellent magnitude characteristics, low computational complexity, and low numerical sensitivity [1]. These filters, that we will now discuss, are based on allpass polyphase filters [22], [14], [17]. By this we mean the polyphase filters $P_0(z)$ and $P_1(z)$ have the form:

$$P_0(z) = \prod_{\ell=1}^{N} \frac{z^{-1} - \alpha_\ell}{1 - \alpha_\ell z^{-1}} \qquad (3.23)$$

$$P_1(z) = \prod_{m=1}^{N} \frac{z^{-1} - \beta_m}{1 - \beta_m z^{-1}} \quad . \qquad (3.24)$$

These filters are efficient because the coefficient symmetry of the allpass difference equation enables an Nth-order filter to be realized with only N multiplies and 2N adds. Such filters may be designed by an optimization procedure as we will discuss later or by classical methods which we will now discuss.

Digital IIR filter design based on applying the bilinear transformation to classical analog filters is perhaps the most popular method for recursive filter design. Such procedures can be used to design lowpass and highpass filters,

$$H_0(z) = \frac{N_0(z)}{D(z^2)} = P_0(z^2) + z^{-1} P_1(z^2) \qquad (3.25)$$

$$H_1(z) = \frac{N_0(-z)}{D(z^2)} = P_0(z^2) - z^{-1} P_1(z^2) \qquad (3.26)$$

which can be represented in terms of the allpass polyphase filters $P_0(z)$ and $P_1(z)$ under certain conditions. Primarily, the filters $H_0(z)$ and $H_1(z)$ must be *complementary* in the following sense:

$$F_0(z) + F_1(z) = 1 \qquad (3.27)$$

where

$$F_0(z) = H_0(z)H_0(z^{-1}) = \frac{N_0(z)}{D(z^2)}\frac{N_0(z^{-1})}{D(z^{-2})} \qquad (3.28)$$

and

$$F_1(z) = H_0(-z)H_0(-z^{-1}) = \frac{N_0(-z)}{D(-z^2)}\frac{N_0(-z^{-1})}{D(-z^{-2})}. \qquad (3.29)$$

This is equivalent to saying that the autocorrelation functions of $H_0(z)$ and $H_1(z)$ must sum to unity. This condition is often called "power complementary" because the power spectrums of $H_0(z)$ and $H_1(z)$ also sum to unity. It is important to note that the filters sought here are related by a frequency shift of π; i.e.,

$$H_0(e^{j\omega}) = H_1(e^{j(\omega-\pi)})$$

and thus

$$F_0(z) = F_1(-z).$$

Odd order digital Butterworth and elliptic filters implicitly satisfy this condition if the frequency response of their autocorrelation function has symmetry about $\pi/2$. Figure 3.5 illustrates the autocorrelation symmetry in the Fourier domain for a lowpass elliptic filter.

The passband/stopband symmetry about $\pi/2$ is visible. Thus it is easy to see by symmetry that the addition of $F_0(e^{j(\omega-\pi)})$ would result in a unity frequency response.

It is well-known and can be easily shown that this frequency domain symmetry property in $F_0(e^{j\omega})$ results in the time domain property that $f_0(n) = 0$ for $n = \pm2, \pm4, \pm6, \cdots$ as illustrated in Figure 3.6.

Thus $f_0(n)$ may be conveniently written as the sum of a DC term and a function $v(n)$ which is only non-zero for odd values of n, i.e.,

$$f_0(n) = \frac{1}{2}\delta(n) + v(n)$$

and $v(2n) = 0$. This implies that

$$F_0(z) = 1/2 + V(z^2) = \frac{D(z^2) + N(z^2)}{D(z^2)} \qquad (3.30)$$

$$F_1(z) = 1/2 - V(z^2) = \frac{D(z^2) - N(z^2)}{D(z^2)} \qquad (3.31)$$

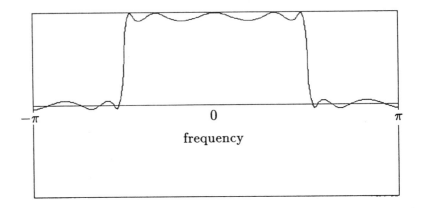

Figure 3.5: Plot of $F_0(e^{j\omega})$, the Fourier transform of the autocorrelation function.

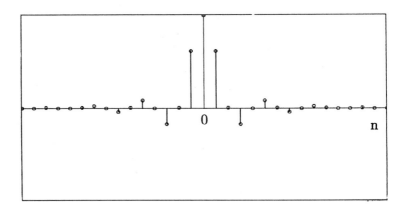

Figure 3.6: Plot of the autocorrelation function, $f_0(n)$.

which leads to three important observations. First, the denominator polynomials for $H_0(z)$ and $H_1(z)$ are identical or, equivalently, both lowpass and highpass filters have the same poles. Second, the denominator polynomial of these filters is a function of z^2. This implies that the poles of the filters can only occur in complex conjugate pairs on the $j\omega$-axis or in complex conjugate quadruples symmetric about this axis[2]. Third, in light of the relationship between $H_0(z)$ and $H_1(z)$ these filters can always be implemented with the polyphase structure of Figure 3.3. The polyphase filters may be obtained from (3.25) and (3.26) resulting in,

$$P_0(z^2) \;=\; \frac{H_0(z) + H_1(z)}{2} \tag{3.32}$$

$$P_1(z^2) \;=\; z\frac{H_0(z) - H_1(z)}{2}. \tag{3.33}$$

Given that the recursive filters $H_0(z)$ and $H_1(z)$ satisfy the condition of (3.27), and thus are expressible in polyphase form, the polyphase filters can be shown to be allpass if the numerator of $H_0(z)$ is linear phase. A proof of this property is given in Appendix A. The requirement that the filter numerator be linear phase is a very mild constraint. Moreover the resulting allpass polyphase representation is a tremendous dividend for meeting this constraint. This property can be used to greatly simplify the filter design process by enabling classical Butterworth and elliptic filters to be employed. It should first be noted that these classical filters have the property that they can be designed to satisfy (3.27), which is the first requirement. In addition, since the zeros of digital lowpass Butterworth filters always occur at $z = -1$ and the zeros of digital lowpass elliptic filters always lie on the unit circle, their numerators are always linear phase. Consequently, these classical filters can be used to design lowpass filters with the allpass polyphase structure. The procedure is simple and consists of the following steps:

1. Using a program for the design of lowpass digital Butterworth and/or elliptic filters based on the bilinear transformation, select an odd filter order.

2. Select a desired stopband deviation, δ_s.

3. Use $\delta_p = \frac{1}{2}(1 - \sqrt{1 - \delta_s^2})$ as the passband deviation.

4. Choose the passband and stopband cutoff frequencies to be symmetric frequencies about $\pi/2$.

The resulting filter will be "complementary" in the sense that (3.27) is true and will have all of its poles on the imaginary axis of the z-plane. To illustrate the

---

[2] For the odd-order symmetric filters discussed here, the poles occur on the $j\omega$-axis.

design procedure, we consider the design of a Butterworth filter, $H_B(z)$, and an elliptic filter, $H_e(z)$, using an IIR design program which maps classical analog filters into the z-plane using the bilinear transformation[3]. For the Butterworth filter, we specified a fifth order filter with passband and stopband cutoffs of $.34\pi$ and $.66\pi$ respectively and a stopband deviation of $.08$. The stopband-passband equation in the design procedure yields $.00160257$ as the passband deviation. These specifications result in the fifth-order Butterworth filter

$$H_B(z) = \frac{a_0 + a_1 z^{-1} + \cdots + a_5 z^{-5}}{1 + b_1 z^{-2} + b_2 z^{-4}} \qquad (3.34)$$

where $a_0 = a_5 = .0527868$, $a_1 = a_4 = .263934$, $a_2 = a_3 = .527868$ and $b_1 = b_3 = 0.0$, $b_2 = .633437$, $b_5 = .0557282$. Conversion of the filters into polyphase form can be done by using (3.32) and (3.33). After factoring the polynomials and canceling the roots we arrive at first-order polyphase filters of the form of (3.23) and (3.24) with $\alpha_1 = -.1055728$ and $\beta_1 = -.5278641$.

The same procedure can be applied to elliptic filters. In the case of our example, $H_e(z)$, we specified $\delta_s = .08$, $\delta_p = .00160257$, $\omega_p = .45\pi$, $\omega_s = .55\pi$ which resulted in a fifth-order elliptic filter with the same form as shown in (3.34) but with coefficients $a_0 = a_5 = .166133$, $a_1 = a_4 = .401607$, $a_2 = a_3 = .633182$ and $b_1 = b_3 = 0.0$, $b_2 = 1.13628$, $b_5 = .267258$. The polyphase filters expressed in the form of (3.23) and (3.24) are also first order with $\alpha_1 = -.332502$ and $\beta_1 = -.803778$. Alternatively $H_0(z)$ and $H_1(z)$ can be designed by optimizing the polyphase coefficients to minimize an error function. A set of recursive filters which was designed in this way specifically for subband image coding is given in Appendix B to this Chapter. We will discuss the advantages of this form of filter design in the last section of this chapter.

If we examine these filters in terms of frequency characteristics and filter order, we find that the filters based on allpass polyphase sections fare very well. For example, consider the design of a lowpass filter with transition width of $.126\pi$, maximum stopband ripple of $.0065$, and maximum overall system frequency response ripple of $.000115$ or less. A ninth order elliptic filter falls well within these tolerances. However, the FIR filters required to meet these specifications are of significantly higher order – approximately 57 in the case of FIR QMFs designed using Johnston's method [12] and 47 in the case of the exactly reconstructing optimal equiripple FIR filters [24]. This example is typical of the kind of reduction in filter order achievable by using IIR filters.

Before concluding our discussion of IIR filter banks, it is noteworthy that this allpass polyphase principle can be extended to uniform multiband filter banks. One such structure which has been considered previously [31], [36] is based on a baseband lowpass filter which is modulated to N uniformly spaced

---

[3]Programs for the design of classical IIR filters are available through the IEEE.

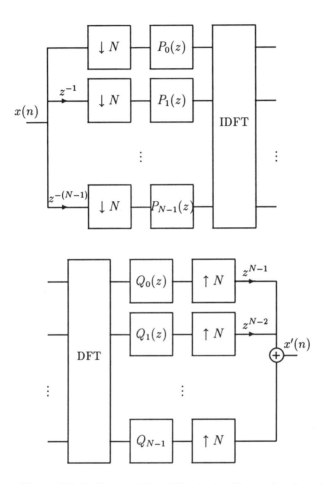

Figure 3.7: Uniform multiband filterbank with complex channels.

center frequencies by a DFT. This structure is shown in Figure 3.7 and has recently been applied to image coding by Husoy and Ramstad [29].

Several observations can be made here. First, the analysis and synthesis sections of the structure in the figure form an identity system. Specifically note that the DFT and inverse DFT are identity systems, the analysis and synthesis polyphase filters are inverses and hence yield an identity system, and the network of delays and downsample-upsample operators also form an identity system. Thus by inspection we see that the system reconstructs exactly. Second, the polyphase filters are assumed to be allpass. By proper selection of the allpass filter coefficients, a baseband lowpass filter with nominal cutoff frequency of π/N can be constructed. Thus by using allpass filters and an FFT the structure is very computationally efficient. Third, the channel outputs are complex. This means that both the real and imaginary parts of the channels must be represented in the coding. Figure 3.8 shows a plot of typical bandpass filters that are implicit in this filter bank. There are many variations of this multiband filter bank extension which can involve the use of FIR and/or IIR filters, real channel or complex channel filter banks, and exact reconstruction or non-exact reconstruction solutions.

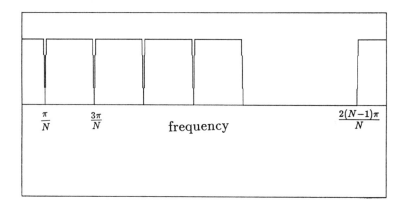

Figure 3.8: Complex channels due to the multiband filter bank.

3.2 Analysis-Synthesis Systems

Thus far various filter banks for splitting and merging of images have been considered. In these discussions, efficient implementations and minimizing the overall distortion due to filtering, decimation and interpolation were of primary concern. In this section another important issue is considered, that of achieving a minimum pixel subband image representation. Since low bit rate coding is the objective, the subband splitting operations should not increase the total number of pixels to be coded (at least not in a significant way). This would only serve to increase the bit rate. Rather the number of pixels in the original image and the total number of pixels in the subband images should be the same. This property, however, does not result naturally from the procedures discussed thus far. If we apply our filters directly to images by linearly convolving the rows and columns and decimating as discussed we end up increasing the overall number of pixels. The problem is that linear convolution of an $M \times M$ image and an $N \times N$ filter results in a larger image of size $(N + M - 1) \times (N + M - 1)$.

Thus the subband images, even though maximally decimated, have a larger aggregate number of pixels. This attribute of linear convolution is contrary to the mission of data compression and thus should be addressed in some way. In order to reduce the impact of this increase, the subband image edges can be cropped or spatially truncated so that they are the proper fractional size of the original image. However, when such truncated images are merged together in the filter bank synthesis section distortion is introduced and occurs primarily at the image boundaries. The distortion can be reduced to negligible (or unnoticeable) amounts by allowing a limited overrun, (perhaps 10 pixels or so), on the sides of the subband image. This approach is based on reaching a compromise between image size increase and reconstruction distortion.

There are solutions to this problem, however, which permit exact reconstruction and do not increase the aggregate number of pixels at all. A detailed discussion of these issues may be found in [21] and [13]. Here, however, we will only focus on two periodic extension methods. We call these methods "periodic extensions" because they are based on periodically extending the image so that all resulting subband images can be viewed as a single period of a periodic 2-D signal.

To illustrate this size enlargement issue and to put these various approaches in perspective, Table 3.1 is given. It shows examples of the total subband size for several different filter lengths and implementations. The two most undesirable approaches are implementations based on "no truncation" and implementations based on "full truncation." In the case of the former, we incur a significant percent increase in data while for the latter case we introduce significant distortion into the system. Clearly the most desirable case shown here is the one which neither expands the data nor introduces distortion – the periodic extension methods.

Image Size	System Type	QMF Filter Length	Number of Pixels to Be Coded	Percent Increase
256 x 256	4-Band	8 taps	69169	5.5%
	4-Band	16 taps	73441	12.1%
	4-Band	32 taps	82369	25.7%
	16-Band	32 taps (Level 1) 32 taps (Level 2)	121801	85.9%
	16-Band	32 taps (Level 1) 16 taps (Level 2)	100489	53.3%
512 x 512	4-Band	8 taps	269361	2.7%
	4-Band	16 taps	277729	5.9%
	4-Band	32 taps	294849	12.5%
	16-Band	32 taps (Level 1) 32 taps (Level 2)	366025	39.6%
	16-Band	32 taps (Level 1) 16 taps (Level 2)	328329	25.2%

Table 3.1:   Increase in pixels due to linear convolution

Circular Extension Method

The simplest and most straightforward approach is to periodically replicate the rows and columns of the image. Since we will be using separable filters we can once again present our treatment in terms of 1-D filtering. For a given row or column, $x(n)$, the replicated or periodic signal, $\tilde{x}(n)$, is defined as $x(n \text{ modulo } N)$ where N is the length of $x(n)$. Figure 3.9 shows a two-band analysis/synthesis system based on this idea.

Note that $\tilde{v}_0(n)$ and $\tilde{v}_1(n)$ are both periodic with period N, because the convolution of a periodic sequence with an aperiodic sequence, which in this case is the filter impulse response, is also periodic. When N is even, $\tilde{y}_0(n)$ and $\tilde{y}_1(n)$ are both periodic with period $N/2$ and so can be unambiguously represented with $N/2$ samples each. We see that the output of the analysis section consists of the two N/2-point sequences $w_0(n)$ and $w_1(n)$. This procedure is nothing more than a conventional filter bank where circular convolution is used in place of linear convolution.

In the synthesis section, a "circular" periodic extension of $w_0(n)$ and $w_1(n)$ restores $\tilde{y}_0(n)$ and $\tilde{y}_1(n)$. Therefore, the relationship between the output $\tilde{X}'(e^{j\omega})$ and the input $\tilde{X}(e^{j\omega})$ is identical to the familiar relationship for two-

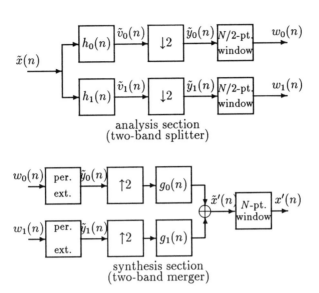

Figure 3.9: Analysis/synthesis system with periodic extension.

band QMF systems:

$$\tilde{X}'(e^{j\omega}) = \frac{1}{2}\tilde{X}(e^{j\omega})[H_0(e^{j\omega})G_0(e^{j\omega}) + H_1(e^{j\omega})G_1(e^{j\omega})] \ . \qquad (3.35)$$

The reconstruction properties are therefore determined solely by the design of the analysis and synthesis filters and are unaffected by the windowing at the analysis section outputs. Thus we have achieved our objective of no data expansion and at the same time preserved the reconstruction properties associated with our filter bank.

Symmetric Extension Method

The symmetric extension method is another solution which, in some cases, yields improved performance in the presence of coding noise. The approach we will take in introducing this method is to examine the procedure in the time domain and to use symmetry arguments. In the symmetric extension approach, the periodically extended signal $\tilde{x}(n)$ has a period $M = 2N$ (and N is assumed to be a positive even integer). The extended signal is formed by periodically replicating the symmetric extension of $\bar{x}(n)$, called $x_{se}(n)$, where

$$x_{se}(n) = \begin{cases} x(n), & 0 \leq n \leq N-1 \\ x(2N - n - 1), & N \leq n \leq 2N - 1 \\ 0, & \text{otherwise.} \end{cases}$$

This replication results in the periodic sequence,

$$\tilde{x}(n) = x_{se}(n \text{ modulo } 2N).$$

Although $\tilde{x}(n)$ is periodic with period $2N$, its symmetry allows it to be uniquely represented by only N samples. When linear phase filters are used in the analysis/synthesis system of Figure 3.9, the sequences $\tilde{v}_0(n)$ and $\tilde{v}_1(n)$ will also have symmetric structure. If N is even, the decimated sequences $\tilde{y}_0(n) = \tilde{v}_0(2n)$ and $\tilde{y}_1(n) = \tilde{v}_1(2n)$ can be represented by $N/2$ samples each, as required by the image size constraint.

This process can be conveniently illustrated in the time domain by way of the example in Figure 3.10. The input $x(n)$ shown in Figure 3.10a is symmetrically extended to form $\tilde{x}(n)$, shown in Figure 3.10b. This is the input to the two-band system in Figure 3.9. If the analysis filter $h_0(n)$ is a symmetric function, then $h_0(n)$ convolved with $\tilde{x}(n)$ is also a symmetric function – a well-known property in signal theory [9]. Similarly, if $h_1(n)$ is an anti-symmetric function, then $h_1(n)$ convolved with $\tilde{x}(n)$ is necessarily anti-symmetric. This point is illustrated in Figures 3.10c-f where Figures 3.10c and 3.10d depict the analysis filters and Figures 3.10e and 3.10f show the symmetries in the outputs from the convolution operations. These signals are then decimated to form the outputs of the analysis section. Since the outputs (Figures 3.10g and 3.10h) exhibit a symmetric structure, they may be rectangularly windowed without losing any information. The actual outputs are the $N/2$-point sequences shown in Figures 3.10i and 3.10j.

With the length constraint satisfied, the remaining problem is to reconstruct the input with near perfect accuracy. The synthesis section in Figure 3.9 computes $\tilde{y}_0(n)$ and $\tilde{y}_1(n)$ by extending $w_0(n)$ and $w_1(n)$ according to their respective symmetries. The problem is thus reduced to the conventional two-band analysis/synthesis problem where the input is just a periodic signal with even period and the analysis/synthesis filters meet the symmetry requirements. The original N-point sequence is obtained trivially by windowing the output sequence with a rectangular window.

We can also think of these periodic extension procedures in the frequency domain. The periodic replication implies circular convolution in the filtering process. Thus in the case of the circular periodic extension, filtering can be viewed as the product of the N-point DFTs of the filter and $x(n)$. In the case of the symmetric extension it is convenient to view the midpoints of the symmetrically extended rows of the image and the midpoints of the filters as being centered at zero. In this way, the symmetric extension method can be thought of in terms of a kind of 2N-point DCT where

$$X_{DCT}(k) = 2 \sum_{n=0}^{N-1} x(n) \cos\left(\frac{\pi k(n + \frac{1}{2})}{N}\right). \tag{3.36}$$

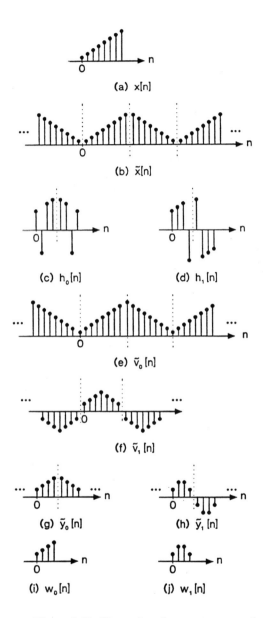

Figure 3.10: Illustration of symmetry properties.

Thus from a frequency domain perspective, the circular extension method can be modeled in terms of the product of DFT and the symmetric extension method in terms of DCTs.

Viewing size-limiting filter banks as systems operating on periodic sequences conveniently shows that all the reconstruction properties of conventional two-band analysis/synthesis systems are preserved. However, the practical implementation of the symmetric extension method does not require processing an infinite duration periodic sequence but rather processing one half period. It should be noted that two-band QMF systems based on circular extension and symmetric extension both have the same computational complexity. This is because each system produces a total of N samples at the analysis output and each system requires the same number of multiplies and adds per output sample. To illustrate the impact of these methods on the performance of a coding system, two coded images of size 256×256 are shown in Figure 3.11.

Figure 3.11: Building image of size 256×256 coded with SBC at .6 bits/pixel. a) Circular Convolution Method, b) Symmetric Extension Method.

Both images were obtained by using a 16-band uniform QMF tree-structured subband image coder. Subband images were coded with DPCM quantization and adaptive bit allocation [27] at a bit rate of .6 bits/pixel. Figure 3.11a was implemented with the circular extension method while Figure 3.11b employed the symmetric extension technique. The ringing which can be seen at the top of Figure 3.11a is due to the amplitude discontinuity across the bottom and top boundaries. This distortion is not present in Figure 3.11b, which was coded with the symmetric extension method. In addition, it is not uncommon for the symmetric extension method to yield a slight improvement in terms of peak SNR. It is important to emphasize that the performance difference between these two methods is a function of the image. For images with no appreciable boundary amplitude differences, both methods essentially yield the same subjective and objective performance quality. However, in regions where this boundary difference does exist, the distortion from using a circular extension is noticeable at low bit rates.

3.2.1 IIR Implementations

It is clear at this point how one might implement one of the analysis/synthesis systems just described by using FIR filters. However, several questions remain regarding implementation of IIR filters in this framework. In this section we discuss two kinds of exact reconstruction IIR analysis/synthesis systems that may be used for coding images. Stability is usually a major obstacle preventing exactly reconstructing recursive synthesis filters from being realized in many 1-D applications. But as stated earlier, stability restrictions that exist in causal 1-D systems are not present here; the only exception is that poles are not permitted on the unit circle. In general, the new proposed class of exact reconstruction filters may have transfer functions with regions of convergence that are either interior, exterior, or annular. For discussion purposes, we consider these filters to be composed of a minimum phase part (implying an exterior region of convergence) and a maximum phase part (implying an interior region of convergence). A filter $F(z)$ with an annular region of convergence is represented as a product of its minimum and maximum phase components:

$$F(z) = F_{\min}(z)F_{\max}(z) \tag{3.37}$$

where $F_{\min}(z)$ and $F_{\max}(z)$ are the minimum and maximum phase components respectively. These filters can be implemented by appropriately using forward and backward difference equations.

In addition to exact reconstruction and filter implementation, the issue of the image size constraint must be addressed. Specifically, we want the sum of the sizes of the subband images to equal the original image size. At first glance, IIR filters may seem inappropriate because the resulting filtered subband images will be infinite in extent. However, this problem may be resolved by applying the periodic extension concept discussed in the previous section.

We proceed by considering the filtering of the periodically-extended sequence $\tilde{x}(n)$ with an IIR filter, where the filter impulse response is not assumed to be symmetric. If the impulse response is indeed asymmetric, the symmetric extension method cannot be used. Therefore, we will develop the general case using the circular convolution method.

The periodic sequence $\tilde{x}(n)$ is formed by periodically extending the N-point sequence $x(n)$ as before:

$$\tilde{x}(n) = \sum_{\ell=-\infty}^{\infty} x(n - \ell N) \quad .$$

The immediate task is filtering this periodic sequence $\tilde{x}(n)$ with a stable causal IIR filter $F_{\min}(z)$ and calculating one period of the output, which is denoted $y(n)$. This can be done efficiently by solving analytically for the appropriate initial conditions of $y(n)$ and implementing the filter as a difference equation.

A reasonably general expression for the initial conditions can be derived for the filter $F_{\min}(z)$ with P zeros and Q poles of the form

$$F_{\min}(z) = \frac{\displaystyle\sum_{m=0}^{P} a_m z^{-m}}{1 + \displaystyle\sum_{\ell=1}^{Q} b_\ell z^{-\ell}} \quad . \tag{3.38}$$

For simplicity and convenience we will assume that $P \leq Q$. The impulse response is then of the form

$$f_{\min}(n) = \beta\delta(n) + \sum_{i=1}^{Q} C_i \alpha_i^n u(n) \quad . \tag{3.39}$$

Convolution of $\tilde{x}(n)$ and $f_{\min}(n)$ results in

$$\begin{aligned} y(n) &= \tilde{x}(n) * f_{\min}(n) \\ &= \sum_{m=-\infty}^{\infty} \tilde{x}(n - m) f_{\min}(m) \\ &= \beta\tilde{x}(n) + \sum_{m=0}^{\infty} \tilde{x}(n - m) \sum_{i=1}^{Q} C_i \alpha_i^m \quad . \end{aligned} \tag{3.40}$$

The output can be expressed in terms of the N-point sequence $x(n)$ by substituting $m = k + \ell N$, where $k = 0, 1, \cdots, N - 1$ and $\ell = 0, 1, \cdots, \infty$, resulting in

$$\begin{aligned} y(n) &= \beta\tilde{x}(n) + \sum_{k=0}^{N-1}\sum_{\ell=0}^{\infty} \tilde{x}(n - (k + \ell N)) \sum_{i=1}^{Q} C_i \alpha_i^{k+\ell N} \\ &= \beta x(n) + \sum_{k=0}^{N-1} x(n - k) \sum_{i=1}^{Q} \frac{C_i \alpha_i^k}{1 - \alpha_i^N} \quad . \end{aligned} \tag{3.41}$$

This formula may be used to compute the initial conditions $y(-1), y(-2), \ldots, y(-Q)$, thereby allowing the difference equation

$$y(n) = \sum_{m=0}^{P} a_m x(n - m) - \sum_{\ell=1}^{Q} b_\ell y(n - \ell) \tag{3.42}$$

to be used for filtering. Similarly, for a stable anti-causal filter,

$$F_{\max}(z) = \frac{\displaystyle\sum_{m=0}^{P} c_m z^m}{1 + \displaystyle\sum_{\ell=1}^{Q} d_\ell z^\ell}, \tag{3.43}$$

with impulse response

$$f_{\max}(n) = \Gamma\delta(n) + \sum_{i=1}^{Q} D_i \gamma_i^n u(-n) \tag{3.44}$$

the initial conditions can be obtained from the equation

$$y(n) = \Gamma\tilde{x}(n) + \sum_{k=0}^{N-1} \tilde{x}(n+k) \sum_{i=0}^{Q} \frac{D_i \gamma_i^{-k}}{1 - \gamma_i^{-N}}. \tag{3.45}$$

The filter may be implemented efficiently via the difference equation

$$y(n) = \sum_{m=0}^{P} c_m x(n+m) - \sum_{\ell=0}^{Q} d_\ell y(n+\ell). \tag{3.46}$$

When N is even, $y(2n)$ will be periodic with period $N/2$ and so can be represented unambiguously with $N/2$ samples. Thus analysis/synthesis systems that employ recursive filters can be used in conjunction with the circular extension method to satisfy the image size constraint.

High quality IIR analysis/synthesis systems can be designed with first-order allpass filters. The computational efficiency possible with these filters is quite remarkable. We can best illustrate this point with an example. Consider the analysis and synthesis filters composed of first-order allpass polyphase filters,

$$P_0(z) = \frac{z^{-1} - a}{1 - az^{-1}}, \quad |a| < 1 \tag{3.47}$$

$$P_1(z) = \frac{z^{-1} - b}{1 - bz^{-1}}, \quad |b| < 1. \tag{3.48}$$

The polyphase realization shown in Figure 3.3 requires two filtering operations in the analysis, which may be performed very efficiently using the difference equations

$$\tilde{s}_0(n) = a(\tilde{s}_0(n-1) - \tilde{t}_0(n)) + \tilde{t}_0(n-1) \tag{3.49}$$

$$\tilde{s}_1(n) = b(\tilde{s}_1(n-1) - \tilde{t}_1(n)) + \tilde{t}_1(n-1). \tag{3.50}$$

The initial conditions are obtained from (3.40) by evaluating the simple summation

$$\tilde{s}_0(-1) = -\frac{1}{a}\tilde{t}_0(-1) + \sum_{k=0}^{N/2-1} \frac{\tilde{t}_0(-1-k)(\frac{1}{a}-a)a^k}{1-a^N} \qquad (3.51)$$

$$\tilde{s}_1(-1) = -\frac{1}{b}\tilde{t}_1(-1) + \sum_{k=0}^{N/2-1} \frac{\tilde{t}_1(-1-k)(\frac{1}{b}-b)b^k}{1-b^N} \qquad . \qquad (3.52)$$

It appears that computing the initial conditions for each row and column adds an additional 2N multiplies to the arithmetic for each row and column that is filtered. However, upon examining the actual coefficient values (see, for example, the filter coefficients given in Appendix B to this Chapter), we see that the convergence of the summation is very rapid. Consequently, accurate initial conditions are obtainable by evaluating only a few terms in the summation, often less than 40.

3.2.2   IIR Linear Phase Analysis Filters

From the discussion in Section 3.2, we would expect quality improvement if we implemented an IIR analysis/synthesis system with symmetric extension. However, the IIR filters we discussed cannot be used in a symmetric extension implementation because the analysis and synthesis filters are required to be linear phase. In this section, we discuss an exact-reconstruction symmetric-extension system based on IIR linear phase filters. In other words, we now consider the design of suitable linear phase IIR filters.

Efficient recursive systems of this type require several constraints in the design process. As before, we restrict the filters to have only even powers of z^{-1} in the denominator, thereby satisfying the conditions for the polyphase form:

$$H_0(z) = P_0(z^2) + z^{-1}P_1(z^2). \qquad (3.53)$$

In addition, however, the proposed class of filters is constrained to have poles in complex conjugate reciprocal quadruples on the imaginary axis and zeros in complex conjugate pairs on the unit circle and at $z = -1$. Consequently, the numerator order of $H_0(z)$ is odd and the denominator order is even. Moreover, we see that both the numerator and denominator are linear phase polynomials in z. Hence, the IIR filter also has linear phase. A typical pole/zero plot is shown in Figure 3.12.

It is a simple task to design these filters. The procedure we consider here consists of minimizing a cost function (as in [12]) using an iterative optimization algorithm. The parameter set is formulated to explicitly constrain the poles to be on the imaginary axis in reciprocal complex conjugate quadruples and to restrict the zeros to conjugate pairs on the unit circle. Thus only one parameter is needed to specify a set of quadruple poles and only one parameter

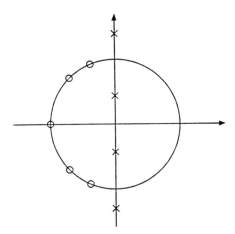

Figure 3.12: Pole/zero plot of exact reconstruction linear phase IIR filters.

for a pair of conjugate zeros. This results in minimization over a relatively small parameter space.

Because these filters are restricted to have linear phase, higher filter order is required to achieve a magnitude response comparable to that obtained when $P_0(z)$ and $P_1(z)$ were allpass. As part of our treatment of this topic we include a variety of filters in Appendix B to this Chapter. The filter coefficients and characteristics are provided for easy comparison as well as for practical use. To illustrate the filter quality obtainable, consider the linear phase IIR filter, LPIIR1 (listed in Appendix B), which has four poles and five zeros.

Figure 3.13 shows the magnitude response for this filter, an allpass polyphase filter, and a QMF filter (24D) designed by Johnston [12]. The magnitude characteristics of LPIIR1 are very close to the Johnston QMF filters, 32D and 24D, and are comparable to the allpass filter, a11. The computational efficiency achieved by LPIIR1 is not as great as that of the IIR allpass-based filter, a11. However, significant efficiency improvement is obtained over QMF filters with comparable magnitude characteristics.

Recursive linear phase filters may be implemented in several ways. A convenient representation for implementation is as a parallel combination of minimum and maximum phase filters. For example, recognize that the filter LPIIR1 has the form

$$\frac{b_0 + b_1 z^{-1} + b_2 z^{-2} + b_2 z^{-3} + b_1 z^{-4} + b_0 z^{-5}}{a_0 + a_1 z^{-2} + a_0 z^{-4}} \ . \tag{3.54}$$

The polyphase filters $P_0(z)$ and $P_1(z)$ may then be implemented in parallel

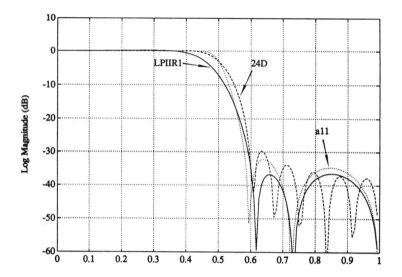

Figure 3.13: Magnitude response plots for the allpass polyphase lowpass filter, (a11), the recursive linear phase IIR filter, (LPIIR1), and the Johnston QMF filter, (24D).

form:

$$P_0(z) = \frac{b_0 + b_2 z^{-1} + b_1 z^{-2}}{a_0 + a_1 z^{-1} + a_0 z^{-2}}$$

$$= A_1 + \frac{A_2}{1 - \alpha z^{-1}} + \frac{A_3}{1 - \frac{1}{\alpha} z^{-1}} \qquad (3.55)$$

$$P_1(z) = \frac{b_1 + b_2 z^{-1} + b_0 z^{-2}}{a_0 + a_1 z^{-1} + a_0 z^{-2}}$$

$$= B_1 + \frac{B_2}{1 - \beta z^{-1}} + \frac{B_3}{1 - \frac{1}{\beta} z^{-1}} \qquad (3.56)$$

where one term in the parallel expansion is a constant, one term is minimum phase, and one term is maximum phase. These polyphase filters $P_0(z)$ and $P_1(z)$ have an annular region of convergence and therefore may be implemented with difference equations having forward and backward terms.

A subband analysis system can be implemented using the LPIIR1 filter with approximately 5.1 multiplies and 5.6 adds per output sample. The savings in multiplies over the 32D analysis system is a factor of 3.1; the savings in total operations is a factor of 2.5.

3.3 Comparisons

We have considered several types of filter banks and several ways of applying these filter banks to form an analysis/synthesis system. In this section we try to draw some conclusions with respect to the obvious questions: how do the various properties of the filter bank and analysis/synthesis system impact the overall performance of the subband image coder; which analysis/synthesis system is best in a given sense and what are the tradeoffs; and what are the performance issues when we attempt to operate at low bit rates.

Different features and properties of the analysis/synthesis system can be associated with different types of distortion that appear in the system output at low bit rates. Thus it is useful to understand these relationships. It is virtually impossible to address all of these questions and issues in a precise way due to the subjective nature of measuring quality and the dependence of the performance on the input. We can, however, make some general statements that can serve as guidelines.

We begin by examining coding quality differences among the various methods we discussed for handling the image size constraint. At medium to high bit rates (1.5 bits/pixel and higher) there is generally negligible difference in terms of performance. Differences occur when the bit rates are lowered. Noticeable distortions can often be observed in the vicinity of .5 to 1 bits/pixel and below. The distortions occur because the aliasing cancellation and frequency response properties break down due to the rather severe quantization that takes place between analysis and synthesis operations. Some methods of analysis/synthesis are more sensitive than others with respect to the breakdown in the reconstruction properties.

With the circular extension method, lowpass and highpass filtering is performed across the opposite boundaries of the picture, which can lead to degradation at the image boundaries. If the left sequence boundary has values at the low end of the dynamic range while the right sequence boundary has values at the high end, for example, signal energy is "smeared" across the boundaries. In addition, the response of the highpass filter to the "step edge" from one boundary to the other can be quite sharp. Consequently, accurate coding of this information is difficult at low bit rates and results in ringing near the edge boundaries. Low bit-rate quantizers generally attempt to exploit the slow amplitude variations of the pixels to improve the accuracy in selecting the quantization step size. Regions of sharp discontinuity in pixel intensity often represent the worst case condition for coding. The symmetric extension method attempts to reduce the occurrence of these sharp step edges by inhibiting their occurrence at the picture boundaries. (ref. Figure 3.11)

Karlsson and Vetterli [13] have investigated other extensions which avoid these discontinuities and have reported that they too perform well. The important point is that methods which tend to avoid discontinuities at the boundaries seem to perform better.

Partially truncating the decimated subbands was another approach to the size expansion issue which we discussed in Section 3.2. This approach does not artificially introduce boundary discontinuities either. However, it does have the disadvantage of increasing the number of pixels which must be coded as shown in Table 3.1. For systems where the ratio of image size to the effective filter length is very large, the increase in number of pixels is relatively small. In such cases this truncation approach has also been shown to work well in practice.

The next issue we consider is the performance difference between linear phase and non-linear phase filters. In particular, we consider the difference between FIR QMFs, which are linear phase, and IIR filter banks based on all-pass polyphase filters, which have non-linear phase. These may be considered to represent the two extremes in terms of phase. In general subjective preference is given to the FIR QMFs at low bit rates. The distortion often takes the form of pixel intensity oscillations (or ringing) in the vicinity of image edges. This distortion can be seen clearly in Figure 3.14, particularly in 3.14b and 3.14d.

There are several ways we can view the source of this distortion. Here we will examine the impact of the filters in terms of their step response. Bear in mind that at low bit rates the fine structure in the high frequency bands is often lost or, at best, severely distorted. Thus a step edge being received by the synthesis filter in the lowpass channel produces the step response of the lowpass synthesis filter at the output. The lowpass step response inherently contains ripples, commonly associated with the Gibbs phenomenon. At higher bit rates, the fine grain structure in the highpass channel would also produce ripples which would tend to cancel the ripples in the lowpass synthesis channel resulting in a smooth image edge. The ringing distortion that we observe in Figure 3.14 is directly related to the step response of the synthesis filters. If we reduce the ripple amplitudes in the synthesis filter step response, we also reduce the amplitude of the ripples in the coding image.

Recursive filters designed using the methods we discussed in the last section tend to have step responses with pronounced overshoots followed by decaying ripples that converge to unity as shown in Figure 3.15a. The rate of ripple decay is related to the proximity of the poles to the unit circle. By contrast, the step response of the FIR QMFs tend to have ripples with somewhat smaller amplitude but distributed on both sides of the step transition as shown in Figure 3.15b. These smaller ripple amplitudes tend to somewhat mollify the subjectively displeasing effects we see in the output.

In light of the Gibbs phenomenon, which implies the necessity of step response ripples in order to achieve good magnitude response characteristics,

Figure 3.14: Examples of original and coded 256×256 images at 0.7 bits/pixel using various systems: a) original, b) circular convolution method with FIR QMF 32D, c) symmetric extension method with FIR QMF 32D, d) circular convolution method using IIR allpass polyphase filter a11, e) symmetric extension method using recursive linear phase filter LPIIR1.

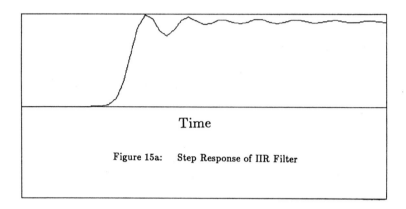

Time

Figure 15a: Step Response of IIR Filter

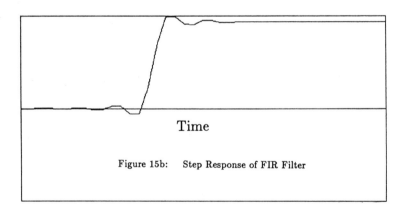

Time

Figure 15b: Step Response of FIR Filter

Figure 3.15: Illustration of unit step response for a) IIR filters, b) FIR filters.

we see that our design objectives are diametrically opposed. In other words, it is impossible to achieve small step response ripples and good lowpass frequency characteristics in a single filter. Nonetheless, filters can be designed to achieve some desired balance between frequency domain and step response characteristics.

Such an approach was undertaken by Jeanrenaud and Smith, [44], in which three parameters were used for the time domain characterization: **os**, the value of the maximum overshoot; **us**, the magnitude of the maximum undershoot; and **Rp**, the sum of the absolute values of the ripples before the undershoot and after the overshoot. Frequency domain properties where characterized by the stopband and passband error energy. A set of filters based on allpass polyphase sections where designed by minimizing a weighted step-response-frequency-response error function. Several observations in this regard are worthy of mention. First, as the transition widths of the recursive filters are reduced, the aliasing effects diminish but, simultaneously, the poles move closer to the unit circle forcing the step response effects to be more dominant. Thus one seems to pay a price for filters with very sharp transition cutoffs when operating at low bit rates. Second, giving strong emphasis to the reduction of the step response ripples leads to a substantial broadening of the transition width. For low bit rate operation, this translates into aliasing. In terms of quality, this often leads to loss of edge definition and can result in a roughness along normally smooth edges of the image. Thus for very low bit rate operation, subjective performance can often be improved by properly selecting a suitable compromise between ringing and aliasing distortion. Third, the impact of the filter bank on the overall system performance clearly depends on the way in which the coding is performed. This suggests that changing the bit allocation strategy will also change the impact of the ringing distortion. This could make some other tradeoff between the step response and magnitude response characteristics more desirable. Finally, some authors have reported successful systems using short length FIR filters [47]. These tend to have small amplitude step response characteristics. Consequently they do not display the ringing distortion one observes in using narrow transition width FIR and IIR filters. They do, however, contain aliasing distortion.

We can gain some intuition regarding the impact of our various analysis/synthesis systems on the performance of a coding system by examining Figure 3.14. These coded images were produced by a uniform 16-band subband image coders for 256×256 images. The coding systems, in each case, employ DPCM coders with the dynamic bit allocation scheme as described in [27] at the overall rate of 0.7 bits/pixel.

Symmetric extension systems with IIR and FIR filters have comparable subjective quality, as seen in Figure 3.14c and 3.14e, although the SNR is slightly higher for the FIR QMF case. In comparing FIR QMFs with IIR allpass-based filters using the circular convolution method, the QMF system displays higher quality at bit rates below 1.0 bits/pixel. Less ringing is observed at the image boundaries and in other regions with sharp discontinuities in pixel intensity. We attribute the higher amplitude ringing observed when some of the IIR allpass polyphase filter are used to the higher step response overshoot of the filter. This ringing distortion can be reduced by using filters with low step response ripples (see Appendix B to this Chapter).

Filter	Adds per output point	Multiplies per output point	Transition width (radians)	Passband ripple	Stopband ripple
8-tap QMF (8A)	4	4	2.50	.0033	.026
12-tap QMF (12B)	6	6	1.25	.0023	.020
16-tap QMF (16B)	8	8	1.15	.0001	.006
24-tap QMF (24D)	12	12	0.51	.0041	.033
32-tap QMF (32D)	16	16	0.44	.0017	.014
5 zero 4 pole (LPIIR1)	≈ 5.6	≈ 5.1	0.69	.0241	.015
Allpass Polyphase (a11)	≈ 3.1	≈ 1.1	0.52	.0003	.024

Table 3.2:   Comparison of FIR and IIR filters in terms of computational complexity and filter characteristics.

The most dramatic difference between these approaches is in terms of computational complexity. Table 3.2 compares the computational complexities of the different filtering schemes. The IIR allpass-based filters are by far the most efficient. Consider the coded images shown in Figure 3.14b and 3.14d. The subjective quality difference between these two is not dramatic. The difference in peak SNR is also very close: 29.95 dB for the FIR system and 29.98 dB for the IIR system. The advantage of the IIR system is its lower computational complexity. The IIR analysis system requires approximately 1 multiply and 3 adds per output sample, while the 32-tap FIR analysis system requires 16 multiplies and 16 adds per output sample. The savings in number of multiplies is a factor of 16; the savings in total operations is a factor of 8 in the case of this example. In practice, however, people now typically use FIR filters of length 16 without incurring noticeable subjective distortion. Nonetheless, the computational benefits of IIR filters are still substantial.

Appendix A: Derivation of the Allpass Polyphase Property

Consider the IIR lowpass and highpass filters of the form

$$H_0(z) = \frac{N_0(z)}{D(z^2)} \qquad (3.57)$$

and

$$H_0(-z) = \frac{N_0(-z)}{D(z^2)} \qquad (3.58)$$

respectively, that satisfy the condition

$$H_0(z)H_0(z^{-1}) + H_1(z)H_1(z^{-1}) = 1. \qquad (3.59)$$

The numerator polynomial, $N_0(z)$, is assumed to have an even length, M, while the denominator polynomial, $D(z^2)$, is assumed to have length $M-1$. If $N_0(z)$ has linear phase then the filters are expressible in terms of allpass polyphase components and thus can be implemented with tremendous efficiency.

To see that an allpass polyphase structure is guaranteed, observe that expressing (3.59) in terms of $N_0(z)$ and $D(z)$ results in

$$N_0(z)N_0(z^{-1}) + N_0(-z)N_0(-z^{-1}) = D(z^2)D(z^{-2}) \quad . \qquad (3.60)$$

The key to this derivation is to impose the linear phase constraint that $N_0(z)$ be a symmetric polynomial of length M which results in the condition

$$N_0(z) = z^{-(M-1)}N_0(z^{-1}) \quad . \qquad (3.61)$$

This in turn implies that $N_0(-z)$ is an anti-symmetric polynomial of the form

$$N_0(-z) = -z^{-(M-1)}N_0(-z^{-1}) \quad . \qquad (3.62)$$

With these conditions, (3.60) may be written as

$$z^{M-1}N_0(z)N_0(z) - z^{M-1}N_0(-z)N_0(-z) = D(z^2)D(z^{-2}) \qquad (3.63)$$

which may be factored to yield

$$\overbrace{[N_0(z) + N_0(-z)]}^{B(z^2)}\overbrace{[N_0(z) - N_0(-z)]z^{M-1}}^{B(z^{-2})} = D(z^2)D(z^{-2}) \quad . \qquad (3.64)$$

Notice that the left side of the equation has been factored into two expressions $B(z^2)$ and $B(z^{-2})$ and that these expressions are time reversed versions of each other. The implicit time reversal can be seen by observing that

$$B(z^{-2}) = [N_0(z) - N_0(-z)]z^{M-1} = [N_0(z^{-1}) + N_0(-z^{-1})] \qquad (3.65)$$

which comes from (3.61) and (3.62). Time reverse relationship between these terms implies that each root of $B(z)$ will have a reciprocal root in $B(z^{-1})$. Clearly the roots of $B(z)B(z^{-1})$ are identical to the roots of $D(z)D(z^{-1})$ since these terms are equal. However, $B(z)$ and $D(z)$ are not equal in general.

Without loss of generality, we can express the denominator polynomials in terms of their roots,

$$D(z^2) = A \prod_{\ell=1}^{(M-2)/2} (1 - \alpha_\ell z^{-2}) \tag{3.66}$$

$$D(z^{-2}) = A \prod_{\ell=1}^{(M-2)/2} (1 - \alpha_\ell z^2) \tag{3.67}$$

where A is the constant gain term. Similarly the polynomials $B(z^2)$ and $B(z^{-2})$ can be written as a function of their roots,

$$B(z^2) = \prod_{\ell=1}^{m} (1 - \alpha_\ell z^{-2}) \prod_{\ell=m+1}^{(M-2)/2} (z^{-2} - \alpha_\ell) \tag{3.68}$$

$$B(z^{-2}) = \prod_{\ell=1}^{m} (1 - \alpha_\ell z^2) \prod_{\ell=m+1}^{(M-2)/2} (z^2 - \alpha_\ell) . \tag{3.69}$$

The polyphase structure is seen by expressing the polyphase filters in terms of these roots. Owing to (3.17) and (3.18) the polyphase filters $P_0(z^2)$ and $P_1(z^2)$ can be expressed in terms of $H_0(z)$ and $H_1(z)$ and also in terms of $N_0(z)$ and $D(z)$ (from (3.57) and (3.58)) to yield

$$P_0(z^2) = \frac{H_0(z) + H_1(z)}{2} = \frac{N_0(z) + N_0(-z)}{2D(z^2)} \tag{3.70}$$

$$P_1(z^2) = z\frac{H_0(z) - H_1(z)}{2} = z\frac{N_0(z) - N_0(-z)}{2D(z^2)}. \tag{3.71}$$

Using (3.66 – 3.71), we can substitute and express the polyphase filters in terms of allpass sections:

$$P_0(z^2) = \frac{1}{2A} \prod_{\ell=m+1}^{(M-2)/2} \frac{z^{-2} - \alpha_\ell}{1 - \alpha_\ell z^{-2}} \tag{3.72}$$

$$P_1(z^2) = \frac{1}{2A} \prod_{\ell=1}^{m} \frac{z^{-2} - \alpha_\ell}{1 - \alpha_\ell z^{-2}}. \tag{3.73}$$

Thus we see that lowpass/highpass filters which satisfy these modest conditions may be implemented using allpass polyphase filter and consequently can achieve great efficiency gains.

filter name	poles of $P_0(z)$	poles of $P_1(z)$	stopband attenuation	transition width	overshoot undershoot	Rp
a10	-0.4		-33 dB	1.50	12%	0.28
b10	-0.465		-25 dB	1.00	12%	0.34
c10	-0.5		-22 dB	0.81	13%	0.38
d10	-0.58		-17 dB	0.52	12%	0.46
a11	-0.141279	-0.589817	-60 dB	1.47	18%	0.47
b11	-0.178561	-0.645045	-45 dB	1.01	19%	0.53
c11	-0.201240	-0.675826	-39 dB	0.80	19%	0.57
d11	-0.253721	-0.736267	-32 dB	0.52	19%	0.64
e11	-0.315235	-0.796538	-26 dB	0.32	19%	0.71
f11	-0.369129	-0.839389	-22 dB	0.22	18%	0.76
g11	-0.397013	-0.869436	-18 dB	0.12	18%	0.78
a21	-0.112867 -0.749637	-0.381317	-55 dB	1.45	21%	0.70
b21	-0.113363 -0.750005	-0.381055	-53 dB	1.00	21%	0.70
c21	-0.103923 -0.752617	-0.372313	-58 dB	0.83	21%	0.70
d21	-0.141226 -0.812351	-0.461468	-45 dB	0.47	20%	0.77
f21	-0.147158 -0.864082	-0.517388	-28 dB	0.22	18%	0.83
g21	-0.233335 -0.919366	-0.652248	-24 dB	0.12	17%	1.03
e22	-0.123120 -0.669745	-0.397433 -0.892960	-51 dB	0.32	21%	1.11
f22	-0.116383 -0.683303	-0.392720 -0.905316	-39 dB	0.22	21%	1.10
g22	-0.101035 -0.720961	-0.391873 -0.930456	-28 dB	0.12	20%	1.07
h22	-0.026014 0.059748	0.147565 -0.567475	-39 dB	1.05	9% 6%	0.33

Table 3.3: Allpass polyphase filter coefficients and characteristics

Appendix B: Set of Recursive Filters

The filters given here in Table 3.3 were designed by a gradient search algorithm which minimized a weighted time-domain-frequency-domain error function [44]. The time domain characterization was composed of three parameters: **os**, the value of the maximum step response overshoot; **us**, the magnitude of the maximum step response undershoot; and **Rp**, the sum of the absolute values of the step response ripples before the undershoot and after the overshoot. The frequency domain characteristics were represented by the squared error in the stopband and passband regions.

The filters are labeled to show the orders of the respective allpass polyphase component filters. For example, filter a11 is a recursive lowpass filter where $P_0(z)$ and $P_1(z)$ are both first order allpass polyphase filters. Note that the first four filters only contain one allpass filter. The filter $P_1(z)$ is just a direct connection. These represent the lowest order filters of this type. Filters composed of second order allpass filters are also included. They represent the highest order filters included here of this type.

The filters in Table 3.3 are specified in terms of their polyphase roots; i.e. they are expressed in the form of (3.23) and (3.24). In addition, the stopband attenuation in dB, the transition width, the maximum step response overshoot and undershoot, and (**Rp**) the sum of the absolute step response ripple values are given with each filter. Table 3.4 provides coefficient for recursive linear phase filters. These filters have the form of (3.55) and (3.56) with coefficients A_1, A_2, A_3, B_1, B_2, B_3, α, and β.

filter name	coef. of $P_0(z)$ A_1, A_2, A_3, α	coef. of $P_1(z)$ B_1, B_2, B_3, β	stopband attenuation	transition width	overshoot undershoot	Rp
LPIIR0	2.55085 -1.02187 -0.52898 .711746	1.0 .528984 1.02187 .711746	29 dB	0.39	10.4 %	0.62
LPIIR1	3.046848 -1.37448 -0.672367 .64611	1.0 .672367 1.37448 .64611	37 dB	0.69	9.8 %	0.52
LPIIR2	3.59776 -1.773942 -0.82381 .588897	1.0 0.82381 1.773942 .588897	46 dB	1.06	9.0 %	0.46

Table 3.4: Recursive linear phase filter coefficients and characteristics

Acknowledgments

The author wishes to thank Steven Eddins and Philippe Jeanrenaud for their research contributions, coding simulations, and help in preparing the tables and figures. The author is also very grateful for the many informative discussions with Drs. P. P. Vaidyanathan and T. Ramstad on this topic.

Bibliography

[1] R. Ansari and B. Lui, "A class of low-noise computationally efficient recursive digital filters with applications to sampling rate alterations," *IEEE Trans. on Acoustics, Speech, and Signal Process.*, vol. ASSP-33, pp. 90-97, 1985.

[2] T. P. Barnwell III, "Sub-band coder design incorporating recursive quadrature filters and optimum ADPCM coders," *IEEE Trans. Acoust., Speech, and Signal Process.*, vol. ASSP-30, pp. 751-765, October 1982.

[3] R. E. Crochiere and L. R.Rabiner, *Multirate Digital Signal Processing*, Prentice Hall, Englewood Cliffs, NJ, 1983, Ch. 7.7, pp. 376-395.

[4] R. E.Crochiere, S. A.Webber, J. L.Flanagan, "Digital coding of speech in sub-bands," *Bell Syst. Tech. J.*, vol 55, pp. 1069-1085, October 1976.

[5] A. Croisier, D. Esteban, C. Galand, "Perfect channel splitting by use of interpolation, decimation and tree decomposition techniques," *Proc. of Int. Conf. on Information-Sciences/Systems*, Patras, Greece, August 1976, pp. 443-446.

[6] D. Esteban and C. Galand, "Application of quadrature mirror filters to split band voice coding schemes," *Proc. Int Conf. on Acoustics, Speech, and Sig. Process.*, May 1977, pp. 191-195.

[7] M. J. T. Smith and S. L. Eddins, "Analysis/synthesis techniques for subband image coding" *IEEE Trans. on Acoustics, Speech, and Signal Process.*, August 1990.

[8] H. Gharavi, A. Tabatabai " Sub-band coding of monochrome and color images," *IEEE Trans. Acoust., Speech, Signal Process.*, vol. ASSP-35, pp. 207-214, February 1988.

[9] A. V. Oppenheim and R. W. Schafer, *Digital Signal Processing*, Prentice Hall, Englewood Cliffs, New Jersey, 1975, Ch. 5.5, pp. 239-241.

[10] R.M. Gray, "Vector quantization," IEEE, *ASSP Magazine*, vol.1, pp. 4-29, April 1984.

[11] P. Jeanrenaud, *Subband Coding of Images with Recursive Allpass Filters, using Vector Quantization*, Master's thesis, Georgia Institute of Technology, Atlanta, November 1988

[12] J.D. Johnston, "A filter family designed for use in quadrature mirror filter banks," *Proc.Int. Conf. Acoust. Speech and Signal Process.*, April 1980, pp. 291-294.

[13] G. Karlsson, and M. Vetterli, "Extension of finite length signals for subband coding," *Signal Processing*, no. 17,pp. 161-168, 1989.

[14] T. Kronander, "A new approach to recursive mirror filters with a special application in subband coding of images," *IEEE Trans. on Acoustics, Speech, and Signal Process.*," vol. 36, pp. 1496-1500, September 1988.

[15] V.J. Mathews, R.W. Waite, T.D. Tran, "Image compression using vector quantization of linear (one-step) prediction errors" *Proc. Int. Conf. Acoust. Speech Signal Process.* Dallas, Texas, April 1987, pp. 733-736.

[16] T.A.Ramstad and O.Foss, "Sub-band coder design using recursive quadrature mirror filters," *Signal Processing: Theories and Applications*, 1980, pp. 747-752.

[17] T.A.Ramstad, "IIR filterbank for subband coding of images," *Proc. of Int. Symposium on Cir. and Systems*, June, 1988, Espoo, Finland, pp. 827-834.

[18] M.R. Schroeder and B.S. Atal, "Code-excited linear prediction (CELP): High quality speech at very low bit rates" *Proc. Int. Conf. Acoust. Speech Signal Process.*, Tampa, Florida, March 1985, pp. 937-940.

[19] C. R. Galand and H. J. Nussbaumer, "Quadrature mirror filters with perfect reconstruction and reduced computational complexity," *Proc. Int. Conf. Acoust. Speech Signal Process.*, April 1985, pp. 525-528.

[20] M.J.T. Smith, T.P.Barnwell III, "Exact reconstruction techniques for tree structured subband coders," *IEEE Trans. Acoust., Speech, and Signal Process.*, vol. ASSP-34, pp. 434-441, June 1986.

[21] M.J.T. Smith, S.L.Eddins, "Sub-band coding of images with octave band tree structures," *Proc. Int. Conf. Acoust. Speech and Signal Process.* Dallas, Texas, April 6-9, pp. 1382-1385.

[22] M.J.T. Smith, R.M. Mersereau, and T. P. Barnwell "Exact reconstruction recursive filter banks for sub-band image coding," *Proc. of IEEE Miami Technicon 87*, Miami, Florida, October 1987, pp. 121-124.

[23] P.P. Vaidyanathan, "Quadrature mirror filter banks, M-band extensions and perfect reconstruction techniques," *ASSP Magazine*, vol.4, pp. 4-20, July 1987.

[24] M. Smith and T. Barnwell, "A new filter bank theory for time-frequency representation," *IEEE Trans. on Acoustics, Speech, and Signal Process.*, vol. 35, pp. 314-327, March 1987.

[25] P.P. Vaidyanathan, S. K. Mitra, and Y. Neuvo, "A new approach to the realization of low-sensitivity IIR digital filters," *IEEE Trans. on Acoustics, Speech, and Signal Process.*, vol. ASSP-34, pp. 350-361, April 1986.

[26] M. Vetterli, "Multidimensional sub-band coding: some theory and algorithms," *Signal Processing*, vol. 6, pp. 97-112, April 84.

[27] P. H. Westerink, J. Biemond, and D. Boeckee, "An optimal bit allocation algorithm for sub-band coding" *Proc. Int. Conf. Acoust. Speech Signal Process.*, New York, April 1988, pp. 757-760.

[28] J.W. Woods and S.D. O'Neill, "Sub-band coding of images," *IEEE Trans. Acoust., Speech, and Signal Process.*, vol. ASSP-34, pp.1278-1288, October 1986.

[29] J. H. Husoy, T. A. Ramstad, "Application of an efficient parallel IIR filterbank to image subband coding," submitted to *Signal Processing*.

[30] A. Fettweis, J. A. Nossek, and K. Meerkotter, "Reconstruction of signals after filtering and sampling-rate reduction," *Proc. Int. Conf. Acoustics, Speech, and Signal Process.*, San Diego, CA, March 1984, pp. 11.7.1-11.7.4.

[31] A. Constantinides and R. Valenzuela, "An efficient and modular transmultiplexer design," *IEEE Trans. on Communications*, vol. COM-30, pp. 1629-1641, July 1982.

[32] R. Ansari and S. H. Lee, "Two-dimensional multirate processing on non-rectangular grids: Theory and filtering procedures," to appear in *IEEE Trans. on Circuits and Systems*, 1989.

[33] M. Vetterli, "A theory of multirate filter banks," *IEEE Trans. on Acoustics, Speech, and Signal Process.*, vol. ASSP-35, pp. 356-372, March 1987.

[34] J. H. Rothweiler, "Polyphase quadrature filters, a new sub-band coding technique," *Proc. Int. Conf on Acoust., Speech, and Signal Process.*, Boston MA, April 1983.

[35] P. L. Chu, "Quadrature mirror filter design for an arbitrary number of equal bandwidth channels," *IEEE Trans. on Acoust. Speech, and Sig. Process.*, vol. 33, pp. 203-328, February, 1985.

[36] K. Swaminathan and P.P. Vaidyanathan, "Theory and design of uniform DFT, parallel, quadrature mirror filter banks," *IEEE Trans. on Circuits and Systems*, vol. CAS-33, pp. 1170-1191, December 1986.

[37] P.P. Vaidyanathan and P. Q. Hoang, "Lattice structures for optimal design and robust implementation of two-channel prefect-reconstruction QMF banks," *IEEE Trans. on Acoustics, Speech, and Sig. Process.*, vol. ASSP-36, pp. 81-94, January 1988.

[38] F. Mintzer, "Filters for distortion free two-band multirate filter banks," *IEEE Trans. on Acoustics, Speech and Sig. Process.*, vol. ASSP-33, pp. 626-630, June 1985.

[39] C. Galand and H. Nussbaumer, "New quadrature mirror filter design in the time domain," *IEEE Trans. on Acoustics, Speech and Sig. Process.*, Vol. ASSP-32, pp. 522-531, June 1984.

[40] H. Göckler, "Design of recursive polyphase networks with optimum magnitude and minimum phase," *Signal Processing 3*, 1981, pp. 365-376, North-Holland Publishing Company.

[41] H. Martinez and T. Parks, "Design of recursive digital filters with optimum magnitude and attenuation poles on the unit circle," *IEEE Trans. on Acoustics, Speech and Signal Process.*, vol. ASSP-36, April 1978.

[42] M. Smith and T. Barnwell, "A unifying framework for maximally decimated analysis/synthesis systems," *Proc. Int. Conf. on Acoustics, Speech, and Signal Process.*, March 1985, pp. 521-524.

[43] M. Smith and K. Tracy, "Multi-dimensional frequency domain coding", *Proceedings of IEEE Miami Technicon*, October 1987, pp. 70-73.

[44] P. Jeanrenaud and M. Smith, "Recursive subband image coding using adaptive prediction and finite state vector quantization," *Signal Processing*, May 1990.

[45] M. Smith and T. Barnwell, "A procedure for designing exact reconstruction filter banks for tree-structured subband coders," *Proc. Int. Conf. on Acoustics, Speech, and Signal Process.*, March 1984, pp. 27.1.1-27.1.4.

[46] T. Nguyen and P.P. Vaidyanathan, "Manual: PRQMFs Two Channel PR FIR QMF Bank Design Package," Dept. of Electrical Engineering, California Institute of Technology, Pasadena, CA 91125.

[47] D. Le Gall and A. Tabatabai, "Subband coding of digital images using symmetric short kernel filters and arithmetic coding techniques", *Proc. Int. Conf. on Acoustics, Speech, and Signal Process.*, April 1988, New York, pp. 761-764.

[48] M. Bellanger, G. Bonnerot and M. Coudreuse, "Digital filtering by polyphase network: Application to sample rate alteration and filter banks," *IEEE Trans. on Acoust. Speech and Signal Process.*, vol. 24, pp. 109-114, April 1976.

[49] T. Nguyen and P.P. Vaidyanathan, "Two-channel perfect reconstruction FIR QMF structures which yield linear phase FIR analysis and synthesis filters," *IEEE Trans. on Acoustics, Speech and Signal Process.*, vol. ASSP-37, pp. 676-690, May 1989.

[50] P. P. Vaidyanathan, "Perfect reconstruction QMF banks for two dimensional applications", *IEEE Trans. on Circuits and Systems*, vol. 34, pp. 976-978, August 1987

[51] P. P. Vaidyanathan, "Theory and design of M-channel maximally decimated quadrature mirror filters with arbitrary M, having the perfect reconstruction property," *IEEE Trans. on Acoustics, Speech and Signal Process.*, vol. ASSP-35, pp. 476-492, April 1987.

[52] R. H. Bamberger and M. Smith, " Filter banks for the directional decomposition of images: Theory and design," submitted to *IEEE Trans. on Acoustics, Speech, and Signal Process.*, December, 1989.

[53] P. P. Vaidyanathan, "Multirate digital filters, filter banks, polyphase networks, and applications: A tutorial," *Proc. of the IEEE*, December 1989.

Chapter 4

Subband Transforms

by: Eero P. Simoncelli[†] and Edward H. Adelson[‡]
Vision Science Group, The Media Laboratory, and
[†]Department of Electrical Engineering and Computer Science
[‡]Department of Brain and Cognitive Science
Massachusetts Institute of Technology
Cambridge, Massachusetts 02139

Linear transforms are the basis for many techniques used in image processing, image analysis, and image coding. Subband transforms are a subclass of linear transforms which offer useful properties for these applications. In this chapter, we discuss a variety of subband decompositions and illustrate their use in image coding. Traditionally, coders based on linear transforms are divided into two categories: transform coders and subband coders. This distinction is due in part to the nature of the computational methods used for the two types of representation.

Transform coding techniques are usually based on *orthogonal* linear transforms. The classic example of such a transform is the discrete Fourier transform (DFT), which decomposes a signal into sinusoidal frequency components. Two other examples are the discrete cosine transform (DCT) and the Karhunen-Loeve transform (KLT). Conceptually, these transforms are computed by taking the inner product of the finite-length signal with a set of basis functions. This produces a set of coefficients, which are then passed on to the

This work was supported by contracts with the IBM Corporation (agreement dated 1/1/84) and DARPA (Rome Airforce F30602-89-C-0022) and a grant from the NSF (IRI 871-939-4). The opinions expressed are those of the authors and do not necessarily represent those of the sponsors.

quantization stage of the coder. In practice, many of these transforms have efficient implementations as cascades of "butterfly" computations. Furthermore, these transforms are usually applied independently to non-overlapping sub-blocks of the signal.

Subband transforms are generally computed by convolving the input signal with a set of bandpass filters and decimating the results. Each decimated subband signal encodes a particular portion of the frequency spectrum, corresponding to information occurring at a particular spatial scale. To reconstruct the signal, the subband signals are upsampled, filtered, and then combined additively. For purposes of coding, subband transforms can be used to control the relative amounts of error in different parts of the frequency spectrum. Most filter designs for subband coders attempt to minimize the "aliasing" resulting from the subsampling process. In the spatial domain, this aliasing appears as evidence of the sampling structure in the output image. An ideal subband system incorporates "brick-wall" bandpass filters which avoid aliasing altogether. Such filters, however, produce ringing (Gibbs phenomenon) in the spatial domain which is perceptually undesirable.

Although coders are usually classified in one of these two categories, there is a significant amount of overlap between the two. In fact, the latter part of this chapter will focus on transforms which may be classified under either category. As an example, consider the block discrete cosine transform (DCT), in which the signal (image) is divided into non-overlapping blocks, and each block is decomposed into sinusoidal functions. Several of these sinusoidal functions are depicted in Figure 4.1. The basis functions are orthogonal, since the DCT is orthogonal and the blocks are chosen so that they do not overlap. Coders employing the block DCT are typically classified as transform coders.

We may also view the block DCT as a subband transform. Computing a DCT on non-overlapping blocks is equivalent to convolving the image with each of the block DCT basis functions and then subsampling by a factor equal to the block spacing. The Fourier transform of the basis functions (also shown in Figure 4.1) indicates that each of the DCT functions is selective for a particular frequency subband, although it is clear that the subband localization is rather poor. Thus, the DCT also qualifies as a subband transform.

4.1 Subband Transform Properties

Given the overlap between the categories of transform and subband coders, what criteria should be used in choosing a linear transformation for coding purposes? We will consider a set of properties which are relevant to the problem

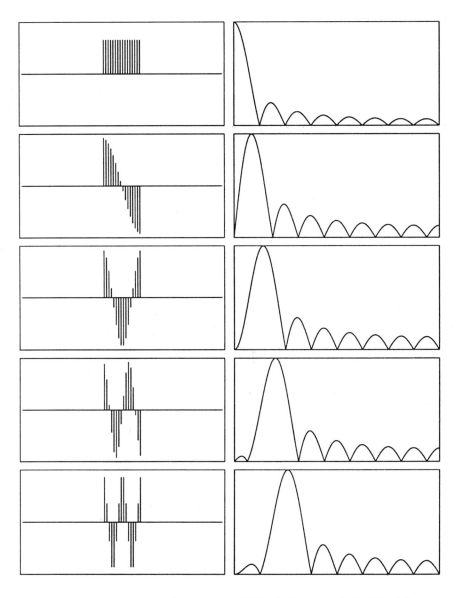

Figure 4.1: Several of the 16-point DCT basis functions (left) with their corresponding Fourier transforms (right). The Fourier transforms are plotted on a linear scale over the range from 0 to π.

of image coding.

Scale and Orientation

An explicit representation of scale is widely accepted as being important for effective image representation [1, 2, 3, 4, 5, 6]. Images contain objects and features of many different sizes which may be viewed over a large range of distances, and therefore, a transformation should analyze the image simultaneously (and independently) at different scales. Several authors have argued that the correct partition in terms of scale is one in which the scales are related by a fixed constant of proportionality. In the frequency domain, this corresponds to a decomposition into localized subbands with equal widths on a logarithmic scale.

For two-dimensional signals, a localized region in the frequency plane corresponds spatially to a particular scale and *orientation*. Orientation specificity allows the transform to extract higher order oriented structures typically found in images, such as edges and lines. Thus, it is useful to construct transformations which partition the input signal into localized patches in the frequency domain.

Spatial localization

In addition to localization in frequency, it is advantageous for the basis functions to be spatially localized; that is, the transform should encode positional information. The necessity of spatial localization is particularly apparent in machine vision systems, where information about the location of features in the image is critical. This localization should not, however, occur abruptly as in the block DCT example given earlier – abrupt transitions lead to poor localization in the frequency domain.

The concept of joint localization in the spatial and spatial-frequency domains may be contrasted with the two most common representations used for the analysis of linear systems: the sampled or *discrete* signal, and its Fourier transform. The first of these utilizes the standard basis set for discrete signals consisting of impulses located at each sample location. These basis functions are maximally localized in space, but convey no information about scale. On the other hand, the Fourier basis set is composed of even and odd phase sinusoidal sequences, whose usefulness is primarily due to the fact that they are the eigenfunctions of the class of linear shift-invariant systems. Although they are maximally localized in the frequency domain, each one covers the entire spatial extent of the signal.

It is clear that representation in the space or frequency domains is extremely useful for purposes of system analysis, but this does *not* imply that impulses or sinusoids are the best way to encode signal information. In a number of recent papers [7, 8, 9], the importance of this issue is addressed and related to a 1946 paper by Dennis Gabor [10], who showed that the class of linear transformations may be considered to span a range of joint localization with the impulse basis set and the Fourier basis set at the two extremes. He demonstrated that one-dimensional signals can be represented in terms of basis functions which are localized both in space *and* frequency. We will return to Gabor's basis set in Section 4.3.

Orthogonality

A final property to be considered is orthogonality. The justification usually given for the orthogonality constraint is in terms of decorrelation. Given a signal with prescribed second order statistics (i.e. a covariance matrix), there is an orthogonal transform (the Karhunen-Loeve transform) which will decorrelate the signal (i.e. diagonalize the covariance matrix). In other words, the second order correlations of the transform coefficients will be zero. Orthogonality is usually not discussed in the context of subband transforms, although many such transforms are orthogonal. The examples in the next section will demonstrate that although orthogonality is not strictly necessary, a transform that is strongly non-orthogonal may be undesirable for coding.

4.2 Linear Transformations on Finite Images

The results presented in this chapter are based on analysis in both the spatial and the frequency domains, and thus rely on two separate notational frameworks: the standard matrix notation used in linear algebra, and the Fourier domain representations commonly used in digital signal processing. In this Section, we describe the two types of notation and make explicit the connection between them. For simplicity, we will restrict the discussion to analysis of one-dimensional systems, although the notation may be easily extended to multiple dimensions.

4.2.1 Analysis/Synthesis Filter Bank Formulation

We will be interested in linear transformations on images of a *finite* size which may be expressed in terms of convolutions with finite impulse response (FIR) filters. The schematic diagram in Figure 4.2 depicts a convolution-based system known as an analysis/synthesis (A/S) filter bank [11]. The notation in the diagram is standard for digital signal processing [12], except that for the purposes of this paper, the boxes $\boxed{H_i(\omega)}$ indicate *circular* convolution of a finite input image of size N with a filter with impulse response $h_i(n)$ and Fourier transform

$$H_i(\omega) = \sum_n h_i(n)e^{-j\omega n}$$

We do not place a causality constraint on the filter impulse responses, since they are meant for application to images. We do, however, assume that the region of support of the filter is smaller than the image size. The boxes $\boxed{k_i \downarrow}$ indicate that the sequence is subsampled by a factor of k_i where k_i is an integer for all i. The boxes $\boxed{k_i \uparrow}$ indicate that the sequence should be upsampled by inserting $k_i - 1$ zeros between each sample. We will assume that the integers k_i are divisors of N.

The analysis section of the A/S system takes an input sequence $x(n)$ of length N and performs a linear transformation to decompose it into M sequences $y_i(n)$ of length N/k_i. The synthesis section performs the inverse operation of the analysis transformation. Here the M sequences $y_i(n)$ are upsampled and, after filtering with filters $g_i(n)$, are combined additively to give an approximation $\hat{x}(n)$ to the original sequence, $x(n)$. Note that although one-dimensional signals are indicated in the diagram, the system is equally valid for multidimensional signals if we replace occurences of the scalars n, ω, k_i with vectors $\boldsymbol{n}, \boldsymbol{\omega}$, and a matrix $\mathbf{K}_i$, respectively.

The use of the A/S formulation emphasizes the computation of the transform coefficients through convolution. This is intuitively desirable since different regions of the image should be processed in the same manner. Furthermore, the expression of the problem in the frequency domain allows us to easily separate the error $e(n) = \hat{x}(n) - x(n)$ into two parts: an aliasing component and a shift-invariant component. To see this, we write the contents of the intermediate signals $y_i(n)$ in the frequency domain as

$$Y_i(\omega) = \frac{1}{k} \sum_{j=0}^{k-1} H_i\left(\frac{\omega}{k} + \frac{2\pi j}{k}\right) X\left(\frac{\omega}{k} + \frac{2\pi j}{k}\right) \tag{4.1}$$

and the A/S system output as

$$\hat{X}(\omega) = \sum_{i=0}^{M-1} Y_i(k\omega)G_i(\omega)$$

where we have used well-known facts about the effects of upsampling and downsampling in the frequency domain [12]. Combining the two gives

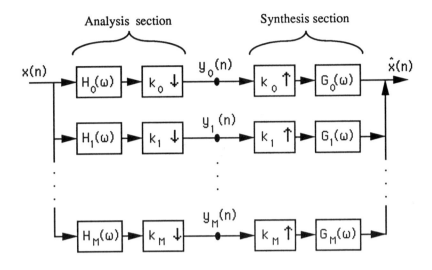

Figure 4.2: An analysis/synthesis filter bank.

$$\hat{X}(\omega) = \frac{1}{k} \sum_{i=0}^{M-1} \left[\sum_{j=0}^{k-1} H_i \left(\omega + \frac{2\pi j}{k} \right) X \left(\omega + \frac{2\pi j}{k} \right) \right] G_i(\omega)$$

$$= \frac{1}{k} \sum_{i=0}^{M-1} H_i(\omega) G_i(\omega) X(\omega)$$

$$+ \frac{1}{k} \sum_{j=1}^{k-1} X \left(\omega + \frac{2\pi j}{k} \right) \sum_{i=0}^{M-1} H_i \left(\omega + \frac{2\pi j}{k} \right) G_i(\omega) . \qquad (4.2)$$

The first sum corresponds to a linear shift-invariant system response, and the second contains the system aliasing.

4.2.2 Cascaded Systems

A further advantage of the A/S system is that it allows explicit depiction and analysis of hierarchically constructed transformations. If we assume that we are dealing with A/S systems with perfect response (that is, $\hat{x}(n) = x(n)$), then any intermediate signal $y_i(n)$ of an A/S system may be further decomposed by application of any other A/S system. To make this notion more precise, an example is given in the diagram of Figure 4.3 in which an A/S system has been re-applied to its own intermediate signal $y_0(n)$. If the original A/S system (as shown in Figure 4.2) had a perfect response then it is clear that the two-stage system shown in Figure 4.3 will also have a perfect response. If the cascading is applied to each of the M intermediate signals $y_i(n)$, we will call the system a *uniform* cascade system. Otherwise, it will be termed a *non-uniform* or *pyramid* cascade. A system which we will discuss in Section 4.3 is based on pyramid cascades of two-band A/S systems. Such a cascade produces an octave-width subband decomposition, as illustrated in the idealized frequency diagram in Figure 4.4.

4.2.3   Matrix Formulation

An alternative to the frequency domain notation associated with the A/S filter bank is the matrix notation of linear algebra. An image of finite extent which has been sampled on a discrete lattice may be written as a finite length column vector x which corresponds to a point in $\mathbf{R}^N$, the set of all real N-tuples. The value of each component of x is simply the corresponding sample value in the image. Multi-dimensional images are converted to this vector format by imposing an arbitrary but fixed order on the lattice positions. If we let N be the length of the vector x, a linear transformation on the image corresponds to multiplication of x by some matrix $\mathbf{M}$ with N columns.

Since the analysis and synthesis stages of the system in Figure 4.2 each correspond to linear transformations, we may represent the same transformations using matrix notation. Using the definition of convolution, and assuming (for simplicity) a one-dimensional system, we may write

$$y_i(m) \;=\; \sum_{l=0}^{N-1} x(l) h_i(k_i m - l)$$

and

$$\hat{x}(n) \;=\; \sum_{i=0}^{M-1} \sum_{m=0}^{\frac{N}{k_i}-1} y_i(m) g_i(n - k_i m)$$

where the filter and image sample locations $(k_i m - l)$ and $(n - k_i m)$ are computed modulo N. These expressions may be formulated as matrix-vector products

$$y \;=\; \mathbf{H}^t x$$

and

$$\hat{x} \;=\; \mathbf{G} y$$

or combining these two equations

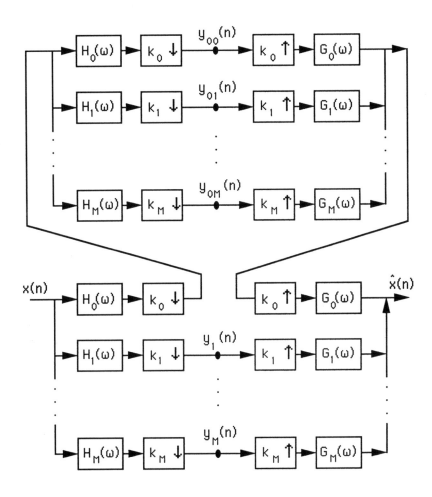

Figure 4.3: A non-uniformly cascaded analysis/synthesis filter bank.

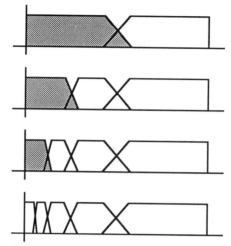

Figure 4.4: Octave band splitting produced by a four-level pyramid cascade of a two-band A/S system. The top picture represents the splitting of the two-band A/S system. Each successive picture shows the effect of re-applying the system to the lowpass subband (indicated in grey) of the previous picture. The bottom picture gives the final four-level partition of the frequency domain. All frequency axes cover the range from 0 to π.

$$\hat{x} \quad = \quad \mathbf{G}\mathbf{H}^t x \tag{4.3}$$

where y and $\hat{x}$ are N-vectors, the superscript t indicates matrix transposition, and

$$
\mathbf{H} \;=\; \left[
\begin{array}{cccccc}
h_0(0) & h_0(k_0) & & h_1(0) & h_1(k_1) & \\
h_0(-1) & h_0(k_0-1) & & h_1(-1) & h_1(k_1-1) & \\
h_0(-2) & h_0(k_0-2) & & h_1(-2) & h_1(k_1-2) & \\
\vdots & h_0(k_0-3) & \cdots & \vdots & h_1(k_1-3) & \cdots \\
& h_0(k_0-4) & & & h_1(k_1-4) & \\
h_0(2) & \vdots & & h_1(2) & \vdots & \\
h_0(1) & & & h_1(1) & &
\end{array}
\right] \tag{4.4}
$$

and

$$
\mathbf{G} \;=\; \left[
\begin{array}{cccccc}
g_0(0) & g_0(k_0) & & g_1(0) & g_1(k_1) & \\
g_0(1) & g_0(k_0+1) & & g_1(1) & g_1(k_1+1) & \\
g_0(2) & g_0(k_0+2) & & g_1(2) & g_1(k_1+2) & \\
\vdots & g_0(k_0+3) & \cdots & \vdots & g_1(k_1+3) & \cdots \\
& g_0(k_0+4) & & & g_1(k_1+4) & \\
g_0(-2) & \vdots & & g_1(-2) & \vdots & \\
g_0(-1) & & & g_1(-1) & &
\end{array}
\right] \tag{4.5}
$$

The columns of $\mathbf{G}$, composed of copies of the filter kernels shifted by increments of k_i and imbedded in vectors of length N, are known as the *basis* functions of the transformation, and the columns of $\mathbf{H}$, composed of copies of the *time-inverted* filters $h_i(-n)$ shifted by increments of k_i, are the *sampling* functions of the transformation.

From the discussion above, it is clear that we can express any linear A/S system in matrix form. The converse of this result is also true: there is an A/S system corresponding to the linear transformation and inverse transformation defined by any invertible matrix $\mathbf{M}$. Given a transformation matrix $\mathbf{M}$ with l rows, we trivially create an analysis filter bank with $k_i = N$ for each i, containing l different filters, each defined by a row of the matrix $\mathbf{M}$.

4.2.4 Inverse Transforms

A primary advantage of the matrix notation is the ease with which it can express the conditions for transform invertibility. From (4.3), we see that in order for the A/S system to perfectly reconstruct the original signal $x(n)$, the corresponding matrices must obey

$$\mathbf{G}\mathbf{H}^t = \mathbf{I} \tag{4.6}$$

where $\mathbf{I}$ is the identity matrix. If $\mathbf{H}$ has rank N and is square, we may choose a synthesis matrix

$$\mathbf{G} = (\mathbf{H}^{-1})^t \tag{4.7}$$

which will also be square with rank N. Thus, transform inversion in the spatial domain is a conceptually simple procedure and we will find it useful in the analysis of A/S systems. Furthermore, it should be clear that $\mathbf{H}$ and $\mathbf{G}$ may be interchanged, thus using the basis functions as sampling functions and vice versa.

If the matrix $\mathbf{H}$ is of rank N but is *not* square (that is, the representation is *overcomplete*), we may always build a perfect reconstruction system by choosing $\mathbf{G}$ to be the generalized inverse or pseudo-inverse [13] of $\mathbf{H}$:

$$\mathbf{G} = (\mathbf{H}\mathbf{H}^t)^{-1}\mathbf{H} \ . \tag{4.8}$$

If $\mathbf{H}$ is square, (4.8) reduces to the solution given in (4.7). Similarly, if we start with a (possibly non-square) matrix $\mathbf{G}$ of rank N, we may choose $\mathbf{H} = (\mathbf{G}\mathbf{G}^t)^{-1}\mathbf{G}$.

4.2.5   Orthogonal Transforms

As mentioned in the Introduction, the issue of orthogonality is usually not considered when discussing subband filters. It is, however, a property which is relevant to image coding, as we will discuss in the next section. A matrix $\mathbf{M}$ corresponding to an orthogonal transformation is a square matrix with the property that

$$\mathbf{MM}^t = \mathbf{M}^t\mathbf{M} = \mathbf{I} \ . \tag{4.9}$$

In terms of the columns or basis functions of M, this means that the inner product of any two distinct columns must be zero, and the inner product of a column with itself must be unity.

The orthogonality condition places a number of restrictions on the corresponding A/S system. Since the transformation matrix must be square, the number of samples in the transformed signal must be equal to N, the number of samples in the original image. For the A/S system, this means that

$$\sum_{i=0}^{M-1} \frac{1}{k_i} = 1$$

where we have assumed that N is divisible by all of the k_i. Such a system has been termed a *maximally decimated* or *critically sampled* filter bank [11].

A second, more important constraint is placed on the A/S system by orthogonality. Combining the perfect reconstruction requirement in (4.6) with the orthogonality constraint in (4.9) gives

$$\mathbf{G} = \mathbf{H} \ .$$

If we consider the relationships between the A/S filters h and g and the matrices H and G described by equations (4.4) and (4.5), this means that the filters must obey

$$g_i(n) = h_i(-n), \qquad \text{for all } i \ . \tag{4.10}$$

In other words, the synthesis filters of an orthogonal transform are *time-inverted* versions of the analysis filters.

4.3   Some Example Transforms

In this Section, we will briefly discuss three one-dimensional (1-D) transforms to illustrate some of the points made in the previous sections. Each transform will demonstrate both advantageous and disadvantageous properties for coding.

The Gabor Transform

In the introduction to this chapter, we argued that the basis functions of a useful decomposition should be localized in both the spatial and the spatial-frequency domains. One solution to the problem of spatially localized subband decomposition is that proposed by Dennis Gabor [10]. Gabor introduced a 1-D transform in which the basis functions are sinusoids weighted by Gaussian windows. The Gabor transform can be considered to perform a localized frequency decomposition in a set of overlapping windows. The resulting basis functions are localized in both space and spatial frequency; in fact, Gabor showed that this joint localization was optimal with respect to a measure that he chose (although Lerner [14] later noted that altering the measure of joint localization produces different optimal functions). The first five basis functions of a Gabor transform are shown in Figure 4.5, along with their frequency spectra. Both the basis functions and their transforms are smooth and compact. In two dimensions, the Gabor basis functions are directional sinusoids weighted by gaussian windows. Daugman [15, 16] has used two-dimensional (2-D) Gabor transforms for image compression.

The primary difficulty with the Gabor transform is that it is strongly non-orthogonal (i.e. the sampling functions are drastically different from the basis functions). The sampling functions corresponding to the Gabor transform (computed by inverting the transformation matrix) are depicted in Figure 4.6. These functions are extremely poorly behaved, both in the spatial and spatial-frequency domains. In a coding application, errors introduced by quantization of the coefficients will be distributed throughout the spatial and frequency domains, even though the coefficient values are computed based on information in localized spatial and frequency regions.

It is interesting to note that the localization of the inverse Gabor functions can be substantially improved if one uses an overcomplete Gabor basis set. This can be accomplished by spacing the Gaussian windows more closely than is required, or by dividing each window into more frequency bands. This results in an increase in the number of coefficients, however, which may be disadvantageous for coding systems. The use of overcomplete Gabor sets for coding remains a topic for further research.

Several authors have discussed related overcomplete oriented transforms for use in image coding. Kunt [17] advocated the use of directional (i.e. orientation) subdivision for image coding, and used an oriented decomposition for this purpose. Watson [18] developed the Cortex transform, an overcomplete transform which decomposes the image into oriented octave-bandwidth subbands, and used it to compress image data.

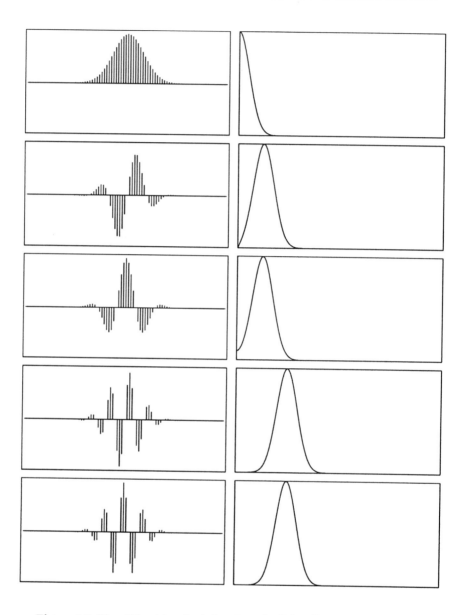

Figure 4.5: Five of the sixteen basis functions of a Gabor filter set, with their corresponding Fourier transforms. The transforms are plotted on a linear scale over the range from 0 to π.

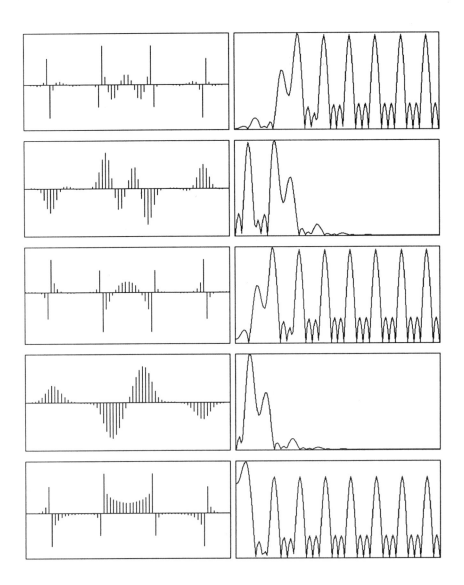

Figure 4.6: The inverse (sampling) functions of the Gabor filter set given in Figure 4.5

The DCT and LOT Transforms

The use of the discrete cosine transform (DCT) in image coding systems is often justified with the statement that it approximates the optimal transform for a signal with first-order Gauss-Markov statistics [19]. In practice, the transform is usually not computed globally, but is applied independently to non-overlapping sub-blocks of the image. As illustrated in Figure 4.1, the resulting block DCT basis functions constitute a subband transform, but the subbands are not very well localized. Considered in the framework of the A/S system, the subsampled subband images will contain severe amounts of aliasing. Since the transform is invertible (in fact, orthogonal), it should be clear that this subband aliasing is cancelled in the synthesis stage. However, if the transform coefficients are quantized or discarded (e.g. in a coding system), the aliasing no longer cancels, and the errors appear as block edge artifacts in the reconstructed image.

Recent work by Cassereau et. al. [20] describes an elegant technique for reducing the aliasing of the block DCT. They perform an orthogonal transformation on the block DCT coefficients which combines coefficients computed from adjacent blocks. In the resulting transform, which they have called a *Lapped Orthogonal Transform* (LOT), the basis functions from adjacent blocks overlap each other, and their impulse responses are tapered at the edges. Malvar [21] has implemented an efficient version of this transform, known as the fast LOT, in which the additional orthogonal transformation is computed using a butter-fly network of simple rotation transformations. Several of the even-symmetric basis functions of the fast LOT are shown in Figure 4.7. One limitation which applies to both the DCT and the LOT is that the transforms are limited to equal-sized subbands. As discussed previously, it may be advantageous to subdivide the spectrum into equal log-width subbands.

The Laplacian Pyramid

One of the first techniques for octave subband decomposition was developed by Burt [22] and applied to image coding by Burt and Adelson [23]. They used a pyramid cascade of small Gaussian-like filters to create an overcomplete sub-band representation which they called a *Laplacian pyramid*. A system for constructing one level of this pyramid (in one dimension) is illustrated in Figure 4.8. The signal is blurred with a lowpass filter, $B(\omega)$, and then subsampled to produce a lowpass subband $W_0(\omega)$. A highpass subband, $W_1(\omega)$, is formed by upsampling $W_0(\omega)$, convolving with an interpolation filter $A(\omega)$, and subtracting from the original signal. The signal is reconstructed by upsampling and filtering $W_0(\omega)$ with $A(\omega)$ and adding it to $W_1(\omega)$. This reconstruction is

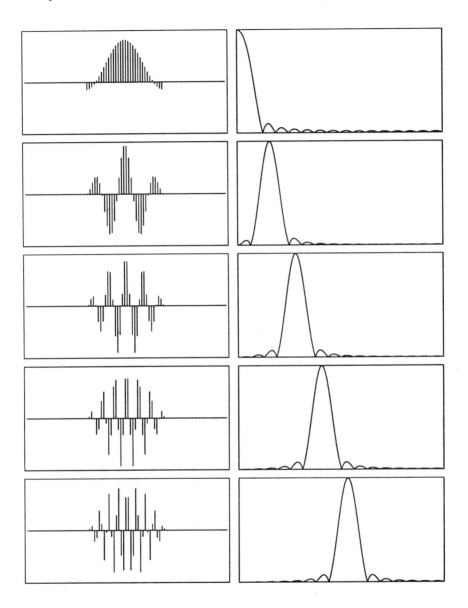

Figure 4.7: Five of the eight even-symmetric basis functions of a LOT. The basis functions are illustrated on the left, and their Fourier transforms on the right. The transforms are plotted on linear axes and cover the range from 0 to π.

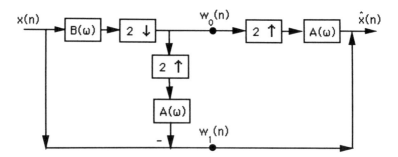

Figure 4.8: Signal processing diagram depicting the standard construction technique for one level of the Laplacian pyramid. A full pyramid is built by non-uniformly cascading this system. This transformation may also be described as an A/S filter bank (see text).

exact, regardless of the choice of the filters $B(\omega)$ and $A(\omega)$. The full pyramid is constructed recursively, by re-applying the system to the lowpass subband. Typically, the filters $A(\omega)$ and $B(\omega)$ are set to some common, compact lowpass filter, although better coding results are obtained by choosing the two filters independently. Some example basis and sampling functions (with $A(\omega) = B(\omega)$) are plotted in Figures 4.9 and 4.10, respectively.

In addition to its suitability for data compression, the multi-scale nature of the pyramid makes it particularly useful for the task of progressive transmission. Progressive transmission is a process by which an image is sent through a low-capacity channel so that a low resolution or blurred version of the image becomes available quickly, and higher resolution information is added in a gradual manner. In the case of a pyramid, this is easily accomplished by sending the transform coefficients in order from lowest to highest resolution.

For comparison to other subband transforms, we have re-formulated the Laplacian pyramid scheme as a three-band A/S system (see diagram in Figure 4.2) by separating $W_1(\omega)$ into two subsignals: $Y_1(\omega)$ contains the even-numbered samples, and $Y_2(\omega)$ contains the odd-numbered samples. The subsampling factors are $k = 2$ for all three A/S branches, thus producing a representation that is overcomplete by a factor of $3/2$. The appropriate filters for the A/S system are defined in terms of the original filters $A(\omega)$ and $B(\omega)$ as follows:

$$
\begin{aligned}
H_0(\omega) &= B(\omega), & G_0(\omega) &= A(\omega) \\
H_1(\omega) &= \tfrac{1}{2}\big[1 - B(\omega)A(\omega) - B(\omega)A(\omega + \pi)\big], & G_1(\omega) &= 1 \\
H_2(\omega) &= \tfrac{e^{j\omega}}{2}\big[1 - B(\omega)A(\omega) + B(\omega)A(\omega + \pi)\big], & G_2(\omega) &= e^{-j\omega}.
\end{aligned}
$$

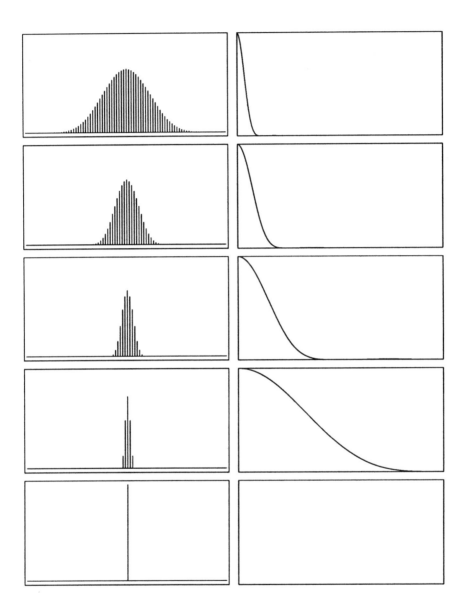

Figure 4.9: Five example basis functions of a four level Laplacian pyramid.

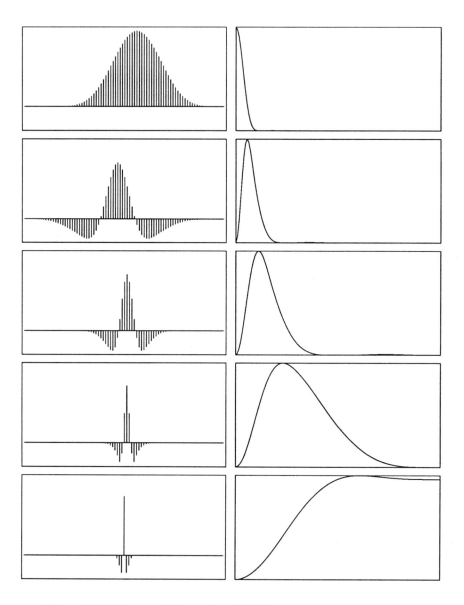

Figure 4.10: Five example inverse (sampling) functions of the Laplacian pyramid.

Notice that when $A(\omega)$ and $B(\omega)$ are lowpass filters, the sampling functions are bandpass and the basis functions are broadband. Since the resulting A/S system violates the constraint in (4.10), the transform is clearly not orthogonal. In two dimensions, Burt and Adelson constructed Laplacian pyramids using the same algorithm, but with a separable 2-D blurring filter. The two dimensional Laplacian pyramid may be re-formulated as a 5-band A/S system with each band subsampled by a factor of two both horizontally and vertically.

As mentioned already in Chapter 2, the Laplacian pyramid has certain disadvantages for image coding. The most serious of these is the fact that quantization errors from highpass subbands do not remain in these subbands. Instead, they appear in the reconstructed image as broadband noise. As with the Gabor transform, the non-orthogonality of the transform is the source of the difficulty. Furthermore, the basis set is overcomplete, requiring an increase (in two dimensions) by a factor of $\frac{4}{3}$ in the number of sample points over the original image. Finally, the 2-D basis functions are not oriented, and thus will not extract the oriented structural redundancy typically found in natural images. Despite these disadvantages for still-image coding, the Laplacian pyramid has been effectively used for motion-compensated video coding, where its overcompleteness makes it robust in to motion-compensation errors [24].

4.4 Quadrature Mirror Filters

In the previous Section, we described three example transforms, each demonstrating useful properties for coding. Now we consider a transform which captures the advantages of the previous examples, while avoiding the disadvantages.

As was illustrated with the Laplacian pyramid, an octave subband transform may be constructed by cascading a two-band A/S system in a non-uniform manner. A useful two-band subband transform which was developed for speech coding is based on banks of quadrature mirror filters (QMF), developed by Croiser et. al. [25, 26]. They discovered a class of non-ideal FIR bandpass filters that could be used in an A/S system while still avoid aliasing in the overall system output. Although they did not describe them as such, these filters form an *orthogonal* subband transform, as was discussed by Adelson et al. [27] and Mallat [3, 28]. Mallat related QMFs to the mathematical theory of wavelets. Vetterli [29] was the first to suggest the the application of QMFs to two-dimensional images. In this Section, we give a brief review of QMFs in one dimension. A more thorough review may be found in [30] or [31].

The original QMF problem was formulated as a two-band critically sampled analysis/synthesis filter bank problem. The overall system response of the filter bank is given by (4.2), with subsampling factor on each branch set to $k = 2$:

$$\hat{X}(\omega) = \frac{1}{2}\Big[H_0(\omega)G_0(\omega) + H_1(\omega)G_1(\omega)\Big]X(\omega)$$

$$+ \frac{1}{2}\Big[H_0(\omega + \pi)G_0(\omega) + H_1(\omega + \pi)G_1(\omega)\Big]X(\omega + \pi). \quad (4.11)$$

The first term is a linear shift-invariant (LSI) system response, and the second is the system aliasing.

The term QMF refers to a particular choice of filters that are related by spatial shifting and frequency modulation. We define

$$\begin{array}{rcl} H_0(\omega) & = & G_0(-\omega) = F(\omega) \\ H_1(\omega) & = & G_1(-\omega) = e^{j\omega}F(-\omega + \pi) \end{array} \quad (4.12)$$

for $F(\omega)$ an arbitrary function of ω. This is a more general definition than that originally provided by Croisier et. al., and makes explicit the orthogonality of the transform (see Section 4.2.5). In particular, the analysis and synthesis filters satisfy the relationship in (4.10), and the relationship between the filters in the two branches (i.e. H_0 and H_1) ensures that the corresponding basis functions are orthogonal.

With the choice of filters given in (4.12), (4.11) becomes

$$\hat{X}(\omega) = \frac{1}{2}\Big[H(\omega)H(-\omega) + H(-\omega + \pi)H(\omega + \pi)\Big]X(\omega)$$

$$+ \frac{1}{2}\Big[H(\omega + \pi)H(-\omega) + e^{j\pi}H(-\omega)H(\omega + \pi)\Big]X(\omega + \pi).$$

The second (aliasing) term cancels, and the remaining LSI system response is

$$\hat{X}(\omega) = \frac{1}{2}\Big[H(\omega)H(-\omega) + H(-\omega + \pi)H(\omega + \pi)\Big]X(\omega). \quad (4.13)$$

Note that the aliasing cancellation is exact, independent of the choice of the function $F(\omega)$. We should emphasize, however, that it is the overall system aliasing that cancels — the individual subbands *do* contain aliasing.

4.4.1   QMF Design

The design problem is now reduced to finding a lowpass filter with Fourier transform $H(\omega)$ that satisfies the constraint

$$\frac{1}{2}\Big[H(\omega)H(-\omega) + H(-\omega + \pi)H(\omega + \pi)\Big] = 1$$

or

$$\left|H(\omega)\right|^2 + \left|H(\omega + \pi)\right|^2 = 2. \tag{4.14}$$

Several authors have studied the design and implementation of these filters [32, 33, 34, 35, 36]. Johnston [32] designed a set of widely used even-length filters by minimizing an error function containing a shift-invariant error term and a weighted stopband ripple term for a fixed number of filter taps. Jain and Crochiere [33, 34] used a similar error criterion in the time domain, and formulated an iterative design scheme which in which each iteration required the constrained minimization of a quadratic function.

A technique for design of perfect reconstruction filter sets is given by Smith and Barnwell in [37]. They first design a lowpass *product filter* $F(\omega)$ which is factorable as

$$F(\omega) = \left|H(\omega)\right|^2$$

and which satisfies

$$f(n) \cdot \frac{\left(1 + (-1)^n\right)}{2} = \delta(n) \ .$$

The resulting $F(\omega)$ is factored to get $h(n)$, the lowpass filter. Wackersreuther [35] independently arrived at a similar design method in the time domain. The problem with these design methods is the somewhat arbitrary choice of the product filter.

Simoncelli [36] proposed an exploratory design method utilizing an iterative matrix averaging technique, and designed a set of odd-length filters using a frequency-sampling method with error criteria similar to Johnston. The design constraints for QMFs do not necessitate sharp transitions and thus frequency-sampling designs perform quite well. Furthermore, it was found that odd-length filters could be made smaller for a given transition band width. The basis functions for a four-level QMF pyramid based on a 9-tap kernel are shown in Figure 4.11. A set of example QMF kernels and a more detailed description of this design technique are given in the appendix to this chapter.

QMFs are typically applied to images in a separable manner. In order to compute a multi-scale pyramid, the transform is applied recursively to the lowpass subimage. Such a cascaded transformation partitions the frequency domain into octave-spaced oriented subbands, as illustrated in the idealized frequency diagram of Figure 4.12. Thus, the QMF pyramid satisfies the properties described in the introduction to this chapter: it is multi-scale and oriented, it is spatially localized, and it is an orthogonal transformation, and so constrains quantization errors to remain within subbands. One unfortunate aspect of the transform is that the orientation decomposition is incomplete.

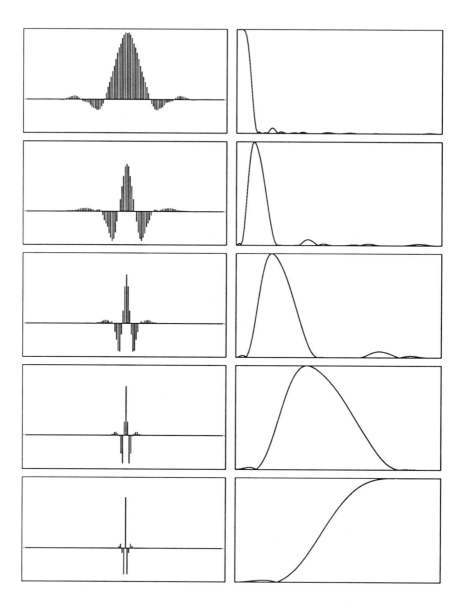

Figure 4.11: Five of basis functions of a 9-tap QMF pyramid transform.

The two diagonal orientations are lumped together in a single subband. We will address this problem in Section 4.5.

4.4.2 An Asymmetrical System

Thus far, we have ignored the issue of computational efficiency. For many applications, this is relatively unimportant due to the steady increase in the speed of signal processing hardware. There are, however, situations where it is desirable to quickly decode or encode a coded image using conventional or general-purpose hardware. For example, an image data base that will be accessed by millions of users with personal computers should be quickly decodable on standard hardware; the cost of encoding these images is of relatively minor importance as long as the decoding is simple. At the other extreme, a remotely piloted vehicle demands a very simple encoding scheme in order to minimize weight and power requirements.

For these situations, it is advantageous to develop asymmetric coding techniques in which simplicity is emphasized at one end at the expense of complexity at the other end. For a QMF transform, the computational complexity is directly proportional to the size of the filters employed. Thus, we wish to relax the orthogonality constraint which forces the synthesis filters to be time-reversed copies of the analysis filters. Consider the situation in which we require efficient decoding. The increase in efficiency can be accomplished by using a very compact filter pair in the synthesis stage of an A/S system [27, 38]. In particular, one can choose the 3-tap lowpass filter $g_0(n) = [1, 2, 1]$, with a highpass counterpart $g_1(n) = [-1, 2, -1]$. Convolutions with these filters may be performed using only arithmetic shifting and addition operations.

The relationship $G_1(\omega) = e^{j\omega}G_0(-\omega + \pi)$, as in (4.12) ensures that the linear subspaces spanned by the basis functions corresponding to each filter will be orthogonal. Conceptually, a set of inverse filters $h_i(n)$ is found by forming a square matrix of the $g_i(n)$ as in (4.5) and inverting it. The size of the matrix determines the size of the resulting inverse filters. In practice, a better design technique is to minimize an error function for a given kernel size. We have designed a set of inverse filters (given in the appendex) by minimizing the maximal reconstruction error for a step edge input signal. These kernels are given in the appendix.

Another highly efficient A/S system was proposed by LeGall. He derived the following set of simple filters for use in an A/S filter bank:

$$H_0(\omega) \;=\; A(\omega), \qquad\qquad G_0(\omega) \;=\; B(\omega)$$

$$H_1(\omega) \;=\; e^{j\omega}B(\omega + \pi), \qquad G_1(\omega) \;=\; e^{-j\omega}A(\omega + \pi)$$

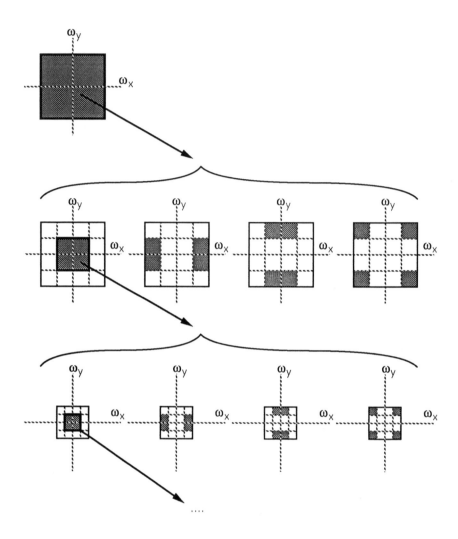

Figure 4.12: Idealized diagram of the partition of the frequency plane resulting from a 4-level pyramid cascade of separable 2-band filters. The top
plot represents the frequency spectrum of the original image, with axes ranging
from $-\pi$ to π. This is divided into four subbands at the next level. On each
subsequent level, the lowpass subband (outlined in bold) is subdivided further.

where the filter kernels (impulse responses) corresponding to $A(\omega)$ and $B(\omega)$ are

$$
\begin{aligned}
a(n) &= [1, 2, 1] \\
b(n) &= [-1, 2, 6, 2, -1] \ .
\end{aligned}
$$

Note that the definitions of $a(n)$ and $b(n)$ may be interchanged. These filters allow efficient encoding *and* decoding, and provide exact reconstruction (with perfect aliasing cancellation).

4.5 Non-separable QMF Transforms

In the previous section, we described the separable QMF pyramid transform. Most two-dimensional work with QMFs has employed separable filters or non-oriented non-separable filters [29]. As discussed in the previous section, separable application of one-dimensional QMFs produces a representation in which one of the subbands contains a mixture of two orientations. This problem is inherent in the rectangular sampling scheme. Rectangular sampling of a signal in the spatial domain corresponds to summing aliased or modulated copies of the spectrum in the frequency domain. Thus, the frequency response of any rectangularly sampled function has the same value at the points (π, π), $(-\pi, \pi)$, $(\pi, -\pi)$, and $(-\pi, -\pi)$ (i.e. this point corresponds to two opposing orientations). Splitting the frequencies in the neighborhood of this point into different orientation bands requires the use of very large filters. In general, the high-frequency diagonal regions of the spectra of natural images are relatively insignificant. But if the filter bank is cascaded to form a pyramid, then the lower frequency diagonals (where there *is* significant power) will also be mixed.

4.5.1 Hexagonal Systems

In this Section, we will discuss the use of hexagonal sampling systems and filters. We will show that the mixed orientation problem discussed above can be avoided by using hexagonally symmetric filters. This non-separable extension of the QMF concept was first described by Adelson et al. [27] and improved and generalized in later work [36, 39]. Other authors have also explored the use of hexagonal sampling systems for image representation. Crettez and Simon [40] and Watson [41] describe decompositions on hexagonal lattices using non-overlapping basis functions. The blocked nature of these functions suggests that they are unlikely to offer efficient image compression.

Figure 4.13 shows a hexagonal sampling lattice and its Fourier transform. The sampling lattice is defined by a pair of *sampling vectors* in the plane:

$$v_0 = \begin{pmatrix} \sqrt{3}/2 \\ 1/2 \end{pmatrix}, \qquad v_1 = \begin{pmatrix} 0 \\ 1 \end{pmatrix}.$$

The locations of the lattice points consist of all linear combinations of these vectors with integer coefficients. In the frequency domain, the effect of this sampling is to convolve the original frequency spectrum of the image with a *modulation* or *reciprocal* lattice which is the Fourier transform of the sampling lattice. The modulation lattice is defined by a pair of *modulation vectors* in the frequency plane:

$$\tilde{v_0} = \begin{pmatrix} 4\pi/\sqrt{3} \\ 0 \end{pmatrix}, \qquad \tilde{v_1} = \begin{pmatrix} -2\pi/\sqrt{3} \\ 2\pi \end{pmatrix}.$$

Thus if $F(\omega)$ is the Fourier transform of a hexagonally sampled signal (image) then it is invariant to translations by multiples of the vectors $\tilde{v_i}$:

$$F(\omega) = F(\omega + n_0\tilde{v_0} + n_1\tilde{v_1}) \tag{4.15}$$

for n_0 and n_1 any two integers.

In general, the relationship between the sampling vectors and modulation vectors is easily described in terms of matrices [42, 43]. If we consider the *sampling matrix* $\mathbf{V}$ with columns containing the vectors v_i and the *modulation matrix* $\tilde{\mathbf{V}}$ with columns containing the vectors $\tilde{v_i}$, then the two matrices are related by the equation

$$\tilde{\mathbf{V}} = 2\pi(\mathbf{V}^{-1})^t. \tag{4.16}$$

Note that we know $\mathbf{V}$ is invertible since we assume that the sampling vectors span the space (i.e. they are linearly independent).

As stated in Section 4.2.1, the A/S system depicted in Figure 4.2 is valid for two-dimensional signals, but the filtering and subsampling is done in two dimensions: ω is now a two-dimensional vector, and the subsampling is parameterized by a non-singular two-by-two *subsampling matrix*, $\mathbf{K}$, with integer entries. Figure 4.14 illustrates two-dimensional subsampling in both the spatial and frequency domains.

In order to write a general expression for the output of a multidimensional analysis/synthesis system, we need a frequency-domain equation analogous to that given in (4.1) relating the subsampled signal to the sampled signal. We also need an equation relating an upsampled signal to the original sampled signal. For rectangular sampling lattices in d dimensions, the relationship is simple. The sampling matrix $\mathbf{K}$ generates a sublattice defined by

$$\{n : n = \mathbf{K}m, m \in \mathbf{Z}^d\},$$

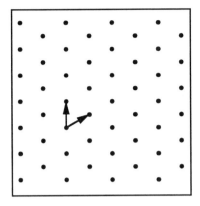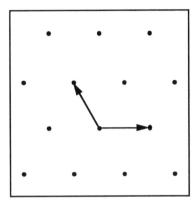

Figure 4.13: Relationship between hex sampling lattices in the spatial and spatial-frequency domains. On the left is the lattice defined by the sampling vectors. On the right is the Fourier transform of this lattice, defined by the modulation vectors.

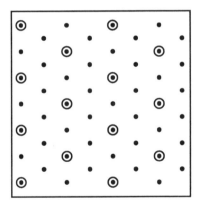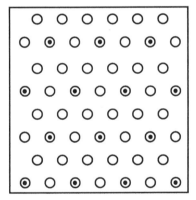

Figure 4.14: Illustration of subsampling on a hexagonal lattice. The points in the diagram on the left represent the original sampling lattice and the circles represent the subsampled lattice points. The picture on the right shows the Fourier transform of the lattice (points) and the Fourier transform of the subsampled lattice (circles).

where $\mathbf{Z}^d$ is the set of all d-dimensional vectors with integer components. The sublattice has $|\mathbf{K}|$ distinct cosets, each coset being a copy of the sublattice translated by an integer vector, and the union of the cosets is the original sampling lattice [44]. Consider two signals related by subsampling: $s(\mathbf{n}) = r(\mathbf{K}\mathbf{n})$. Then their Fourier transforms are related by the expression

$$S(\boldsymbol{\omega}) = \frac{1}{|\mathbf{K}|} \sum_{i=0}^{|\mathbf{K}|-1} R\left((\mathbf{K}^{-1})^t(\boldsymbol{\omega} - 2\pi \mathbf{k}_i)\right)$$

where $S(\boldsymbol{\omega})$ and $R(\boldsymbol{\omega})$ are the Fourier transforms of $s(\mathbf{n})$ and $r(\mathbf{n})$ respectively, and the $\mathbf{k}_i$ are a set of polyphase shift vectors corresponding to each of the $|\mathbf{K}|$ sublattice cosets [44]. A simple example of a set of shift vectors is the following:

$$\left\{\mathbf{k} : (\mathbf{K}^{-1})^t\mathbf{k} \in [0,1)^d, \mathbf{k} \in \mathbf{Z}^d\right\},$$

where $[0,1)^d$ is the half-open unit interval in d dimensions.

The corresponding expression for non-rectangular sampling lattices is obtained by mapping from the rectangular case. The result of subsampling in the analysis/synthesis system may then be written as a convolution of the sampled spectrum with a set of *subsampling modulation vectors* $\tilde{\boldsymbol{\kappa}}_j$:

$$Y_i(\boldsymbol{\omega}) = \frac{1}{|\mathbf{K}|} \sum_{j=0}^{|\mathbf{K}|-1} H_i\left((\mathbf{K}^{-1})^t\boldsymbol{\omega} + \tilde{\boldsymbol{\kappa}}_j\right) X\left((\mathbf{K}^{-1})^t\boldsymbol{\omega} + \tilde{\boldsymbol{\kappa}}_j\right), \qquad (4.17)$$

where the $\tilde{\boldsymbol{\kappa}}_j$ are defined as

$$\left\{\tilde{\boldsymbol{\kappa}}_j : j = 0, 1, \ldots, |\mathbf{K}| - 1\right\}$$
$$= \left\{\tilde{\mathbf{V}}(\mathbf{K}^{-1})^t\mathbf{n} : (\mathbf{K}^{-1})^t\mathbf{n} \in [0,1)^d, \mathbf{n} \in \mathbf{Z}^d\right\}. \quad (4.18)$$

The effect of upsampling in the frequency domain is the same as for the rectangular case [43]. Combining (4.17) with the frequency domain upsampling relationship gives an expression for the overall filter bank response:

$$\begin{aligned}
\hat{X}(\boldsymbol{\omega}) &= \frac{1}{|\mathbf{K}|} \sum_{i=0}^{|\mathbf{K}|-1} G_i(\boldsymbol{\omega}) Y_i(\mathbf{K}^t\boldsymbol{\omega}) \\
&= \frac{1}{|\mathbf{K}|} \sum_{i=0}^{|\mathbf{K}|-1} G_i(\boldsymbol{\omega}) \sum_{j=0}^{|\mathbf{K}|-1} H_i(\boldsymbol{\omega} + \tilde{\boldsymbol{\kappa}}_j) X(\boldsymbol{\omega} + \tilde{\boldsymbol{\kappa}}_j) \\
&= \frac{1}{|\mathbf{K}|} \sum_{j=0}^{|\mathbf{K}|-1} X(\boldsymbol{\omega} + \tilde{\boldsymbol{\kappa}}_j) \left[\sum_{i=0}^{|\mathbf{K}|-1} G_i(\boldsymbol{\omega}) H_i(\boldsymbol{\omega} + \tilde{\boldsymbol{\kappa}}_j)\right]. \quad (4.19)
\end{aligned}$$

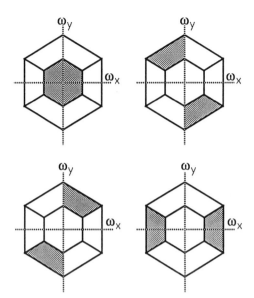

Figure 4.15: Illustration of the modulating effect of subsampling in the frequency domain. Assume that the sampled image has a spectrum bandlimited to the gray region in the upper left frequency diagram. Subsampling will modulate the spectrum to the gray regions in the other three diagrams. The resulting spectrum will be the sum of the four spectra.

As in (4.11), the first term of the sum ($j = 0$) corresponds to the LSI system response, and the remaining terms are the system aliasing.

Returning now to the specific case of the hexagonal sampling lattice, we describe a system obtained by using a specific sampling matrix $\mathbf{K}$. Since we want to be able to apply the transform recursively, we choose a subsampling scheme which preserves the geometry of the original sampling lattice:

$$\mathbf{K} = \begin{bmatrix} 2 & 0 \\ 0 & 2 \end{bmatrix}.$$

On the hexagonal sampling lattice with this subsampling scheme, the definition given in (4.18) produces the following modulation vectors:

$$\tilde{\kappa}_0 = \begin{pmatrix} 0 \\ 0 \end{pmatrix}, \quad \tilde{\kappa}_1 = \begin{pmatrix} 2\pi/\sqrt{3} \\ 0 \end{pmatrix}, \quad \tilde{\kappa}_2 = \begin{pmatrix} \pi/\sqrt{3} \\ \pi \end{pmatrix}, \quad \tilde{\kappa}_3 = \begin{pmatrix} -\pi/\sqrt{3} \\ \pi \end{pmatrix}.$$

Figure 4.15 offers an idealized picture of this modulation.

Analogous to the one-dimensional case, we can choose the filters to eliminate the aliasing terms in (4.19):

$$H_0(\omega) = G_0(-\omega) = F(\omega) = F(-\omega)$$

$$
\begin{aligned}
H_1(\omega) &= G_1(-\omega) = e^{j\omega \cdot s_1} F(\omega + \tilde{\kappa}_1) \\
H_2(\omega) &= G_2(-\omega) = e^{j\omega \cdot s_2} F(\omega + \tilde{\kappa}_2) \\
H_3(\omega) &= G_3(-\omega) = e^{j\omega \cdot s_3} F(\omega + \tilde{\kappa}_3)
\end{aligned}
\tag{4.20}
$$

where H is a function that is invariant under negation of its argument, the s_i are a set of spatial shift vectors (defined in the next paragraph), and the expressions $\omega \cdot s_i$ indicates an inner product of the two vectors. As in (4.12), the filters are related by spatial shifting and frequency modulation. For the subsampling matrix we are using here, there are four sublattice cosets and therefore four distinct shifting vectors (including the zero vector). Two assignments of the s_i lead to system aliasing cancellation, and these two assignments are related by reflection through the origin. So without loss of generality, we choose the shifting vectors to be

$$
s_1 = \begin{pmatrix} \sqrt{3}/2 \\ 1/2 \end{pmatrix}, \qquad
s_2 = \begin{pmatrix} 0 \\ 1 \end{pmatrix}, \qquad
s_3 = \begin{pmatrix} -\sqrt{3}/2 \\ 1/2 \end{pmatrix}.
$$

After cancellation of the aliasing terms in (4.19), the remaining LSI system response is

$$
\begin{aligned}
\hat{X}(\omega) &= \frac{1}{4} X(\omega) \sum_{i=0}^{3} G_i(\omega) H_i(\omega) \\
&= \frac{1}{4} X(\omega) \sum_{i=0}^{3} F(-\omega + \tilde{\kappa}_i) F(\omega + \tilde{\kappa}_i) \\
&= \frac{1}{4} X(\omega) \sum_{i=0}^{3} |F(\omega + \tilde{\kappa}_i)|^2.
\end{aligned}
\tag{4.21}
$$

As in one dimension, the aliasing cancellation is exact, independent of the choice of $F(\omega)$, and the design problem is reduced to finding a filter with Fourier transform $F(\omega)$ satisfying the constraint

$$
\sum_{i=0}^{3} |F(\omega + \tilde{\kappa}_i)|^2 = 4.
\tag{4.22}
$$

This is analogous to the one-dimensional equation (4.14). Again, a lowpass solution will produce a band-splitting system which may be cascaded hierarchically to produce an octave-bandwidth decomposition in two dimensions. An idealized illustration of this is given in Figure 4.16. Finer frequency and orientation subdivisions may be achieved by recursively applying the filter bank to some of the high frequency subbands, as illustrated in Figure 4.17.

Filters may be designed using the methods described in Section 4.4.1 [39]. Several example filter sets are given in the appendix to this chapter. The

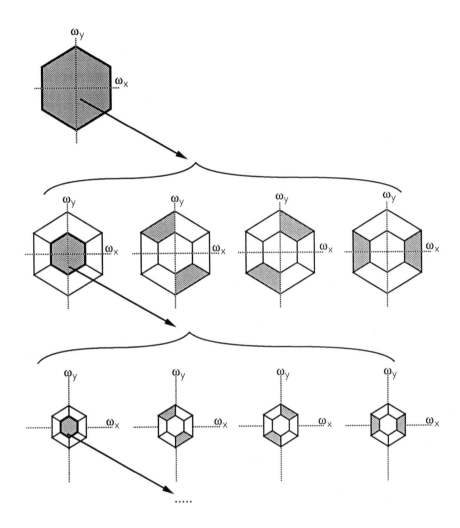

Figure 4.16: Idealized diagram of the partition of the frequency plane resulting from a four-level pyramid cascade of hexagonal filters. The top plot represents the frequency spectrum of the original image. This is divided into four subbands at the next level. On each subsequent level, the lowpass subband (outlined in bold) is sub-divided further.

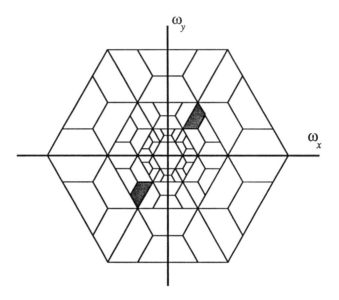

Figure 4.17: An example of a frequency-domain partition which could be computed using a non-pyramid cascade of the hexagonal filter bank transform described in the text. The shaded region indicates the frequency region associated with one of the transform subbands.

power spectra of an example set of filters (the "4-ring" filters) are plotted in Figure 4.18. These filters are extremely compact, requiring only nine multiplications per convolution point (assuming one takes advantage of the twelve-fold hexagonal symmetry). Figure 4.19 shows the results of applying this bank of filters recursively to an image of a disk. Examples of images coded using these filters will be given in Section 4.6.

4.5.2 Rhombic Dodecahedral Systems

The extension of the concepts developed in the previous section to three-dimensional signal processing is fairly straightforward. Such systems are useful for applications such as compression of medical images or video motion sequences. Analogous to the two-dimensional hexagonal case, one can choose a periodic sampling lattice which corresponds to the densest packing of spheres

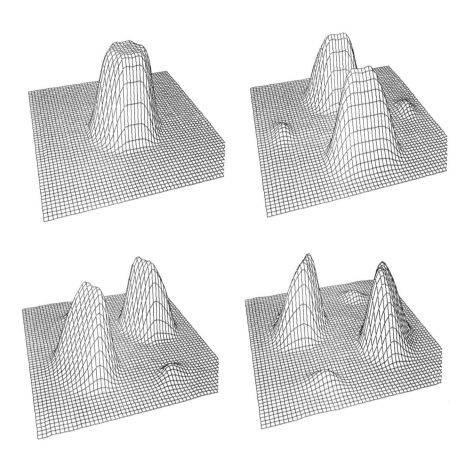

Figure 4.18: The power spectra for the "4-ring" set of hexagonal QMF filters. The filter kernels are given in the appendix.

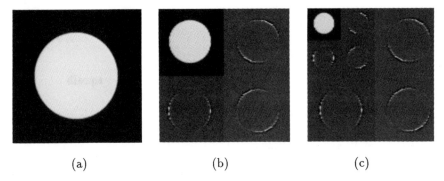

(a)   (b)   (c)

Figure 4.19: Results of applying a hexagonal QMF bank to an image of a disk. (a) The original image. (b) The result after one application of the analysis section of the filter bank. The image has been decomposed into a lowpass and three oriented high-pass images at 1/4 density. (c) The result of applying the filter bank recursively to the lowpass image to produce a two-level pyramid decomposition.

in three dimensions. This packing corresponds to the crystal structure of garnet. We choose as a band limiting region the *Voronoi* region of this lattice (a rhombic dodecahedron) which is illustrated in Figure 4.20. The sampling matrix for the lattice is

$$\mathbf{V} = \begin{bmatrix} 2 & 1 & 1 \\ 0 & 1 & 0 \\ 0 & 0 & 1/\sqrt{2} \end{bmatrix}.$$

Using (4.16), the modulation matrix is then

$$\tilde{\mathbf{V}} = \begin{bmatrix} \pi & 0 & 0 \\ -\pi & 2\pi & 0 \\ -\sqrt{2}\pi & 0 & 2\sqrt{2}\pi \end{bmatrix}.$$

To preserve the geometry of the original sampling lattice, we choose an eight-band A/S system with subsampling matrix

$$\mathbf{K} = \begin{bmatrix} 2 & 0 & 0 \\ 0 & 2 & 0 \\ 0 & 0 & 2 \end{bmatrix}.$$

This produces the following subsampling modulation points, as determined by (4.18):

$$\tilde{\boldsymbol{\kappa}}_0 = \begin{pmatrix} 0 \\ 0 \\ 0 \end{pmatrix}, \quad \tilde{\boldsymbol{\kappa}}_1 = \begin{pmatrix} 0 \\ \pi \\ \sqrt{2}\pi \end{pmatrix}, \quad \tilde{\boldsymbol{\kappa}}_2 = \begin{pmatrix} 0 \\ \pi \\ 0 \end{pmatrix}, \quad \tilde{\boldsymbol{\kappa}}_3 = \begin{pmatrix} \pi/2 \\ \pi/2 \\ -\pi/\sqrt{2} \end{pmatrix},$$

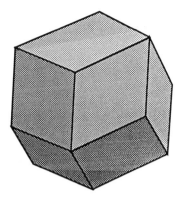

Figure 4.20: A rhombic dodecahedron. This is the shape of the bandlimiting frequency region for the "garnet" filter.

$$\tilde{\kappa}_4 = \begin{pmatrix} \pi/2 \\ -\pi/2 \\ \pi/\sqrt{2} \end{pmatrix}, \quad \tilde{\kappa}_5 = \begin{pmatrix} \pi/2 \\ -\pi/2 \\ -\pi/\sqrt{2} \end{pmatrix}, \quad \tilde{\kappa}_6 = \begin{pmatrix} \pi/2 \\ \pi/2 \\ \pi/\sqrt{2} \end{pmatrix}, \quad \tilde{\kappa}_7 = \begin{pmatrix} 0 \\ 0 \\ \sqrt{2}\pi \end{pmatrix}.$$

When applied to video motion sequences, these modulation vectors correspond to a decomposition into the following subbands: lowpass, stationary vertical, stationary horizontal, motion up/right, motion up/left, motion down/right, motion down/left, and combined stationary diagonals and full-field flicker. Unfortunately, there seems to be no way to avoid the last filter which contains mixed orientations. The overall system response of the filter bank is

$$\hat{X}(\omega) = \frac{1}{8} \sum_{j=0}^{7} X(\omega + \tilde{\kappa}_j) \left[\sum_{i=0}^{7} G_i(\omega) H_i(\omega + \tilde{\kappa}_j) \right]. \tag{4.23}$$

where the first term is the LSI system response, and the remaining terms are aliasing terms.

Once again, we can choose filters related by shifts and modulations that will cancel the system aliasing terms:

$$\begin{aligned} H_0(\omega) &= G_0(-\omega) = F(\omega) = F(-\omega) \\ H_i(\omega) &= G_i(-\omega) = e^{j\omega \cdot s_i} F(\omega + \tilde{\kappa}_i), \quad i \in \{1, 2, ...7\} \end{aligned}$$

where the shift vectors s_i are defined as

$$s_0 = \begin{pmatrix} 0 \\ 0 \\ 0 \end{pmatrix}, \quad s_1 = \begin{pmatrix} 1 \\ 0 \\ 1/\sqrt{2} \end{pmatrix}, \quad s_2 = \begin{pmatrix} 1 \\ 1 \\ \sqrt{2} \end{pmatrix}, \quad s_3 = \begin{pmatrix} 1 \\ 2 \\ 1/\sqrt{2} \end{pmatrix},$$

$$s_4 = \begin{pmatrix} 2 \\ 1 \\ 1/\sqrt{2} \end{pmatrix}, \quad s_5 = \begin{pmatrix} 2 \\ 0 \\ 0 \end{pmatrix}, \quad s_6 = \begin{pmatrix} 1 \\ 1 \\ 0 \end{pmatrix}, \quad s_7 = \begin{pmatrix} 0 \\ 1 \\ 1\sqrt{2} \end{pmatrix}.$$

Note that as in the hexagonal case, the choice of shift vectors is not unique.

With the choice of filters given above, the aliasing terms in (4.23) cancel and the remaining LSI system response is

$$\hat{X}(\omega) = \frac{1}{8} X(\omega) \sum_{i=0}^{7} |F(\omega + \tilde{\kappa}_i)|^2 ,$$

independent of the choice of the function $F(\omega)$. The design constraint equation is now

$$\sum_{i=0}^{7} |F(\omega + \tilde{\kappa}_i)|^2 = 8.$$

To illustrate the use of the garnet filter, we apply it to an image sequence of a sinusoidal pinwheel rotating in a counterclockwise direction. One frame of the sequence is shown in Figure 4.21(a). The squared responses of the four different motion-selective filters (filters $H_3(\omega)$ through $H_6(\omega)$) are shown in Figure 4.21(b-e).

4.6   Image Coding Examples

Several authors have used QMFs for purposes of image coding. Woods and O'Neill [45] were the first to implement an image coding system using QMFs. They constructed a separable sixteen-band decomposition using a uniform cascade of 32- and 80-tap filters designed by Johnston [32], and then coded the bands using adaptive DPCM. Gharavi and Tabatabai [46] used a pyramid of separable filters and in [47], applied it to color images. Tran et. al. [48] used an extension of Chen and Pratt's [49] combined Huffman and run-length coding scheme to code QMF pyramids. Adelson et al. have used both separable and hexagonal QMF pyramids for image coding [27, 36, 39]. Mallat [3] used filters derived from wavelet theory to code images. Westerink et. al. [50] have used vector quantization for subband coding of images.

In Figures 4.22 and 4.23, we give examples of data compression of the 256×256 "Lenna" image using a separable 9-tap QMF bank, a 3-tap asymmetric filter bank (described in Section 4.4.2), and a hexagonal "4-ring" QMF bank. In all cases, a four-level pyramid transform was computed by recursive application of the analysis portion of the A/S system to the lowpass image. The total bit rate R was fixed and the bit rates assigned to the coefficients

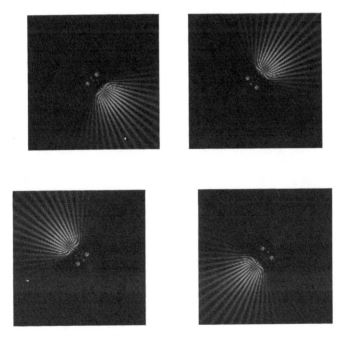

Figure 4.21: At the top is one image from a rotating pinwheel image sequence. The four lower images are the squared result of convolving the sequence with four of the "garnet" filters described in the text. Each filter responds preferentially to one direction of motion.

of the transform were determined using the standard optimal allocation formula [19]:

$$R_k = R + \frac{1}{2} \log_2 \frac{\sigma_k^2}{\left[\displaystyle\prod_{j=0}^{N-1} \sigma_j^2\right]^{1/N}} \tag{4.24}$$

where, as before, σ_k^2 is the variance of the kth coefficient in the transform. Negative values of R_k were set to zero and the other bit rates raised to maintain the correct overall bit rate R.

Note that if we assume stationary image statistics, the σ_k^2 are the same for all coefficients belonging to the same sub-image of the transform. It has been shown [19] that the optimal quantizer for entropy coding is nearly uniform for bit rates which are high enough that the image probability distribution is approximately constant over each bin. Even though the examples shown were compressed to relatively low bit rates, uniform quantization was used due to its simplicity. Each sub-image was quantized with the bin size chosen to give a first order entropy equal to the optimal bit rate R_k for that subimage.

For the hexagonal pyramid, additional pre- and post-processing was necessary to resample the image on a hexagonal grid. Before building the pyramid, we resampled the original image vertically by a factor of 7/4 using sinc interpolation. We then multiplied by the function $f(\boldsymbol{n}) = 1 + (-1)^{(n_x + n_y)}$. This method, which is similar to one suggested in [42], gives a reasonable geometric approximation to a hexagonal sampling lattice. After re-synthesizing the image, we interpolated the zero-valued pixels and vertically resampled by a factor of 4/7.

The hexagonal QMF system generally offers coding performance perceptually superior to that of the separable system, perhaps because the aliasing errors are not as visually disturbing as those of separable QMFs. Of course, the hexagonal system has the disadvantage of being more inconvenient to use in conjunction with standard hardware.

4.7 Conclusion

We have discussed the properties of linear transforms that are relevant to the task of image compression. In particular, we have suggested that the basis and sampling functions of the transform should be localized in both the spatial and the spatial-frequency domains. We have also suggested that it is desirable for the transform to be orthogonal.

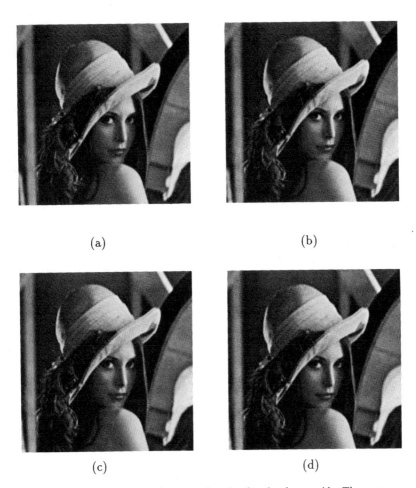

(a)

(b)

(c)

(d)

Figure 4.22: Data compression example using four-level pyramids. The pyramid data was compressed to a total of 65536 bits (i.e. total first-order entropy was 1.0 bit/pixel). (a) Original "Lenna" image at 256×256 pixels. (b) Compressed using 9-tap separable QMF bank. (c) Compressed using 3-tap asymmetrical filter bank. (d) Compressed using "4-ring" hexagonal QMF bank.

(a)   (b)

(c)   (d)

Figure 4.23: Data compression example using four-level pyramids. The pyramid data was compressed to a total of 16384 bits (i.e. total first-order entropy was 0.25 bit/pixel). (a) Original "Lenna" image at 256 × 256 pixels. (b) Compressed using 9-tap separable QMF bank. (c) Compressed using 3-tap asymmetrical filter bank. (d) Compressed using "4-ring" hexagonal QMF bank.

Several examples serve to illustrate these properties. The Gabor basis functions are well-localized, but the severe non-orthogonality of the transform leads to sampling functions which are very poorly localized. The block DCT is an equal-width subband transform with poor frequency localization. The LOT enhancement provides improved frequency localization. The Laplacian pyramid is an example of an octave-width subband transform that is non-orthogonal, non-oriented and overcomplete; its properties are non-optimal for coding still images, but may be advantageous for coding moving images.

Subband transforms based on banks of QMFs are well-localized, orthogonal, and can be applied recursively to form octave-width subbands. Separable application of these transforms offers orientation specificity in some but not all of the subbands. Non-separable orthogonal subband transforms based on hexagonal sampling offer orientation specificity in all of the subbands, although they are more difficult to implement. These orthogonal subband transforms are highly effective in image coding applications, and may also be appropriate for applications in image enhancement and machine vision tasks.

Appendix: Filters

In this appendix, we discuss the design of QMFs, present a set of example filter kernels, and compare their theoretical energy compaction properties to those of the DCT and LOT transforms.

Filter Design

A "good" QMF is one that satisfies the constraint given in (4.14). In addition, one would like the sub-band images to have a minimal amount of aliasing. The objective then is to design filters with small regions of support that satisfy both of these constraints. Assuming symmetric (linear phase) filter designs, a filter of size N is determined by a set of $\lceil N/2 \rceil$ free parameters, where $\lceil \cdot \rceil$ indicates the ceiling function. Therefore, filters may be designed by minimizing an error function defined on the space of these free parameters.

For a fixed filter size, we define a frequency-domain filter bank error function as the maximal deviation of the overall filter bank response given in (4.13) from its ideal value:

$$E_1 = \max_{\omega} \left\{ f_1(\omega) \left| |F(\omega)|^2 + |F(\omega + \pi)|^2 - 2 \right| \right\}$$

where ω ranges over the samples in the frequency spectrum. The function $f_1(\omega)$ is a frequency weighting function roughly matched to the sensitivity of

n	QMF-5	QMF-9	QMF-13
0	0.8593118	0.7973934	0.7737113
1	0.3535534	0.41472545	0.42995453
2	-0.0761025	-0.073386624	-0.057827797
3		-0.060944743	-0.09800052
4		0.02807382	0.039045125
5			0.021651438
6			-0.014556438

Table 4.1: Odd-length QMF kernels. Half of the impulse response sample values are shown for each of the normalized lowpass QMF filters (All filters are symmetric about $n = 0$). The appropriate highpass filters are obtained by delaying by one sample and multiplying with the sequence $(-1)^n$.

the human visual system and the statistics of images:

$$f_1(\omega) = 1/|\omega|.$$

We also define an intra-band aliasing error function:

$$E_2 = \max_{\omega'} \left\{ f_2(\omega') \, |F(-\omega')F(\omega' + \pi)| \right\}$$

where the function $f_2(\omega')$ is defined as

$$f_2(\omega') = 1/|\omega'|^2.$$

The frequency vector ω' ranges over all of the samples in the frequency spectrum, except for the point at $\pi/2$. Aliasing within subbands cannot be eliminated at this point because the overall filter bank response at this point would then be forced to zero, violating the constraint in (4.14).

Finally, we combine the two error functions as a weighted sum:

$$E = \alpha E_1 + (1 - \alpha)E_2, \quad \alpha \in [0, 1].$$

Given a set of values for the free parameters, we can construct a kernel and compute the value of the error function E. To design filters, we used a downhill simplex method to search the space of free parameters for minima in E. The weighting factor α was adjusted to give a filter bank response error E_1 less than a fixed threshold. A set of example odd-length filter kernels are given in Table 4.1.

The same design technique was used for multidimensional non-separable filters. For the hexagonally symmetric filters, the free parameters comprise a wedge-shaped region covering approximately one twelfth of the kernel. The two error functions are defined in the same manner as for the 1-D filters. The frequency vector ω' now ranges over all of the samples in the 2-D frequency spectrum, except for those in a hexagonal boundary containing the point $(0, \frac{\pi}{2})$. A set of kernel values is given in Table 4.3

In Table 4.4 we give several inverse filters for the 3-tap asymmetrical system described in Section 4.4.2. These kernels were designed by minimizing the maximal absolute-value reconstruction error for a step edge input signal.

Filter Compaction Properties

An optimal transform for data compression should minimize the bit rate for a given allowable error in the reconstructed image. If the basis functions of the transform are orthonormal, and if expected mean square difference is used as an error measure, this is equivalent to maximizing the following expression for the gain in coding over PCM [19, 45]:

$$G = \frac{\frac{1}{N} \sum_{j=0}^{N-1} \sigma_j^2}{\left[\prod_{j=0}^{N-1} \sigma_j^2 \right]^{1/N}}$$

where σ_j^2 is the variance of the jth transform coefficient.

This measure was computed for some of the 1-D QMF filters given in this Appendix, as displayed in Table 4.5. Values were computed assuming Markov second order signal statistics, where the autocorrelation matrix $\mathbf{R}_{xx}$ is a symmetric Toeplitz matrix of the form

$$\mathbf{R}_{xx} = \begin{bmatrix} 1 & \rho & \rho^2 & \cdots & \rho^N \\ \rho & 1 & \rho & & \rho^{(N-1)} \\ \rho^2 & \rho & 1 & & \rho^{(N-2)} \\ \vdots & & & \ddots & \\ \rho^N & \rho^{(N-1)} & \rho^{(N-2)} & & 1 \end{bmatrix}$$

and where ρ is the inter-sample correlation coefficient. A value of $\rho = 0.95$ was used to compute the numbers given in Table 4.5. The compaction values are given for a 1-D image of size $N = 256$ with the QMF filter kernels reflected at the edges in a manner that preserves the orthogonality of the basis set [36]. Comparable values for the 16-point LOT (with kernel sizes L=32) and a 16-point block DCT and a 32-point block DCT are also given.

```
                    j       k       l       l       k       j
                k       g       h       i       h       g       k
            l       h       e       f       f       e       h       l
        l       i       f       c       d       c       f       i       l
    k       h       f       d       b       b       d       f       h       k
j       g       e       c       b       a       b       c       e       g       j
    k       h       f       d       b       b       d       f       h       k
        l       i       f       c       d       c       f       i       l
            l       h       e       f       f       e       h       l
                k       g       h       i       h       g       k
                    j       k       l       l       k       j
```

Table 4.2: A hexagonal filter. The letters refer to the free parameters (see text). Only the low-pass filter is shown. The three highpass filters are formed by modulating and shifting the low-pass.

Parameter	3-ring	4-ring	5-ring
a	0.59290695	0.6066799	0.60879886
b	0.32242984	0.3162482	0.31689283
c	-0.016686682	-0.028019974	-0.027267352
d	-0.061579883	-0.0016289932	-0.012790751
e	-0.0020203826	-0.02741341	-0.03874194
f	-0.0038235565	-0.038143888	-0.02383056
g		-0.005958891	0.0008673751
h		0.019682134	0.015554102
i		0.016045252	0.0080001475
j			-0.0009099232
k			0.0022140248
l			-0.0010486352

Table 4.3: Some example hexagonal filter coefficient values. The parameter letters correspond to the diagram shown in Table 4.2

n	15	17	21
0	0.8648855700	0.8662753700	0.8660005000
1	0.3589060300	0.3588442800	0.3586960400
2	-0.1476441600	-0.1488108800	-0.1486006000
3	-0.0618851260	-0.0616580880	-0.0615359620
4	0.0244434030	0.0257062400	0.0255328510
5	0.0106931890	0.0102884290	0.0105768030
6	-0.0030558493	-0.0044906090	-0.0043832410
7	-0.0015278960	-0.0012884160	-0.0017810371
8		0.0006442405	0.0007449251
9			0.0002303323
10			-0.0001151661

Table 4.4: Filter impulse response values for 15, 17, and 21-tap inverses for the 3-tap system described in Section 4.4.2. Half of the impulse response sample values are shown for each of the normalized lowpass filters (All filters are symmetric about $n = 0$). The appropriate highpass filters are obtained by multiplying with the sequence $(-1)^n$ and shifting by one pixel.

The 9-tap subband filter gives slightly better value than the 16-point DCT, and the 13-tap subband filter is substantially better. These comparisons do not necessarily correspond to measurements of subjective quality, however, since they are based on a crude Markov statistical model of images, and since they assume an MSE error measure. We have found that images compressed with a 9-tap subband transform are perceptually superior to the 32-point DCT, primarily because of the absence of the block artifacts. We also find that the 9-tap QMF is preferable to the 13-tap QMF: the 9-tap filter produces more aliasing, but the Gibbs ringing is more noticeable with the 13-tap filter. We have not performed any coding experiments using the LOT, and so cannot comment on its performance.

filter	G_{PCM}
QMF-5	8.07
QMF-9	9.05
QMF-13	9.28
fast-LOT-16	9.32
DCT-16	8.82
DCT-32	9.49

Table 4.5: Theoretical Coding gains over PCM for four-level QMF pyramids, the fast LOT (with $N = 16$ and $L = 32$), and the block DCT. Values were computed assuming first-order Gauss-Markov signal statistics with $\rho = 0.95$, on a 1-D image of size 256.

Bibliography

[1] D. Marr, T. Poggio, and S. Ullman, "Bandpass channels, zero-crossings, and early visual information processing," *J. Opt. Soc. Am.*, vol. 69, pp. 914–916, 1977.

[2] E. H. Adelson, C. H. Anderson, J. R. Bergen, P. J. Burt, and J. M. Ogden, "Pyramid methods in image processing," *RCA Engineer*, vol. 29, pp. 33–41, November/December 1984.

[3] S. G. Mallat, "A theory for multiresolution signal decomposition: the wavelet representation," GRASP Lab Technical Memo MS-CIS-87-22, University of Pennsylvania, Department of Computer and Information Science, 1987.

[4] A. P. Witkin, "Scale-space filtering," in *Proc. Int. Joint Conf. Artificial Intelligence*, pp. 1019–1021, 1985.

[5] J. J. Koenderink, "The structure of images," *Biological Cybernetics*, vol. 50, pp. 363–370, 1984.

[6] A. Pentland, "Fractal based description of natural scenes," *IEEE Trans. PAMI*, vol. 6, no. 6, pp. 661–674, 1984.

[7] S. Marcelja, "Mathematical description of the response of simple cortical cells," *J. Opt. Soc. Am.*, vol. 70, pp. 1297–1300, 1980.

[8] J. G. Daugman, "Uncertainty relation for resolution in space, spatial frequency, and orientation optimized by two-dimensional visual cortical filters," *J. Opt. Soc. Am. A*, vol. 2, pp. 1160–1169, July 1985.

[9] J. G. Daugman and D. M. Kammen, "Pure orientation filtering: A scale-invariant image-processing tool for perception research and data compression," *Behavior Research Methods, Instruments, & Computers*, vol. 18, no. 6, pp. 559–564, 1986.

[10] D. Gabor, "Theory of communication," *J. IEE*, vol. 93, pp. 492–457, 1946.

[11] R. E. Crochiere and L. R. Rabiner, *Multirate Digital Signal Processing*. Signal Processing Series, Englewood Cliffs, NJ: Prentice-Hall, 1983.

[12] A. V. Oppenheim and R. W. Schafer, *Digital Signal Processing*. Englewood Cliffs: Prentice-Hall, Inc., 1975.

[13] G. Strang, *Linear Algebra and Its Applications*. Orlando: Academic Press, 1980.

[14] R. M. Lerner, *Lectures on Communication System Theory*, ch. 10. New York: McGraw-Hill, 1961.

[15] J. G. Daugman, "Complete discrete 2-d gabor transforms by neural networks for image analysis and compression," *IEEE Trans. ASSP*, vol. ASSP-36, pp. 1169–1179, 1988.

[16] J. G. Daugman, "Entropy reduction and decorrelation in visual coding by oriented neural receptive fields," *IEEE Trans. Biomedical Engineering*, vol. 36, no. 1, pp. 107–114, 1989.

[17] M. Kunt, A. Ikonomopoulos, and M. Kocher, "Second generation image-coding techniques," in *Proceedings IEEE*, vol. 73, pp. 549–574, 1985.

[18] A. B. Watson, "Efficiency of a model human image code," *J. Opt. Soc. Am. A*, vol. 12, pp. 2401–2417, 1987.

[19] N. Jayant and P. Noll, *Digital Coding of Waveforms*. Signal Processing Series, Englewood Cliffs, NJ: Prentice-Hall, 1984.

[20] P. M. Cassereau, D. H. Staelin, and G. de Jager, "Encoding of images based on a lapped orthogonal transform," *IEEE Trans. Communications*, vol. 37, no. 2, pp. 189–193, 1988.

[21] H. S. Malvar and D. H. Staelin, "Reduction in blocking effects in image coding with a lapped orthogonal transform," in *Proc. ICASSP*, (New York), pp. 781–784, 1988.

[22] P. J. Burt, "Fast filter transforms for image processing," *Computer Graphics and Image Processing*, vol. 16, pp. 20–51, 1981.

[23] P. J. Burt and E. H. Adelson, "The laplacian pyramid as a compact image code," *IEEE Trans. Communications*, vol. COM-31, pp. 532–540, April 1983.

[24] R. Schäfer, P. Kauff, and U. Gölz, "On the application of spatio-temporal contrast sensitivity functions to HDTV," in *Conference on Applied Vision*, (San Fransico), pp. 118–121, Optical Society of America, July 1989.

[25] A. Croisier, D. Esteban, and C. Galand, "Perfect channel splitting by use of interpolation/decimation/tree decomposition techniques," in *International Conference on Information Sciences and Systems*, (Patras), pp. 443–446, August 1976.

[26] D. Esteban and C. Galand, "Application of quadrature mirror filters to split band voice coding schemes," in *Proceedings ICASSP*, pp. 191–195, 1977.

[27] E. H. Adelson, E. Simoncelli, and R. Hingorani, "Orthogonal pyramid transforms for image coding," in *Proceedings of SPIE*, vol. 845, (Cambridge, MA), pp. 50–58, October 1987.

[28] S. G. Mallat, "A theory for multiresolution signal decomposition: The wavelet representation," *IEEE Trans. PAMI*, vol. 11, pp. 674–693, July 1989.

[29] M. Vetterli, "Multi-dimensional sub-band coding: Some theory and algorithms," *Signal Processing*, vol. 6, pp. 97–112, February 1984.

[30] P. P. Vaidyanathan, "Quadrature mirror filter banks, M-band extensions and perfect-reconstruction techniques," *IEEE ASSP Magazine*, pp. 4–20, July 1987.

[31] M. Vetterli, "A theory of multirate filter banks," *IEEE Trans. ASSP*, vol. ASSP-35, pp. 356–372, March 1987.

[32] J. D. Johnston, "A filter family designed for use in quadrature mirror filter banks," in *Proceedings ICASSP*, vol. 1, pp. 291–294, 1980.

[33] V. K. Jain and R. E. Crochiere, "A novel approach to the design of analysis/synthesis filter banks," in *Proceedings ICASSP*, pp. 5.10–5.10, 1983.

[34] V. K. Jain and R. E. Crochiere, "Quadrature mirror filter design in the time domain," *IEEE Trans. ASSP*, vol. ASSP-32, pp. 353–360, April 1984.

[35] G. Wackersreuther, "On the design of filters for ideal QMF and polyphase filter banks," *Arch. Elekt. Ubertrag*, vol. 39, pp. 123–130, 1985.

[36] E. P. Simoncelli, "Orthogonal sub-band image transforms," Master's thesis, Massachusetts Institute of Technology, Department of Electrical Engineering and Computer Science, Cambridge, MA, May 1988.

[37] M. J. T. Smith and T. P. Barnwell, III, "A procedure for designing exact reconstruction filter banks for tree-structured subband coders," in *Proceedings ICASSP*, vol. 2, pp. 27.1.1–27.1.4, 1984.

[38] E. H. Adelson and E. P. Simoncelli, "Subband image coding with three-tap pyramids," in *Picture Coding Symposium*, (Cambridge, MA), March 1990.

[39] E. P. Simoncelli and E. H. Adelson, "Non-separable extensions of quadrature mirror filters to multiple dimensions," in *Proceedings of the IEEE: Special Issue on Multidimensional Signal Processing*, April 1990.

[40] J. P. Crettez and J. C. Simon, "A model for cell receptive fields in the visual striate cortex," *Computer Graphics and Image Processing*, vol. 20, pp. 299–318, 1982.

[41] A. B. Watson and A. J. Ahumada, "A hexagonal orthogonal-oriented pyramid as a model of image representation in visual cortex," *IEEE Trans. Biomedical Engineering*, vol. 36, pp. 97–106, January 1989.

[42] R. M. Mersereau and T. C. Speake, "The processing of periodically sampled multidimensional signals," *IEEE Trans. ASSP*, vol. ASSP-31, pp. 188–194, February 1983.

[43] D. E. Dudgeon and R. M. Mersereau, *Multidimensional Digital Signal Processing*. Signal Processing Series, Englewood Cliffs, NJ: Prentice-Hall, 1984.

[44] E. Viscito and J. Allebach, "Design of perfect reconstruction multi-dimensional filter banks using cascaded Smith form matrices," in *Proceedings ISCAS*, pp. 831–834, 1988.

[45] J. W. Woods and S. D. O'Neil, "Subband coding of images," *IEEE Trans. ASSP*, vol. ASSP-34, pp. 1278–1288, October 1986.

[46] H. Gharavi and A. Tabatabai, "Sub-band coding of digital images using two-dimensional quadrature mirror filtering," in *Proceedings of SPIE*, vol. 707, pp. 51–61, 1986.

[47] H. Gharavi and A. Tabatabai, "Application of quadrature mirror filters to the coding of monochrome and color images," in *Proceedings ICASSP*, pp. 32.8.1–32.8.4, 1987.

[48] A. Tran, K.-M. Liu, K.-H. Tzou, and E. Vogel, "An efficient pyramid image coding system," in *Proceedings ICASSP*, vol. 2, pp. 18.6.1–18.6.4, 1987.

[49] W. Chen and W. Pratt, "Scene adaptive coder," *IEEE Trans. Communications*, vol. COM-32, pp. 225–232, March 1984.

[50] P. H. Westerink, D. E. Boekee, J. Biemond, and J. W. Woods, "Subband coding of images using vector quantization," *IEEE Trans. COM*, vol. COM-36, pp. 713–719, June 1988.

Chapter 5

Subband Coding of Color Images

by: Peter H. Westerink, Jan Biemond and Dick E. Boekee
    Department of Electrical Engineering
    Delft University of Technology
    2600 GA Delft, The Netherlands

The digital encoding of color images has received considerably less attention than the coding of monochrome images. And although most image coding algorithms indeed concentrate on the case of monochrome images, it is clear that color pictures are generally far more preferred and appreciated than monochrome images by the human observer. Modern image compression schemes therefore will all be suitable for color images, while for typical monochrome applications such as medical image data storage and the transmission of newspaper pictures, they will also be compatible with monochrome image compression. For example, both in present analog color television (NTSC/PAL/SECAM) and in the color television standard for digital television (the CCIR recommendation 601, [8]), the system is compatible with monochrome systems. Thus we will first discuss the encoding of monochrome images using subband coding. Then we will extend the technique to color images.

The general idea of subband coding is that each subband will have a coder and bit rate accurately matched to the properties of that particular band. Especially, an adaptive bit allocation algorithm that distributes the bits among the subbands will make the subband coding scheme suitable for a wide variety of images. It is clear that we need to know the subband statistics to optimally

design a subband coding scheme with separate subband encoders. To that
end, in this chapter we will first consider the general subband coding scheme
as depicted in Figure 5.1. Typically, the input signal is first split into subbands.
Next the subbands are encoded, transmitted (or stored) and decoded. Finally,
the signal is reconstructed again using the decoded subbands. In this case
the channel in Figure 5.1 represents the communication channel including the
channel encoder and decoder.

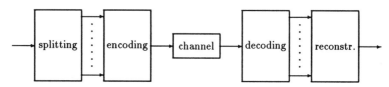

Figure 5.1: General subband coding scheme.

For the separate encoding of the subbands coding techniques such as PCM,
DPCM and vector quantization can be used and it also possible to apply block
DCT encoding on the subbands. The goal of this chapter is to investigate
a number of possible different coding methods for separately encoding the
subbands, extend these techniques to color images and present examples of
subband coded color images.

In Section 5.1 we will first investigate which coding methods are suitable for
the efficient encoding of the separate subbands. Next, in Section 5.2 the actual
quantizers will be designed and examples are given. As a splitting scheme we
will use the scheme as proposed in [46] and the bit allocation algorithm from
[43] will be applied. At that stage we will have designed a subband coder for
monochrome images. Examples of subband encoded monochrome images and
comparisons to other image coding techniques will be given in Section 5.3.

In Section 5.4 the extension to color images will be made. To that end,
first a color domain will be chosen that is suitable for the subband coding of
color images. Next, a color error distortion measure will be introduced, which
is necessary for a bit allocation algorithm to distribute the bits among the
subbands *and* among the color components. Examples of subband coded color
images will be given. Finally, in Section 5.5 some conclusions will be drawn.

5.1   Coding Method

The encoding of subbands in speech coding has been investigated quite inten-
sively, and good results have been obtained in the medium and low bit rate

speech coding. For the encoding of the separate subbands several different coding techniques have been employed. The most commonly applied methods are PCM [14, 16], DPCM [15, 24], vector quantization (VQ) [2, 3, 13, 21, 31, 41] and adaptive predictive coding (APC) [36, 39], while often also combinations and/or modifications of these coding techniques are applied.

Examples of subband coding of images where the subbands are encoded separately can be found, for example, in [23, 29, 48]. In [48] a 256x256 image is split into 16 equally sized subbands, using a splitting tree structure. The subbands are all encoded using adaptive DPCM, each with its own (linear) predictor. In [23] the 512x512 images are split into only 7 subbands, also using a splitting tree structure. In this case the subbands do not all have the same dimensions. The subband containing the lowest frequencies is encoded using DPCM, while the other subbands are encoded with PCM. The block DCT encoding of the lowpass subband is applied to subband video coding in [4], where this lowpass subband is the result of splitting the original image into only 3 subbands. The dimension of this subband is still 640x360. Vector quantization of the separate image subbands has been applied in [29], using finite state vector quantization (FSVQ).

Here, we make a choice between possible coding methods based on subband statistics. For example, based on measurements of the autocovariance matrix of each subband, in [23] it is concluded that there is no need to encode all the subbands using DPCM. In fact, only the lowpass subband has an autocovariance matrix with large coefficients and is DPCM encoded, while the other subbands are encoded using PCM. As an example, in Table 5.1 the 3x3 autocorrelation matrices are shown for the lowpass subband (subband 1) and some other subbands of the Clown image. The subbands that are not shown have similar autocorrelation matrices as the high-pass subbands shown in the table. The splitting scheme used with the subband numbering is displayed in Figure 5.7(b), which is shown in Section 5.3. The total number of subbands is 28. It must be noted here that similar autocovariance matrices for the subbands are obtained if different splitting schemes or other images are used. In Table 5.1 it can be seen that only subband 1 is correlated, while the other subbands appear to be almost uncorrelated.

We will restrict ourselves to the choice of PCM or DPCM. To that end, in Table 5.2 the prediction gain of DPCM over PCM is shown for the same subbands of the Clown image as in the example of Table 5.1. The prediction gain G_P is defined as the ratio σ_x^2/σ_d^2 between the signal variance σ_x^2 and the prediction error variance σ_d^2 [27]. From this table it is clear, that only subband 1 will really benefit from DPCM over PCM encoding. The other remaining subbands (including the subbands not shown) do not show a significant decrease in variance after prediction and it is therefore sufficient to encode these using

Subband	Autocovariance matrix		
1	0.763	0.860	0.771
	0.859	1.000	0.859
	0.771	0.860	0.763
4	-0.018	-0.226	0.052
	-0.147	1.000	-0.147
	-0.052	-0.226	-0.018
8	-0.079	-0.128	-0.088
	0.227	1.000	0.227
	-0.088	-0.128	-0.079
23	0.134	0.410	0.125
	0.282	1.000	0.282
	0.135	0.410	0.134

Table 5.1: Autocovariance matrices (3x3) of some subbands of the Clown image.

Subband	G_P
1	7.44
4	1.08
8	1.07
23	1.29

Table 5.2: Prediction gain G_P of DPCM over PCM for some subbands of the Clown image.

PCM [46]. Note that this is a similar conclusion as was reached in [23].

The (linear) prediction model that was used to fill Table 5.2 is the 3-point "Quarter Plane" (QP) model as shown in Figure 5.2.

$$
\begin{array}{ccccc}
\cdot & \cdot & \cdot & \cdot & \\
\cdot & c_{11} & c_{10} & \cdot & \\
\cdot & c_{01} & \times & \cdot & \\
\cdot & \cdot & \cdot & \cdot &
\end{array}
$$

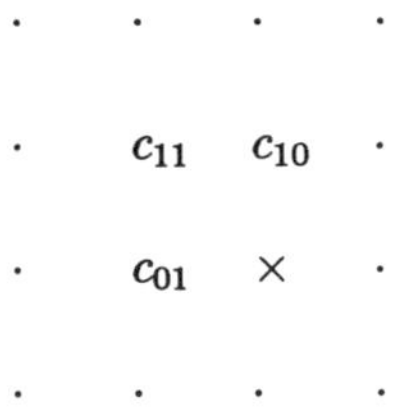

Figure 5.2: Quarter plane linear prediction model for DPCM of the lowpass subband, with the prediction coefficients c_{ij}.

From Table 5.2 it can be seen that this model yields a prediction gain of 7.44 for subband 1. Under the restriction of causality it is also possible to use other prediction models within the DPCM loop. For example, the QP model of Figure 5.2 can be fitted to the separable negative exponential autocorrelation function (acf). This function is used quite often in image processing because of its separability and because it has a reasonably good fit to the autocorrelation matrix of images [27]. The separable negative exponential acf is given by

$$
R_x(k, \ell) = \sigma_x^2 \, \rho_v^{|k|} \, \rho_h^{|\ell|}, \tag{5.1}
$$

where ρ_v and ρ_h are respectively the vertical and horizontal correlation coefficients. It can be shown that for this separable acf the QP model yields the minimum variance causal prediction model with prediction coefficients $c_{01} = \rho_h$, $c_{10} = \rho_v$ and $c_{11} = -\rho_h \rho_v$ [26, 27]. Experiments show that the autocorrelation matrix of the lowpass subband can be fitted very well to the model of (5.1). Also, when the prediction coefficients are forced to fit the separable model, nearly the same values are obtained as in the non-separable case. Indeed, applying the separable model to subband 1 of the Clown image a value of 7.32 is found for the prediction gain G_P. It is therefore not surprising that more complex (causal) prediction models, such as the 4-point "Non-Symmetric Half Plane" (NSHP) model, do not yield a significant increase of the prediction gain. A similar conclusion holds for the prediction of full band images [27]. For the DPCM encoding of subband 1 we therefore choose the QP model as shown in Figure 5.2. Finally, because the prediction coefficients are easier to compute for the non-separable model than for the separable model (which puts constraints on the coefficients), the non-separable QP model will be used.

5.2   Subband Quantizer Design

This Section first covers the modeling of the image data by a histogram-fitting technique. This is then followed by a discussion on Lloyd-Max quantizer design and then a presentation on variable length Huffman coding.

5.2.1   Histogram Fitting

In order to design the quantizers for the DPCM and PCM encoders, we will take a closer look at the histograms of the subband values. For the lowpass subband this means that we use the prediction error values without quantization (assuming that the quantization error is small). In [48], where all subbands are encoded using adaptive DPCM, it is shown that the histograms of the prediction errors of the subbands can all be fitted to the two-sided negative exponential probability density function (pdf). This pdf is also known as the *Laplacian* pdf. The same approach is initially taken by Gharavi *et al.* in [23], where the quantizers for both the lowpass subband DPCM encoder and the PCM encoders are based on the Laplacian pdf. However, after a subjective evaluation they do not find the quantizers to be suitable. Therefore, they design new (symmetric) quantizers by defining a "dead zone" around the zero level of the quantizer. This means that quantizer representation levels close to the zero level are removed. The remaining levels are positioned at uniform distances. These new quantizers are not based on a particular pdf, but in [23] it is motivated that in this manner the noise around the zero level of the quantizer is set to zero and therefore reduced.

In Figure 5.3 the histogram of subband 8 of the Clown image is shown, together with the Laplacian pdf for equal variance. This particular subband histogram has the typical shape that all subband histograms show and is therefore a good subband histogram representative. Just as in the previous section we have used the splitting scheme as depicted in Figure 5.7(b), and once more it must be noted that other splitting schemes and/or images yield similar results. It can easily be seen that the two figures do not match. The subband histogram is more peaked and narrow. Therefore, in this early stage the conclusion can be drawn that *the Laplacian pdf does not fit the subband data well*.

Let us next introduce a class of probability density functions that is known as the Generalized Gaussian pdf. This pdf is given by

$$p(x) = a \exp \left\{ -|bx|^{\gamma} \right\}, \qquad (5.2)$$

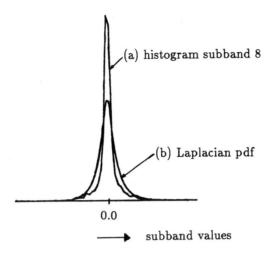

Figure 5.3: Typical subband histogram (a) and the Laplacian pdf (b).

where $\Gamma(.)$ is the Gamma function [1] in

$$a = \frac{b\gamma}{2\Gamma(\frac{1}{\gamma})}, \tag{5.3}$$

and

$$b = \frac{1}{\sigma_x} \sqrt{\frac{\Gamma(\frac{3}{\gamma})}{\Gamma(\frac{1}{\gamma})}}. \tag{5.4}$$

Note that we have two special cases for the parameter γ. For $\gamma = 1.0$ we have the Laplacian pdf and for $\gamma = 2.0$ we have the well-known Gaussian pdf. The Generalized Gaussian pdf is plotted for three different values of the parameter γ in Figure 5.4. The parameter γ clearly determines the shape of the pdf. At first sight the value of $\gamma = 0.5$ seems to yield a pdf that has a closer fit to the subband histogram in Figure 5.3 than for $\gamma = 1.0$, the Laplacian pdf. Therefore, we will next determine the value of the shape parameter γ that best fits the subband histograms.

For the fitting of data to distribution or density functions many statistical tests are known from the literature [12]. The two most commonly known and most frequently used goodness-of-fit tests are the Kolmogorov-Smirnov (KS) test and the chi-squared (χ^2) test. The KS test measures the distance between *distribution* functions, while the χ^2 test measures the (weighted) distance between *density* functions. There is a controversy over which test is the most powerful, but the general feeling seems to be that the Kolmogorov-Smirnov

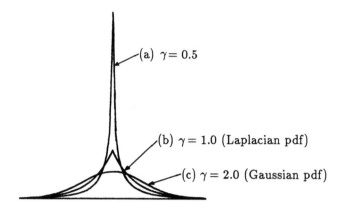

Figure 5.4: Generalized Gaussian probability density function (pdf) for (a) $\gamma = 0.5$, (b) the Laplacian pdf, $\gamma = 1.0$ and (c) the Gaussian pdf, $\gamma = 2.0$.

test is probably more powerful than the chi-squared test in most situations [12]. Further comparisons between the two tests are presented in [38]. To postpone the decision of choosing between the two tests, we will employ both the Kolmogorov-Smirnov goodness-of-fit test and the chi-squared goodness-of-fit test to fit the subband histograms to the Generalized Gaussian pdf for a number of values of the shape parameter γ.

For a given set of data $X = (x_1, x_2, \cdots, x_M)$, the Kolmogorov-Smirnov test compares the sample distribution function $F_X(\cdot)$ to a given distribution function $F(\cdot)$. The KS test statistic t_{KS} [30] is then defined by:

$$t_{KS} = \sqrt{M} \max_{i=1,2,\dots,M} |F_X(x_i) - F(x_i)|. \tag{5.5}$$

The KS test statistic t_{KS} is a measure of the distance between the sample distribution function and the given distribution function. When testing data against several distributions, the distribution that yields the smallest KS test statistic t_{KS} is the best fit to the data.

Using the sample density function $f_X(\cdot)$ and the given density function $f(\cdot)$ the chi-squared test statistic t_{χ^2} [30] is defined as:

$$t_{\chi^2} = M \sum_{i=1}^{M} \frac{[f_X(x_i) - f(x_i)]^2}{f(x_i)}. \tag{5.6}$$

As with the KS test, the lowest value for the chi-squared test statistic t_{χ^2} yields the best fit to a probability density function.

For image block DCT coefficients the statistical distributions have been investigated in [17] and [35]. In [35] model fits were made to the Gaussian and to the Laplacian pdf. For that purpose the Kolmogorov-Smirnov goodness-of-fit test was used. It was found that the DC coefficient can be modeled by a Gaussian pdf, while the other coefficients can be modeled by the Laplacian pdf. The same test was applied in [17], but next to the Laplacian and the Gaussian pdf's the Cauchy distribution was also considered. Again it was found that the Laplacian pdf provided the best fit for the majority of the coefficients.

For fitting the subband data both the KS test and the χ^2 test are applied. For the lowpass subband the tests are applied to the prediction error. The subband distributions (for the KS test) and the subband densities (for the χ^2 test) are fitted to the Generalized Gaussian pdf for different values of the shape parameter γ with a stepsize of 0.05. As an example, the values of the KS test statistic t_{KS} and of the χ^2 test statistic t_{χ^2} are plotted for different values of γ in Figure 5.5, for subband 23 of the Clown image. In both cases a

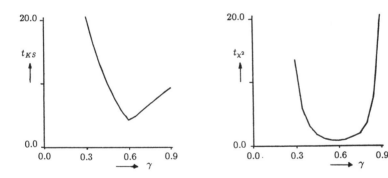

Figure 5.5: Kolmogorov-Smirnov test statistic t_{KS} (a) and chi-squared test statistic t_{χ^2} (b) for different values of γ for subband 23 of the Clown image.

value of $\gamma = 0.60$ yielded the minimum test statistic. It can be seen, that the shape of the KS test results follows the line of the absolute form $|\cdot|$ according to the definition in (5.5). The same holds for the χ^2 test results where the definition of (5.6) results in a parabola according to $(\cdot)^2$.

Finally, the test results for all the subbands of the Clown image are shown in Table 5.3. Also shown are the prediction error variance of subband 1 and the variances of the other subbands.

Subband	Variance	$\gamma_{opt,KS}$	γ_{opt,χ^2}
1	528.08	0.75	0.70
2	77.18	0.65	0.60
3	67.20	0.45	0.50
4	19.19	0.50	0.50
5	7.04	0.50	0.55
6	50.76	0.55	0.55
7	2.94	0.50	0.50
8	7.30	0.50	0.50
9	3.29	0.60	0.60
10	1.73	0.60	0.60
11	13.86	0.50	0.45
12	6.16	0.50	0.50
13	0.66	0.70	0.60
14	1.17	0.60	0.60
15	1.59	0.55	0.50
16	3.57	0.50	0.50
17	1.01	0.60	0.55
18	6.45	0.50	0.50
19	0.47	0.70	0.60
20	1.90	0.60	0.60
21	0.58	0.95	0.65
22	0.45	1.10	0.80
23	3.31	0.60	0.60
24	1.47	0.70	0.65
25	0.22	0.90	0.85
26	0.42	1.10	0.90
27	0.28	0.85	0.65
28	0.76	0.85	0.70

Table 5.3: Kolmogorov-Smirnov (KS) and chi-squared (χ^2) test results for the subband data of the Clown image.

For most of the subbands the KS test and the χ^2 test yield the same or almost the same value of the shape parameter γ. The most common outcome of the tests is for $0.50 \leq \gamma \leq 0.60$. However, from subband number 21 up to number 28 the results of both tests yield much higher values of γ. Moreover, in these cases the results of both tests lie farther apart than with the other subbands. However, the variances of these subbands are very low when compared to the other subbands. In fact, the lower the subband variance, the more the best found value for γ deviates from the range $0.50 \leq \gamma \leq 0.60$. Measurements of the test statistics t_{KS} and t_{χ^2} for other images, such as the Building image, show similar behavior. Hence, there is a close connection between the subband variance and the pdf model fitting results. This can be explained by reasoning that in subbands where the variance is very low the noise variance within that subband is relatively high. The noise can for example be due to the recording system and/or the digitization process. In general, the measured probability density function of the subbands is a combination of the signal pdf and the noise pdf. For example, in the case of additive noise the resulting pdf is the convolution of the noise pdf with the signal pdf. For the high variance subbands we have relatively very little noise, but in the low variance subbands the noise pdf cannot be neglected. Although the actual subband signal (of a natural scene) without noise might very well be modeled by the Generalized Gaussian pdf with $0.50 \leq \gamma \leq 0.60$, in general the presence of the noise will alter this pdf, and the model fitting test results will be influenced by the noise. On the other hand, in image coding we want to encode a given signal as accurately as possible, that is, including noise if present. Fortunately, we do not need to bother much about these subbands, since in most cases they will not be encoded at all. Moreover, coding errors in these bands are negligible when compared to the errors made in other bands (that have much higher variances). Here, we will choose to design the quantizers for *all* the high frequency subbands on the basis of the Generalized Gaussian pdf with $\gamma = 0.50$.

The fitting results to the lowpass subband yield somewhat different results. The histogram of that subband is less peaked than the other subbands and is closer to the Laplacian pdf. However, an exact fit to the Laplacian pdf is also in this case not justified, see for instance Table 5.3. Experiments on several images show that a matching pdf for the lowpass subband is obtained for the Generalized Gaussian pdf with $\gamma = 0.75$.

5.2.2   Lloyd-Max Quantizers

Quantizers that minimize the mean-squared error are also known as Lloyd-Max quantizers. If the quantization error between the input signal x and the quantized signal y is defined as

$$q = y - x \tag{5.7}$$

then the error criterion that is minimized is written as:

$$\sigma_q^2 = \sum_{k=1}^{L} \int_{x_k}^{x_{k+1}} (x - y_k)^2 \, p_x(x) dx, \tag{5.8}$$

where $x_k, k = 1, \ldots, L + 1$ are the decision levels, $y_k, k = 1, \ldots, L$ are the representation levels and L is the number of levels. By partial differentiation it can be shown [27] that the optimum decision levels are given by

$$
\begin{aligned}
x_{1,opt} &= -\infty, \\
x_{k,opt} &= \tfrac{1}{2}\left(y_{k,opt} + y_{k-1,opt}\right), \qquad k = 2, 3, \ldots, L, \\
x_{L+1,opt} &= \infty,
\end{aligned}
\tag{5.9}
$$

and the representation levels are obtained by

$$y_{k,opt} = \frac{\displaystyle\int_{x_{k,opt}}^{x_{k+1,opt}} x\, p_x(x) dx}{\displaystyle\int_{x_{k,opt}}^{x_{k+1,opt}} p_x(x) dx}, \qquad k = 1, 2, \ldots, L. \tag{5.10}$$

Consequently, if we known the probability density function of the data we want to quantize, the Lloyd-Max quantizer is determined by (5.9) and (5.10). Note, that the minimum as described in (5.9) and (5.10) may be a *local* minimum [19].

Basically, there are two design methods for Lloyd-Max quantizers [19]. Since the solution to (5.9) and (5.10 is non-explicit, both procedures are iterative. The first one starts by assigning an initial value to y_1 and solves (5.10) for x_2. Next, the value of y_2 is computed using (5.9). This procedure is repeated until y_L is found. It is then checked whether this value is close enough to the right-hand term of (5.10) and based on this the initial guess of y_1 is changed and the iteration is continued. However, a problem of this algorithm is that it is not obvious with how much the initial guess much be changed after each iteration. Furthermore, solving (5.10) to obtain $x_{k+1,opt}$ can be quite difficult. To avoid these problems, we have used the other quantizer design method. It starts with an inital guess of all representation levels, which can, for instance, be the levels of a uniform quantizer. Next, by means of (5.9) the corresponding decision levels are calculated and with these a new set of representation levels is computed using (5.10). The iteration is repeated until a certain accuracy in the total mean-squared error is reached. Note, that the LBG algorithm for codebook design for vector quantizers is the higher dimensional vector version of this algorithm (although not based on a probability density function). The drawback of the algorithm is its rather slow convergence, but this has not been found a problem.

In this manner quantizers have been designed for the Generalized Gaussian pdf with $\gamma = 0.75$ (for the lowpass subband) and with $\gamma = 0.50$ (for all the other subbands). The number of levels of the quantizers are 1, 2, 3, 4, 5, 6, 7, 8, 15, 16, 31, 32, 63, 64 and 128.

As an example, in Table 5.4 the representation levels are shown for 7 level (symmetric) Lloyd-Max quantizers for the Laplacian pdf and for the Generalized Gaussian (GG) pdf with $\gamma = 0.50$. The corresponding optimal decision levels lie centered between every two representation levels according to (5.9). Both quantizers in Table 5.4 are for zero mean and unity variance. Note that

	Laplacian	GG, $\gamma = 0.50$
1	-2.85	-5.26
2	-1.44	-2.15
3	-0.60	-0.74
4	0.00	0.00
5	0.60	0.74
6	1.44	2.15
7	2.85	5.26

Table 5.4: Representation levels for (symmetric) 7 level Lloyd-Max quantizers for the Laplacian pdf and for the Generalized Gaussian (GG) pdf with $\gamma = 0.50$.

the representation levels of the *smaller* pdf lie *further* apart, which is the result of minimizing the quantization error variance. This is in contrast with maximum output entropy quantization [33] where the interval probabilities are designed to be equal. In that case the representation levels of a narrow pdf will have many levels around the peak. In Section 5.2.1 it was already noted that in [23] the quantizers based on a Laplacian pdf were found not to be suitable and were replaced by quantizers with a larger "dead zone" in the center of the quantizer. In fact, here we see that this effect is automatically created by applying quantizers based on a pdf that fits the data better, i.e. the Generalized Gaussian pdf with $\gamma = 0.50$.

5.2.3 Coding of the Quantizer Outputs

Using the decision levels of a Lloyd-max quantizer it is possible to calculate the occurrence probabilities of the representation levels. For the peaked pdf we are considering, these levels lie relatively far apart and therefore we expect the probabilities to be quite different. For instance, the zero representation level will have a relatively large probability. In that case, it possible to remove further redundancy from the signal by applying "entropy encoding" of the quantizer outputs [7]. In practice this is accomplished with variable length encoding, such as Huffman coding [25, 7] or Shannon coding [7], where quantizer outputs with high probabilities are encoded with shorter codewords than others. As an example, in Table 5.5 these probabilities are given of the representation levels as shown in the example of Table 5.4. Also shown are the assigned codewords when Huffman coding is applied. It can be seen that the zero level, which has the largest probability, has the shortest codeword.

	Level	Probability	Codeword
1	-5.26	0.0063	101010
2	-2.15	0.0402	1011
3	-0.74	0.1551	11
4	0.00	0.5969	0
5	0.74	0.1551	100
6	2.15	0.0402	10100
7	5.26	0.0063	101011

Table 5.5: Codewords and probabilities corresponding to the representation levels of the 7 level Lloyd-Max quantizer for the Generalized Gaussian pdf with $\gamma = 0.50$.

If we assign a codeword of length l_i to the quantizer output i with probability p_i, we can calculate the average codeword length:

$$\bar{l} = \sum_{i=1}^{L} p_i l_i \qquad (5.11)$$

Codes such as the Huffman code or the Shannon code minimize the quantity $\bar{l}$ and thus establish a minimal (average) bit rate to transmit the quantizer representation levels to the receiver. The lower bound of the performance of variable length codes is given by the entropy H [6, 7], which is defined as

$$H = -\sum_{i=1}^{L} p_i \log_2 p_i. \qquad (5.12)$$

For Table 5.5 the entropy is $H = 1.74$, while the average codeword length of the Huffman code is approximately $\bar{l} = 1.81$ bits/sample. This is a considerable gain in bits/sample, since in the case we do not apply Huffman coding to the quantizer outputs, each representation level in Table 5.5 will be assigned a fixed length codeword consisting of 3 bits (leaving one codeword unused). In Figure 5.6 the gain is shown when the Clown image is encoded with or without Huffman coding. It can be seen that the gain in SNR is approximately 1.5 dB.

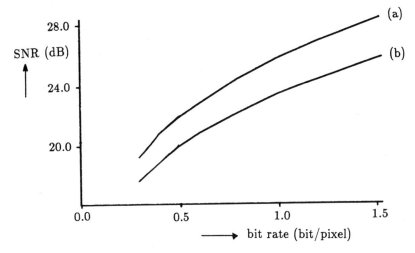

Figure 5.6: Encoding of the Clown image, (a) with Huffman encoding and (b) with fixed length coding of the quantizer outputs.

The described approach to design the quantizer is split into two separate optimization steps. First the average distortion is minimized given a number of representation levels (see Section 5.2.2), and next the average bit rate is minimized by applying variable length encoding. However, the entire quantizer design can also be formulated as a single optimization problem. In that case the distortion is minimized under the constraint of a given average bit rate. The difference from the two-step approach is that now the occurrence probabilities of the representation levels are taken into account while minimizing the distortion, while the constraint on the number of levels is dropped. In [18] Favardin calculates the performances of such quantizers for a number of different probability density functions, among which is the Generalized Gaussian pdf (with $\gamma = 0.50$). He shows that by increasing the number of representation levels quantizers can be obtained that have higher performances than the Lloyd-Max quantizers. However, the actual design is a formidable task. That burden is next shown to be overcome by using uniform threshold quantizers that have nearly the same performance, but it is stated that these will probably have a larger number of levels. The difference in performance between the Lloyd-Max quantizers and the optimal quantizers (with more representation levels) for the Generalized Gaussian pdf with $\gamma = 0.50$ is approximately 0.5 bits/sample at the same distortion.

One of the principal disadvantages of variable length encoding is the sensitivity to channel errors. Since codewords can have different lengths, the decoder must keep track of the beginning of each word. In the case of a channel error, the synchronization of the bit stream may be lost and the decoder then has no means of recovering the remaining codewords. Here, we will neglect the possibility of channel errors and investigate the performance of a subband coder using variable length encoding.

5.3   Coding Results for Monochrome Images

In the previous sections we have given methods for encoding the subbands. Here we will show some examples of subband coded monochrome images. To choose a splitting scheme we adopt the proposition made in [46]. This means, that for 256x256 and 512x512 images we use the splitting schemes that are shown respectively in Figure 5.7(a) and (b). The numbers of the subbands in

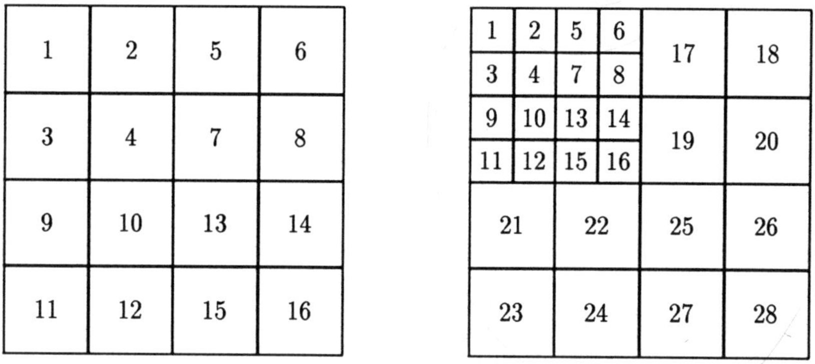

1	2	5	6
3	4	7	8
9	10	13	14
11	12	15	16

(a)

1	2	5	6	17	18
3	4	7	8		
9	10	13	14	19	20
11	12	15	16		
21	22			25	26
23	24			27	28

(b)

Figure 5.7: Best image frequency band divisions for subband coding of (a) 256x256 images and (b) 512x512 images (zero frequency is upper left).

this figure refer to the subband numbering as used earlier in this chapter. For the one-dimensional QMF we have used the filter as denoted by "16B" in [28].

5.3.1   Monochrome Coding Examples

To show what the subbands look like and to demonstrate the effect of the bit allocation algorithm, one would like to make the image subbands visible. The

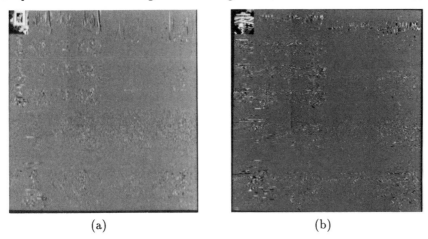

(a)   (b)

Figure 5.8: Visualized subbands of the 512x512 images Clown and Building.

subband containing the low frequencies is a lowpass version of the original image and can be regarded as an image itself. However, the other subbands contain negative values and have an average value of approximately zero. These higher frequency subbands therefore cannot be interpreted as "pictures" and if we want to make these bands visible special measures have to be taken. As an example, in Figure 5.8 the subbands of both the 512x512 images Clown and Building are visualized. This is done by setting the minimum (negative) value of a subband to black and the maximum value to white, while the in-between values are linearly stretched.

In the upper left corner of each picture in Figure 5.8 we have the lowpass subband, which is the only band that has clearly recognizable features of the original image. This is directly related to the amount of correlation that is still left within this subband. All the other subbands have a noisy uncorrelated character with high local variances around the edges and in the high frequency areas of the original images. The directional sensitivity of the higher frequency subbands can also be clearly seen. High frequencies in one direction appear only within subbands in one direction. This fact once again emphasizes the need for a per image adaptive quantization and bit allocation scheme. To demonstrate the effect of the adaptivity of the bit allocation algorithm, in Figure 5.9(a)-(d) the subbands are shown after encoding/decoding at 0.6 and 1.0 bits per pixel, respectively. The uniform gray areas in Figure 5.9 indicate that that particular subband has not been encoded at all. Also, the subbands that were encoded using a 3-level quantizer can be clearly recognized. The lowpass subband is encoded very well and cannot be seen to have any distortion. The numerical values corresponding to Figures 5.9(a) and (b) are listed in Table 5.6, where the variances, number of quantizer levels, bit rates and mean-squared errors are shown for each subband. Note, that the variances as indicated in Table 5.6 are the quantizer input variances, which means for subband 1 that the prediction error variance is listed.

Finally, the reconstructions that correspond to Figure 5.9 are shown in Figure 5.10. The reconstructions in Figure 5.10(a) and (c), which are at 0.6 bits per pixel, do show some minor distortions.

In particular the image Building suffers from the low bit rate coding. However, this image is far more difficult to encode due to the many edges and details. The distortions seem to be confined to flat areas around edges where the loss of higher frequency subbands is clearly noticeable. Also in the busy areas, such as the pencil pots in the image Clown, the images have a lowpass character and look somewhat unsharp. Fortunately, in general these typical subband coding errors are not too annoying, especially not when compared to the well-known blocking effects that occur in block transform coding or vector quantization. We will come to that in the next section. The coding results at 1.0 bits per pixel have a high quality, see Figure 5.10(b) and (d), and in this case coding errors are hard to find. It therefore depends on the application and the type of image, at what bit rate the compressed image is still acceptable. In any case, subband coding seems to be a very good image coding method, both at high and at low bit rates.

5.3.2   Comparison to Other Coding Techniques

The comparison to other coding techniques is carried out for two cases. First, we have implemented a coding scheme based on the 8x8 block DCT, where the complexity of the coding methods is the same as for the subband coding results. The differences in coding results are compared for the 512x512 image Clown, and will be the result of the fundamental differences between subband coding and block DCT. In the second case, for the 256x256 image Lenna, other coding results are taken from literature and are directly compared to subband coding results by means of an SNR measure.

To be able to strictly compare the coding results between the 8x8 block DCT encoding and subband coding we have implemented an 8x8 block DCT coding scheme with the same complexity as the subband coder. First, the image is subdivided into 8x8 blocks to which the 8x8 DCT is applied. Next, the corresponding DCT coefficients from the different blocks are assembled to obtain 64 sets of coefficients. It can be shown, that just as with subband coding, the (2-D) set of DC-coefficients can successfully be encoded using

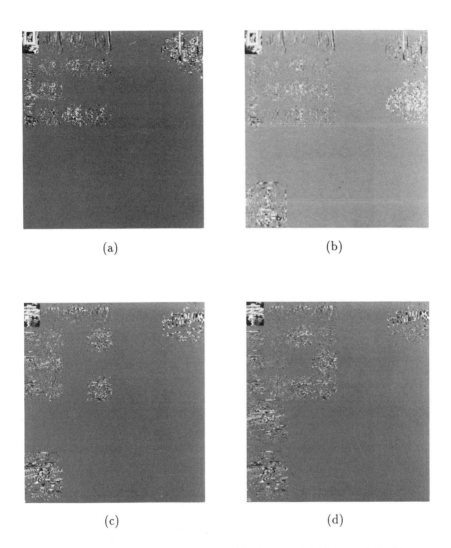

(a)   (b)

(c)   (d)

Figure 5.9: Coded/decoded subbands: (a) Clown at 0.6 bits per pixel, (b) Clown at 1.0 bits per pixel, (c) Building at 0.6 bits per pixel and (d) Building at 1.0 bits per pixel.

Subband	Variance	0.6 bits/pixel			1.0 bits/pixel		
		Levels	Bit rate	MSE	Levels	Bit rate	MSE
1	528.08	128	4.92	0.63	128	4.92	0.63
2	77.18	64	4.01	0.36	128	4.46	0.17
3	67.20	32	3.41	0.60	128	4.32	0.14
4	19.19	15	2.69	0.50	32	3.39	0.16
5	7.04	7	1.83	0.62	15	2.77	0.20
6	50.76	32	3.54	0.47	64	3.97	0.23
7	2.94	5	1.40	0.59	7	1.68	0.32
8	7.30	7	1.75	0.67	15	2.69	0.21
9	3.29	5	1.51	0.52	15	2.86	0.09
10	1.73	5	1.51	0.30	7	1.86	0.18
11	13.86	15	2.65	0.35	32	3.35	0.11
12	6.16	7	1.76	0.60	15	2.71	0.16
13	0.66	1	0.00	0.66	5	1.54	0.11
14	1.17	1	0.00	1.17	7	1.86	0.11
15	1.59	3	1.12	0.63	7	1.77	0.20
16	3.57	7	1.77	0.33	15	2.73	0.09
17	1.01	1	0.00	1.01	1	0.00	0.01
18	6.45	3	1.11	2.96	7	1.72	1.93
19	0.47	1	0.00	0.47	1	0.00	0.47
20	1.90	1	0.00	1.90	3	1.13	0.87
21	0.58	1	0.00	0.58	1	0.00	0.58
22	0.45	1	0.00	0.45	1	0.00	0.45
23	3.31	1	0.00	3.31	5	1.49	0.65
24	1.47	1	0.00	1.47	1	0.00	1.47
25	0.22	1	0.00	0.42	1	0.00	0.22
26	0.42	1	0.00	0.22	1	0.00	0.42
27	0.28	1	0.00	0.28	1	0.00	0.28
28	0.76	1	0.00	0.76	1	0.00	0.76

Table 5.6: Numerical coding values per subband when coding the Clown.

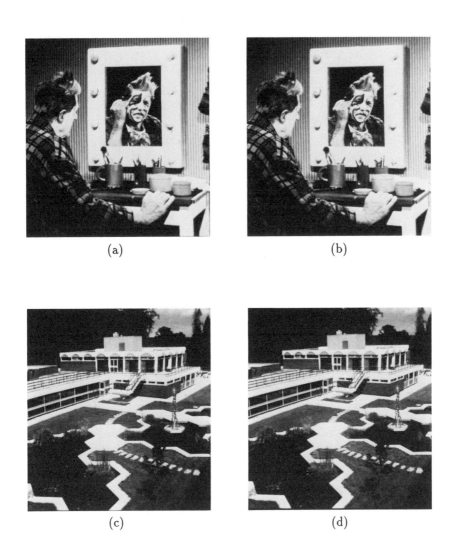

(a)   (b)

(c)   (d)

Figure 5.10: Reconstructions: (a) Clown at 0.6 bits per pixel, (b) Clown at 1.0 bits per pixel, (c) Building at 0.6 bits per pixel and (d) Building at 1.0 bits per pixel.

DPCM, while the other sets of coefficients can be very well encoded with a scalar quantizer. The histograms of respectively the prediction error and the other coefficient values are fitted to the Generalized Gaussian pdf. In contrast to the measurements performed in [35] and [17] the results yielded exactly the same Lloyd-Max quantizers as with subband coding for both the DPCM and the PCM encoders. The quantizer input values are always scaled with respect to their variance. Finally, the bit allocation algorithm as derived in [43] is used. It can be seen that the coding methods of the assembled DCT coefficients are exactly the same as with subband coding.

In Figure 5.11 for the 515x512 image Clown the SNR values are shown at various bit rates for both subband coding and 8x8 block DCT encoding.

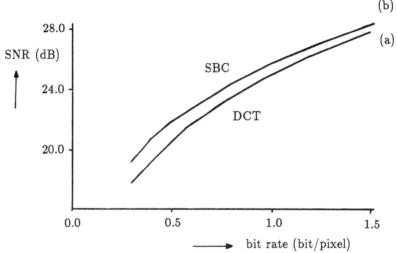

Figure 5.11: SNR versus bit rate for coding the Clown using (a) 8x8 block DCT coding and (b) subband coding.

It is clear, that subband coding is slightly superior to 8x8 block DCT encoding in SNR sense (ranging from 0.6dB to 1.4dB). A subjective evaluation of both coding results shows the differences in coding method even better. In Figure 5.12 two coding results are shown, both at 0.6 bits per pixel. Also shown are close-ups of the face in the mirror to make the differences more pronounced. Clearly, the DCT result has the blocking effects from which such images are known to suffer. The subband coding result does show coding error effects, but they are very different in character. The errors appear to have a slightly "muddy" character around edges and in flat areas. However, in general the blocking effects in the DCT image of Figure 5.12 are less appreciated than the subband coding errors that appear in Figure 5.12. From this comparison we can therefore conclude, that subband coding of images is preferable over

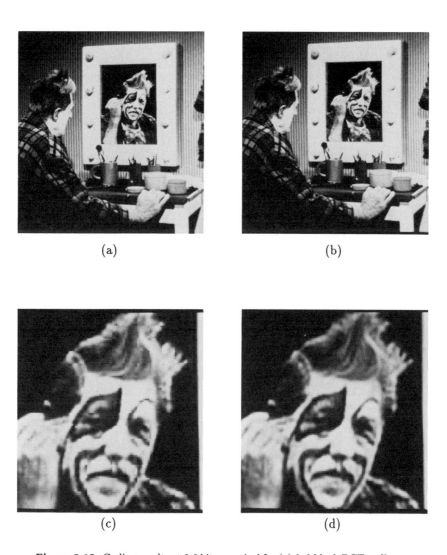

Figure 5.12: Coding results at 0.6 bits per pixel for (a) 8x8 block DCT coding, (b) subband coding, (c) close-up of the DCT result and (d) close-up of the SBC result.

the implemented 8x8 block DCT encoding, both in SNR and in subjective
sense.

To compare subband coding to coding results known from literature, the
256x256 image Lenna is encoded using subband coding at different bit rates.
As a distortion measure the SNR_{255} is used. The results are shown in Figure 5.13. It is apparent, that the subband coding result outperforms all the

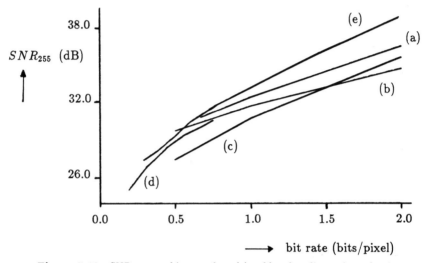

Figure 5.13: SNR versus bit rate for: (a) subband coding using adaptive
DPCM, (b) spatial differential VQ *inside* the training set, (c) adaptive DCT,
(d) subband coding using predictive VQ [42] and (e) subband coding (curves
(a)-(c) are taken from [48]).

other shown coding techniques in SNR_{255} sense. The performance of the
subband coding using adaptive DPCM [48] is also lower, because a number
of sub-optimal coding techniques were applied, such as quantizers based on a
Laplacian probability density function, Huffman encoding only for the quantizers with 3 and 5 levels, and a sub-optimal bit allocation algorithm. Although
the vector quantization result [5] is for coding *inside* the training set, this
is also included here, because it represents the upper bound of this coding
method. The curve for subband coding using predictive VQ as described in
[42] lies also under the curve for separately encoding of the subbands. Finally,
the adaptive DCT coding method [9] follows the curve of the subband coding
result, but with a fairly constant difference in SNR_{255} of about 2.5 dB. Applying "our" 8x8 block DCT encoding scheme on Lenna yielded also a worse
SNR_{255} performance than the subband coding result, with a difference of only
approximately 0.6 dB. This smaller difference is mainly due to the optimal
bit allocation and the Huffman coding of the quantizer outputs.

5.4   Color Images

The main extension from monochrome images to color is the addition of two chrominance components, the three together forming an image which is recognized by a human observer as a color picture. The fact that the color representation needs three components is based on the concept that the retina of the human eye is composed of three sets of receptors ("cones"). Each of these is sensitive to one of three principal colors, respectively corresponding to red, green and blue. This is also known as the *trichromacy* of man. The color perception and representation is therefore basically three-dimensional, and it is in such a three-dimensional color domain in which colors are perceived and evaluated.

In this Section, the topics of color perception and color representation will not be treated, but a review of these matters together with a large number of references on colorimetry and related topics can be found in [32]. Instead, we will concentrate on the aspects of choosing a proper color domain and a corresponding error criterion which is suitable for efficient subband coding. It is apparent that the choice of a particular color domain has its effect on the compressibility and quality of the color image. The color error criterion will determine the distribution of bits between the color components and is therefore also important for the final coder performance. In the CCIR digital television standard the amount of bits allotted to the luminance signal (the monochrome image) and chrominance signals (the 2 additional color images) is fixed to a (4:2:2) or a (3:1:1) ratio. However, since every scene or picture is different, we would rather like to have an adaptive distribution of bits between the luminance and the chrominance signals. For that purpose we will need a proper color distortion measure that can be optimized by the bit assignment to the luminance and chrominance signals.

5.4.1   Color Domain

As is widely known, the encoding of a color image in the RGB domain is not very efficient and it is better to use a color domain such as the NTSC television standard YIQ. The reason for this can be deduced from the differences between these color domains: in the RGB color domain the components are strongly correlated and the component variances are of the same order, while transformation of the RGB signals to the YIQ domain produces nearly decorrelated components and most of the signal energy is contained within the luminance (Y) signal.

Following [20], generally desirable properties of the color component images for efficient (subband-) coding can be formulated as:

1. Signal energy compaction.

2. Uncorrelated components.

3. An equivalence between a relatively simple distortion measure in the color domain (preferably the Euclidian distance) and the perception of color differences.

4. Black and white compatibility (e.g. one component should be directly related to the luminance).

5. Ease of implementation.

Both property 1 and 2 are necessary conditions for efficient data compression of the color image. Property 3 reflects the desire to have a distortion measure that is in accordance with the Human Visual System. Note, that this does *not* guarantee good data compressibility but merely defines the color error criterion. Property 4 is not directly related to image data compression, but, as has been mentioned, in general one wishes to be compatible with monochrome data compression schemes. Finally, property 5 is of importance for simple hardware realization.

Color spaces that are known from literature include the CIE 1974 color domains $L^*a^*b^*$ and $L^*u^*v^*$, which are especially designed for small and large color differences, respectively [11]. Applications of these color domains can be found in, for instance, computer graphics [40] and image display terminals [37]. However, experiments show that the two transformations yield high component cross correlations and a uniform energy distribution, which may not be good for color image compression. Straightforward subband coding of the color components shows very poor coding results. Without further investigations, at the moment, these two color spaces do not seem to be very suitable to be used in color image compression by means of subband coding.

A color domain based on a perception model was introduced by Frei and Baxter in [20]. The color components in this color space are fairly well uncorrelated, while the signal energy compaction is quite good as well. In the model, the neural response of the receptors in the retina is approximated by a logarithmic function, in accordance with the perception of intensity increments. Although it is pointed out that the approximation is incorrect for very low and very high levels of illumination, this is said to be acceptable. However, subband coding of the Clown image within this color domain, yields good coding results except in the areas of high illumination level. In the white areas of the image (the face and the frame of the mirror) color errors occur that are clearly noticeable and therefore unacceptable. Again we must therefore conclude that at this point this color domain cannot be usefully applied together with subband coding, and further investigations are needed.

The two color domains we will be using are the two well-known television standards as they are widely used in analog color television systems, the PAL/SECAM television standard YUV:

$$\begin{pmatrix} Y \\ U \\ V \end{pmatrix} = \begin{pmatrix} 0.299 & 0.587 & 0.114 \\ -0.146 & -0.288 & 0.434 \\ 0.617 & -0.517 & -0.100 \end{pmatrix} \begin{pmatrix} R \\ G \\ B \end{pmatrix}, \qquad (5.13)$$

and the NTSC television standard YIQ:

$$\begin{pmatrix} Y \\ I \\ Q \end{pmatrix} = \begin{pmatrix} 0.299 & 0.587 & 0.114 \\ 0.597 & -0.277 & -0.321 \\ 0.213 & -0.523 & 0.309 \end{pmatrix} \begin{pmatrix} R \\ G \\ B \end{pmatrix}. \qquad (5.14)$$

Both color domains are *linear* transformations of the R-, G- and B color components, and are, of course, invertible. Also, both have a color component corresponding to the luminance, the Y component, and are therefore compatible with monochrome image coding schemes.

In order to illustrate the desired color domain properties 1 and 2, in Table 5.7 the covariance matrix and the signal energy distribution are shown for the color domains RGB, YUV, and YIQ, for the 512x512 Clown image.

Covariance matrix			Color domain	Variances		
1.000	0.920	0.886	R	4419.5	=	33.0%
0.920	1.000	0.977	G	4347.1	=	32.5%
0.886	0.977	1.000	B	4629.3	=	34.6%
1.000	0.066	-0.067	Y	4228.7	=	92.6%
0.066	1.000	-0.685	U	63.0	=	1.4%
-0.067	-0.685	1.000	V	275.9	=	6.0%
1.000	-0.071	-0.026	Y	4228.7	=	92.6%
-0.071	1.000	0.533	I	295.3	=	6.5%
-0.026	0.533	1.000	Q	43.6	=	1.0%

Table 5.7: Covariances and energy distribution for the RGB, YUV and YIQ color domains for the Clown image.

It can be seen that both the YUV and YIQ color domains seem to yield good component decorrelation and signal energy compaction, in contrast to the RGB domain.

5.4.2   Color Error Criterion

In color television, the bit rate per color component is set to a fixed ratio. However, in order to profit from the advantages of subband coding, we would like to have a variable bit rate for each of the color component images, depending on their contents. In subband coding for monochrome images for instance, this is one of the main advantages, and bit allocation algorithms are used that dynamically (that is, per image) distribute the bits over the subbands [43]. To be able to do the same for color images, we need to have a color error criterion that can be optimized by such a bit allocation algorithm. For the YUV and YIQ color domains, however, there is no actual color error distortion measure available.

In order to be able to use the bit allocation algorithm as derived in the previous section, we require a color distortion measure that is additive with respect to the individual color component errors. If the errors per color component are defined as the mean-squared error, a distortion measure for the YIQ domain suitable for bit allocation is given by

$$D_{YIQ} = w_Y E\left[(\hat{Y} - Y)^2\right] + w_I E\left[(\hat{I} - I)^2\right] + w_Q E\left[(\hat{Q} - Q)^2\right], \qquad (5.15)$$

where w_Y, w_I and w_Q are positive weighting coefficients. For compatibility with distortion measures for monochrome images we can immediately set $w_Y = 1$. The choice of a value for the remaining two coefficients, however, is not obvious. In fact, the choice for w_I and w_Q is determined by a trade-off between color errors and luminance errors. It is generally known that it is not necessary to spend many bits on the chrominance signals. Therefore, considering the variances of these signals, see Table 5.7, it seems reasonable to choose equal weighting of all three color components. However, experiments show, that choosing lower values for w_I and w_Q than 1 yields better (subjective) results. Indeed, these results indicate a (subjective) proper division of bits between chrominance and luminance for $w_I = w_Q \approx 0.3$. Taking even lower values for the coefficients, such as $w_I = w_Q \approx 0.1$, yielded to much visible error in color.

Therefore, as a compromise we use for the YIQ domain as a distortion measure

$$D_{YIQ} = E\left[(\hat{Y} - Y)^2\right] + 0.3E\left[(\hat{I} - I)^2\right] + 0.3E\left[(\hat{Q} - Q)^2\right]. \qquad (5.16)$$

The results for the YIQ signals apply also to the YUV domain, where we use the similar distortion measure

$$D_{YUV} = E\left[(\hat{Y} - Y)^2\right] + 0.3E\left[(\hat{U} - U)^2\right] + 0.3E\left[(\hat{V} - V)^2\right]. \qquad (5.17)$$

Note, that these distortion measures are the result of the desire to apply the bit allocation algorithm as derived in [43] to the color domains YUV and YIQ.

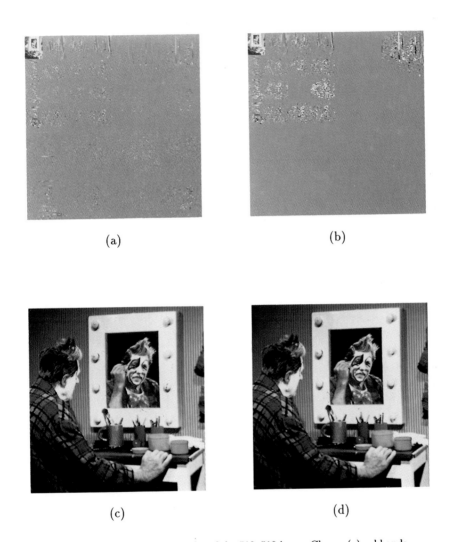

(a)   (b)

(c)   (d)

Figure 5.14: Color coding results of the 512x512 image Clown: (a) subbands, (b) decoded subbands at 0.8 bits/pixel, (c) reconstruction at 0.8 bits/pixel and (d) reconstruction at 1.2 bits/pixel.

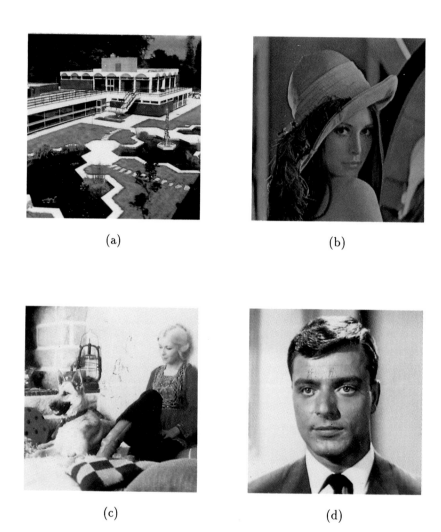

Figure 5.15: Color coding examples of 512x512 images at 0.8 bits/pixel: (a) Building, (b) Lenna, (c) Colette and (d) Jim.

The values of $w_I = w_Q = 0.3$ are experimentally determined to approximately make a good trade-off between color errors and luminance errors.

Finally, for maximum performance of the color subband coder, each (color component) subband must be encoded with an encoder that is optimally adapted to that subband. Statistical measurements from the subbands of monochrome images indicate that correlations within the subbands are very low except for the lowpass subband, see Section 5.1. It can be shown, that this property also holds for each color component, that is, for U, V, I and Q. Also, measurement of the histograms of the subbands of the color components in the two color domains shows that they are not different from the histograms of the monochrome image subbands. In that sense each color component (in YUV and YIQ) acts like a single monochrome image and the subband coders for the different components can be taken similar to the coders for monochrome images. For the bit allocation we again use the algorithm as described in [43], by which the additive distortion measures just derived are optimized.

5.4.3   Coding Results

In Figure 5.14(a) the color subbands are shown for the 512x512 color image Clown. Note that these subbands appear the same for both YUV and YIQ since they are linear transforms. It can be seen, that except for the lowpass subband, all subbands are nearly monochrome. This observation points towards a distribution of bits between luminance and chrominance that will be less than a (4:2:2) ratio. In Figure 5.14(b) the subbands are shown after encoding and decoding at 0.8 bits per pixel. The color domain that is used is YIQ. Most of the subbands are completely monochrome, since their chrominance signals were not encoded. The ratio between luminance (Y) and chrominance signals (I and Q) is approximately (4:1:1), which indeed is far less than the (4:2:2) ratio. Finally, the reconstruction is shown in Figure 5.14(c). The same image, but at a bit rate of 1.2 bits/pixel is shown in Figure 5.14(d). It can be seen that the quality of the images is very high.

Four other examples of subband coding of color images are shown in Figure 5.15 which are all encoded in the YIQ domain at 0.8 bits per pixel. Although we have so few bits for the chrominance components, the reconstructions do not seem to suffer from color errors. Finally, a subjective evaluation between the coding results in both color domains does not show any differences. It can therefore be concluded, that there is no real preference for encoding in either of the two domains, YUV or YIQ.

5.5   Conclusions

Subband coding of images is a very fruitful method for image data compression. It is preferable to other existing coding techniques, such as block DCT encoding and vector quantization, both in the objective (SNR) sense and in the subjective sense. The typical subband coding errors appear mainly in the vicinity of edges, when visible. The character of the errors, however, is better appreciated than the typical blocking effects that occur in, for instance, 8x8 block DCT encoding.

The encoding of color images can be done equally well in both the YIQ domain and the YUV domain, because the results do not show a noticeable difference in performance in the subjective sense. On the average, the bit ratio between the luminance and the chrominance signals is in the order of (4:1:1) which is a surprisingly low number of bits for the chrominance signals. Due to the adaptive bit allocation, however, this distribution of bits can vary dynamically. This is a direct extension of one of the main advantages of subband coding to the subband coding of color images.

Bibliography

[1] M. Abramowitz and I.A. Stegun, *Handbook of Mathematical Functions*, Dover Publications, Inc., New York, 1965.

[2] H. Abut and S.A. Luse, "Vector quantizers for sub-band coded waveforms", *Proc. Int. Conf. on Acous., Speech and Signal Process. (ICASSP)* , San Diego, CA, USA, paper 10.6, March 1984.

[3] H. Abut and S. Ergezinger, "Low Rate Speech Coding Using Vector Quantization and Sub-band Coding", *Proc. of the Int. Conf. on Acous., Speech and Signal Process. (ICASSP)* , Tokyo, Japan, pp. 449-452, April 1986.

[4] R. Ansari, H.P. Gaggioni and D.J. Le Gall, "HDTV Coding Using a Nonrectangular Subband Decomposition", *Proc. SPIE: Visual Communications and Image Processing*, vol. 1001, pp. 821-824, November 1988.

[5] R.L. Baker and R.M. Gray, "Image compression using non-adaptive spatial vector quantization", *Proc. of 16th Asilomar Conf.*, pp. 55-61, November 1982.

[6] T. Berger, *Rate Distortion Theory*, Prentice Hall, Inc., Englewood Cliffs, New Jersey, 1971.

[7] R.E. Blahut, *Principles and Practice of Information Theory*, Addison-Wesley Publishing Company, 1987.

[8] CCIR Recommendation 601, 1982.

[9] W-H. Chen and C.H. Smith, "Adaptive Coding of Monochrome and Color Images", *IEEE Trans. on Communications*, vol. COM-25, pp. 1285-1292, November 1977.

[10] W-H. Chen and W.K. Pratt, "Scene Adaptive Coder", *IEEE Trans. on Communications*, vol. COM-32, pp. 225-232, March 1984.

[11] CIE Colorimetry Committee, "Proposal for Study of Color Spaces and Color-Difference Evaluations", *Journal of the Opt. Soc. of America*, vol. 64, pp. 896-897, June 1974.

[12] W.J. Conover, *Practical Nonparametric Statistics*, John Wiley & Sons, Inc., 1971.

[13] M. Copperi, D. Sereno and L. Bertorello, "16 kbit/s split-band APC coder using vector quantization and dynamic bit allocation", *Proc. Int. Conf. on Acous., Speech and Signal Process. (ICASSP)*, Tokyo, Japan, pp. 845-848, April 1986.

[14] R.E. Crochiere, S.A. Webber and J.L. Flanagan, "Digital Coding of Speech in Sub-bands", *Bell Systems Technical Journal*, vol. 55, pp. 1069–1085, October 1976.

[15] R.E. Crochiere, "Digital signal processor: Sub-band coding", *Bell Systems Technical Journal*, vol. 60, pp. 1633–1653, September 1981.

[16] J.H. Derby and C.R. Galand, "Multirate sub-band coding applied to digital speech interpolation", *Proc. of the Int. Conf. on Acous., Speech and Signal Process. (ICASSP)*, Tampa, Florida, USA, pp. 1680-1683, March 1985.

[17] J.D. Eggerton and M.D. Srinath, "Statistical Distributions of Image DCT Coefficients", *Comput. & Elect. Eng.*, vol. 12, no. 3/4, pp. 137-145, 1986.

[18] N. Farvardin and J.W. Modestino, "Optimum Quantizer Performance for a Class of Non-Gaussian Memoryless Sources", *IEEE Trans. on Information Theory*, vol. IT-30, pp. 485-497, May 1984.

[19] P.E. Fleischer, "Sufficient Conditions for Achieving Minimum Distortion in a Quantizer", *IEEE Int. Convention Rec.*, pt. 1, pp. 104-111, 1964.

[20] W. Frei and B. Baxter, "Rate-Distortion Coding Simulation for Color Images", *IEEE Trans. on Communications*, vol. COM-25, pp. 1385-1392, November 1977.

[21] A. Gersho, T. Ramstad and I. Versvik, "Fully vector-quantized subband coding with adaptive codebook allocation", *Proc. Int. Conf. on Acous., Speech and Sign. Process. (ICASSP)*, San Diego, CA, USA, paper 10.7, March 1984.

[22] H. Gharavi and A. Tabatabai, "Sub-band coding of digital images using two-dimensional quadrature mirror filtering", *Proc. SPIE: Visual Communications and Image Processing*, vol. 707, pp. 51-61, September 1986.

[23] H. Gharavi and A. Tabatabai, "Sub-Band Coding of Monochrome and Color Images", *IEEE Trans. on Circuits and Systems*, vol. 35, pp. 207-214, February 1988.

[24] S. Hayashi, M. Honda and N. Kitawaki, "A backward type band-split adaptive predictive coding system with dynamic bit allocation for wide-band speech and sound signals", *Trans. Inst. Elec. Commun. Eng. in Japan*, vol. J69A, pp. 1234-1242, October 1986.

[25] D.A. Huffman, "A Method for the Construction of Minimum Redundancy Codes", *Proc. IRE*, vol. 40, pp. 1098-1101, September 1952.

[26] A.K. Jain, "Advances in Mathematical Models for Image Processing", *Proc. of the IEEE*, vol. 69, pp. 502-528, May 1981.

[27] N.S. Jayant and P. Noll, *Digital Coding of Waveforms*, Prentice Hall, Inc., Englewood Cliffs, New Jersey, 1984.

[28] J.D. Johnston, "A Filter Family Designed for Use in Quadrature Mirror Filter Banks", *Proc. Int. Conf. on Acous., Speech and Signal Process. (ICASSP)*, Denver, USA, pp. 291-294, April 1980.

[29] C.S. Kim, M.J.T. Smith, J. Bruder and R.M. Mersereau, "Subband coding of color images using finite state vector quantization", *Proc. Int. Conf. on Acous., Speech and Signal Process. (ICASSP)*, New York, USA, pp. 753-756, April 1988.

[30] D.E. Knuth, *The Art of Computer Programming*, vol. 2, "Seminumerical algorithms", Addison-Wesley Publishing Company, 1981.

[31] T. Langlais, J. Masson and Z. Picel, "Real time implementation of 16 kbit/s subband coder with vector quantization", *Proc. of the Third Euro. Signal Process. Conf. (EUSIPCO)*, pp. 419-422, September 1986.

[32] J.O. Limb, C.B. Rubinstein and J.E. Thompson, "Digital coding of color video signals - A review", *IEEE Trans. on Communications*, vol. COM-25, pp. 1349-1385, November 1977.

[33] D.G. Messerschmitt, "Quantizing for maximum output entropy", *IEEE Trans. on Information Theory*, vol. IT-17, pp. 612-612, September 1971.

[34] T.A. Ramstad, "Considerations on quantization and dynamic bit-allocation in subband coders", *Proc. Int. Conf. on Acous., Speech and Signal Process. (ICASSP)*, Tokyo, Japan, pp. 841-844, April 1986.

[35] R.C. Reininger and J.D. Gibson, "Distributions of the two-dimensional DCT coefficients for images", *IEEE Trans. on Communications*, vol. COM-31, pp. 835-839, June 1983.

[36] E.B. Richardson and N.S. Jayant, "Subband coding with adaptive prediction for 56 kbits/s audio", *IEEE Trans. on Acous., Speech and Signal Process.*, vol. ASSP-34, pp. 691-696, August 1986.

[37] A. Santisteban, "The perceptual color space of digital image display terminals", *IBM Journal of Res. Develop.*, vol. 27, pp. 127-132, March 1983.

[38] M.J. Slakter, "A comparison of the Pearson chi-square and Kolmogorov goodness-of-fit tests with respect to validity", *Journal of the American Statis. Assn.*, vol. 60, pp. 854-858, September 1965.

[39] F.K. Soong, R.V. Cox and N.S. Jayant, "A high quality subband speech coder with backward adaptive predictor and optimal time-frequency bit assignment", *Proc. Int. Conf. on Acous., Speech and Signal Process. (ICASSP)*, Tokyo, Japan, pp. 2387-2390, April 1986.

[40] J. Tajima, "Uniform color scale applications to computer graphics", *Computer Vision, Graphics and Image Process.*, vol. 21, pp. 305-325, March 1983.

[41] I. Versvik and H.C. Guren, "Subband coding with vector quantization", *Proc. Int. Conf. on Acous., Speech and Signal Process. (ICASSP)*, Tokyo, Japan, pp. 3099-3102, April 1986.

[42] P.H. Westerink, J. Biemond and D.E. Boekee, "Sub-band coding of images using predictive vector quantization", *Proc. Int. Conf. on Acous., Speech and Sign. Process. (ICASSP)*, Dallas, Texas, pp. 1378-1381, April 1987.

[43] P.H. Westerink, J. Biemond and D.E. Boekee, "An optimal bit allocation algorithm for sub-band coding", *Proc. Int. Conf. on Acous., Speech and Signal Process. (ICASSP)*, New York, pp. 757-760, April 1988.

[44] P.H. Westerink, J. Biemond, D.E. Boekee and J.W. Woods, "Sub-band coding of images using vector quantization", *IEEE Trans. on Communications*, vol. 36, pp. 713-719, June 1988.

[45] P.H. Westerink, J. Biemond and D.E. Boekee, "Quantization error analysis of image sub-band filter banks", *Proc. of the Int. Symp. on Circuits and Systems (ISCAS)*, Helsinki, Finland, pp. 819-822, June 1988.

[46] P.H. Westerink, J. Biemond and D.E. Boekee, "Evaluation of image subband coding schemes", *Proc. Euro. Signal Process. Conf. (EUSIPCO)*, Grenoble, France, pp. 1149-1152, September 1988.

[47] R.C. Wood, "On optimum quantization", *IEEE Trans. on Information Theory*, vol. IT-15, pp. 248-252, March 1969.

[48] J.W. Woods and S.D. O'Neil, "Subband coding of images", *IEEE Trans. on Acous., Speech and Signal Process.*, vol. ASSP-34, pp. 1278-1288, October 1986.

Chapter 6

Subband Coding of Video Signals

by: Hamid Gharavi
Video Systems Technology Research
Bell Communications Research
Red Bank, New Jersey 07701

In video coding applications the main objective is to remove the vast amount of redundancy which normally exists in the spatial domain (within a frame) as well as in the temporal direction (frame-to-frame). Attempts to minimize the temporal redundancies can be accomplished by interframe coding techniques [1]. In addition, many applications of high compression video coding involve the use of hybrid coding [2]. This method, which is a combination of DPCM and transform coding, is presently considered the most effective coding for video teleconferencing applications [3,4]. The main disadvantage of this method however, is the subjective degradation in which the viewers perceive the outlines of the transform blocks. This type of distortion, which appears as discontinuities at the edges of blocks, can be very objectionable to the viewer. As a result, in this Chapter we present a different approach which is not only free of block distortion but also extremely efficient in terms of compression and hardware complexity. The method is based on bandwidth splitting using quadrature mirror filtering (QMF) which has been extensively investigated in recent years for still image applications [5-14].

In Section 6.1, after a brief history of subband image coding, the basic principles of 1-dimensional (1-D) and 2-dimensional (2-D) quadrature mirror filtering (QMF) are presented. This is then followed by the evaluation of the effect of filter impulse response on near-perfect QMF bank system design.

Section 6.2 deals with subband based video coding. This section begins with the input format description of the CCIR recommendation 601 standard and what is known as common intermediate format (CIF). This is then followed by defining two basic models for the interframe/interfield subband coding for video signals. The comparison of the two models, the compatibility consideration, as well as other related issues for video coding are also presented in this section.

Section 6.3 discusses motion detection and estimation techniques for motion compensated (MC) interframe/interfield prediction. A simple and yet efficient criterion is also presented in this section.

Section 6.4 mainly deals with coding algorithms which include quantization and entropy coding. In Sections 6.5 and 6.6 the intraframe coding technique and the coding strategies for color components are described. Simulation results and discussion on the performance of the various subband video coding schemes are presented in Section 6.7. Finally, the conclusion and closing statements are given in Section 6.8.

6.1 Subband Coding: Background

The concept of subband coding and its application to image signals goes back to as early as 1959 [15]. Although this concept seemed very impressive at the time, its practical implication was not followed by other researchers.

Since the introduction of quadrature mirror filtering (QMF) by Esteban and Galand [19], subband coding has been extensively used for bandwidth compression of speech signals. More recently, after the 2-D extension of QMF by Vetterli [5], its application to image coding was first considered by Woods and O'Neil [6,8] and later by Gharavi and Tabatabai [7,9]. In the Woods approach, individual bands were coded using adaptive DPCM, whereas in the Gharavi approach, except for the lowest band, all other bands were coded using a combination of PCM quantization and run-length coding for the transmission of non-zero PCM values. An extension of this method to subband coding of color images was introduced by the same authors [10,11]. In addition, subband coding using predictive vector quantization was also investigated by Westerink, et al [12,13].

6.1.1 Quadrature Mirror Filtering (QMF)

The concept of QMF is based on splitting the bandwidth of the input spectrum into two halves; low band and high band. This process is achieved by designing lowpass $h_0(n)$ and highpass $h_1(n)$ filters. Since the bandwidth of each decomposed band is half the bandwidth of the original signal (in the ideal case), each band is decimated by 2:1. In reality, the lowpass and highpass filters are not ideal and consequently the subbands would generate undesirable aliasing components after reconstruction. To eliminate the aliasing, Esteban and Galand [19] designed a special class of lowpass and highpass filters whose frequency spectra have mirror image symmetry at about the center frequency $(\omega = \frac{\pi}{2})$. This implies the following requirements:

i. For alias-free reconstruction

$$H_0(\omega) \quad = \quad H_1(\omega + \pi)$$

$$G_0(\omega) = 2H_0(\omega) \tag{6.1}$$

$$G_1(\omega) = -2H_1(\omega)$$

ii. For perfect reconstruction

$$|H_0^2(\omega)| + |H_1^2(\omega)| = 1 \tag{6.2}$$

where $H_0(\omega)$, $H_1(\omega)$ and $G_0(\omega)$, $G_1(\omega)$ are the Fourier transforms of the transmitting and receiving pairs of lowpass and highpass filters. The latter requirement can be closely approximated for modest size FIR filters [20]. We shall discuss their performance later in this chapter.

Based on the same principle, the input signal can be divided into smaller frequency bands by cascading 2-band QMF in a tree structural manner. Vetterli [5] applied the above concept to extend 1-D QMF to a 2-D separable QMF filter designed for image application. This was performed by applying 1-D QMF along the rows and then along the columns which would result in a basic four-band decomposition as shown in Figure 6.1.

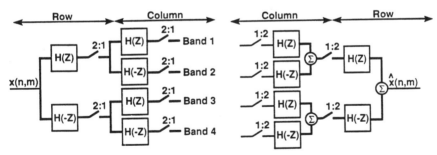

(a) Subband Decomposition   (b) Subband Reconstruction

Figure 6.1 - Separable 2-D QMF

Similar to the 1-D case, 2-D separable QMF requires the following conditions:

i.   For alias-free reconstruction

$$H_{12}\,(\omega_1,\,\omega_2)\;=\;-\,1/4G_{12}\,(\omega_1,\,\omega_2)\;=\;H_{11}(\omega_1+\pi,\,\omega_2)$$

$$H_{21}\,(\omega_1,\,\omega_2)\;=\;-\,1/4G_{21}\,(\omega_1,\,\omega_2)\;=\;H_{11}\,(\omega_1,\,\omega_2+\pi)$$

$$H_{22}\,(\omega_1,\,\omega_2)\;=\;+\,1/4G_{22}\,(\omega_1,\,\omega_2)\;=\;H_{11}\,(\omega_1+\pi,\,\omega_2+\pi)$$

$$H_{11}\,(\omega_1,\,\omega_2)\;=\;+\,1/4G_{11}\,(\omega_1,\,\omega_2)$$

(6.3)

ii. For perfect reconstruction

$$\sum_{\ell}\sum_{k}|H_{k,\ell}^{2}(\omega_1+(k-1)\pi,\omega_2+(\ell-1)\pi)|\;=\;1\qquad k,\ell\;=\;1,2\qquad(6.4)$$

where $H_{k,\ell}(\omega_1+(k-1)\pi,\omega_2+(\ell-1)\pi)$,   and $G_{k,\ell}(\omega_1+(k-1)\pi,$ $\omega_2+(\ell-1)\pi)$ $k,\ell=1,2$ are the Fourier transforms of the transmitting and receiving 2-D QMF.

For the special case of a two-tap FIR filter, the above condition can be easily met but the resulting filter response is very poor. The analysis/synthesis for perfect reconstruction using a short kernel filter was studied by LeGall and Tabatabai [14]. These types of filters may be suitable for a small number of bands [27]. However, their application to high compression image coding which necessitates the decomposition of large numbers of narrow bands, may affect the coding efficiency due to their poor filter frequency responses. In addition, in the case of high compression, excessive coding noise can have a greater effect on the cancellation of aliasing due to the large overlap between the neighboring bands.

Based on our experiments we found that to improve the coding efficiency and reduce the effect of aliasing, the emphasis should be put on the quality of filter impulse response. This would require the employment of long tap filters such as those developed by Johnston [20]. The Johnston FIR filter however, can only provide a near perfect reconstruction. The reconstruction error can become negligible as the length of the filter increases. At the same time, longer tap filters would increase the hardware complexity and also produce a longer delay especially in multistage QMF filter bank systems. In order to examine the tradeoff between the hardware complexity and possible loss of quality due to the reconstruction error, we carried out a series of experimental tests. These are discussed in the following section.

6.1.2 Evaluation of Near Perfect QMF Filters

To achieve high bandwidth compression using subband coding it is essential to split the image into a large number of small subimages. The basic 2-D separable QMF can provide four bands (a QMF bank depicting such a process is shown in Figure 6.1). The decomposition of the image to more than four bands can be achieved by repeating the above process for each band in a tree structural manner (i.e., 16, 32, 64 bands). In a non-uniform decomposition of the image spectrum the lowest band is further split by 2-D QMF. This process is repeated until a desirable resolution for the lowest band is obtained. The relationship between the total number of sub-images N and the decimation factor n of the lowest band is expressed as,

$$N = 3(Log_2 n - 1) + 4 . \tag{6.5}$$

Note that in this arrangement n can only take values to the power of two.

In our evaluation we have considered the basic four bands as well as non-uniform 7 and 10 bands decomposition of the image spectrum. The latter two would obviously require employing several 1-D QMF banks which affect the accumulation of the filters truncation error as well as the reconstruction noise. To measure this we first put the transmitting and receiving QMF banks back to back in order to eliminate the effect of quantization noise. Then the signal-to-noise ratio SNR (peak-to-peak signal to rms noise) was measured by truncating the output of each 1-D QMF to a bit accuracy ranging from 16 to 8 bits. These experiments were conducted using Johnston's [20] 8-tap and 16-tap FIR filters designated as type A. The results, averaged over four images each with the resolution of 512×512, are plotted in Figure 6.2. In this figure the vertical axis represents the SNR and the horizontal axis corresponds to the bit accuracy at the output of each 1-D QMF. By looking at these results we conclude the following:

1. The QMFs with longer tap FIR filters are more sensitive to truncation error when compared with shorter tap QMF. For a 16-tap filter an accuracy of not less than 12 bits is needed, whereas for the 8-tap, even 9 bits seem to be more than sufficient (see Figure 6.2).

2. As the number of bands increases it causes the SNR to drop substantially. This can be easily compensated by employing longer tap filters. For example, as shown in Figure 6.2, the 16-tap filter with 7 bands decomposition can improve the SNR by about 10 dB over the 7 bands with the 8-tap filter. More importantly, long tap filters can also contribute towards improving the coding performance (see the results Section).

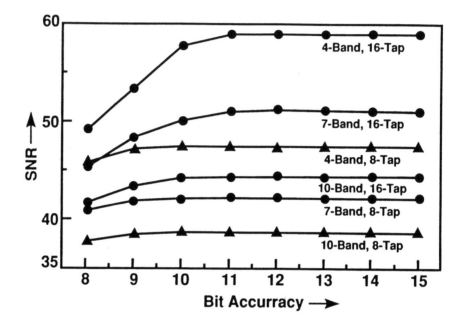

Figure 6.2 - Average SNR Versus Bit Truncation Accuracy

6.2 Subband Video Coding

The television picture is composed of lines and fields (frames), and the correlation between pixels along a line and in successive lines and fields has led to the development of predictors that are said to operate in one, two, and three dimensions, respectively.

Further, areas of the picture in one or more fields (frames) often have correlation coefficients exceeding 0.98. This enables the predictor design to proceed on the basis that the area in which the sample to be predicted resides may be considered as a constant luminance. Of course this is not always true, and the predictor may perform less well unless some form of motion detection and estimation is used.

In this section 3-D subband models for interframe (interfield) prediction are presented. Before we begin let's briefly describe the digital video input standards in which the quality of the video service is dependent.

6.2.1 Digital Video Format

CCIR (International Radio Consoltative Committee) recommendation 601 defines a digital video format for television studios for 525-line and 625-line television systems. This standard is intended to facilitate the development of equipment with many common features and to permit international exchange of programs. It is based, however, on digital component coding which provides one luminance Y and two color-difference components R-Y and B-Y. In one example of the digital coding standard, the luminance and color difference sampling rates are related by the ratio 4:2:2. The sampling frequency is commonly selected as a multiple integer of the line frequency in both the 525-line and 625-line systems

$$f_{luminance} = 858 f_{525-line} = 864 f_{625-line} \qquad (6.6)$$

$$\simeq 13.5 \; MHz$$

$$f_{color} = \frac{f_{luminance}}{2} \simeq 6.75 \; MHz \; . \qquad (6.7)$$

The luminance signal has 220 quantization levels with the black level corresponding to level 16 and the peak white corresponding to level 235. The color difference signals each have 224 quantization levels in the center part of the quantization scale with zero signals corresponding to level 128. The recommendation also defines the number of pixels per active line as 720 pixels for luminance and 360 pixels for each color difference signal. The sampling structure is orthogonal in line, field, and frame repetitively. The R-Y and B-Y pixels are co-sited with odd (1st, 3rd, 5th, etc.) luminance pixels in each line.

Since the CCIR-601 digital picture format could not be considered for low bit rate video applications (due to the high spatial resolution), the CCITT Specialist Group (SGXV) [3] has proposed a new digital format which is called common intermediate format (CIF). This format however, has a simple relationship to the CCIR-601 format. The number of samples per active line is 360 pixels for luminance and 180 pixels for each color difference signal. For the CCITT block hybrid coding these are reduced to 352 pixels for luminance and 176 for each color difference signal so they can provide an integer number of 16× 16 blocks. The number of lines per active picture for luminance is 288 and for each color difference signal 144. The main parameters of the CCIR-601 and CIF are summarized in Table 6.1.

Table 6.1 CCIR-601 and CIR Formats

Parameters	CCIR-601		CIF
	525-line	625-line	
Luminance, Y	720 × 240	720 × 288	360 × 288
Color Difference C_1, C_2	360 × 240	360 × 288	180 × 144
Field/Frame	59.94	50	29.97

6.2.2 Interframe Subband Models

Let us assume that each pixel of an image is undergoing linear translation and appears at other locations at other time instants. If $s(k, \ell, t)$ denotes the intensity of the pixel at location (k, ℓ) at time t and $s(k - dk, \ell - d\ell, t - T)$ is the pixel intensity at the previous location at time $t - T$. Then the motion trajectory can be given by

$$s(k, \ell, t) = s(k - dk, \ell - d\ell, t - T) = constant \qquad (6.8)$$

where dk and $d\ell$ are the horizontal and vertical components of the motion trajectory during a one frame period T.

In this chapter we consider two interframe subband models which are described as follows.

6.2.2.1 Model I

The block diagram of this model is shown in Figure 6.3 [21]. According to this figure the input pixel is first subtracted from the Motion Compensated (MC) estimate to form the prediction error signal. Thus,

$$e(k, \ell, t) = s(k, \ell, t) - \hat{s}(k, \ell, t) \quad . \qquad (6.9)$$

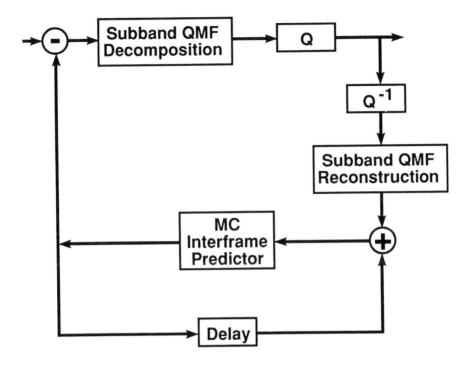

Figure 6.3 - Basic Structure Subband Model I

Assuming that the motion trajectory is estimated piecewise, the MC interframe estimate $\hat{s}(k,\ell,t)$ is given by

$$\hat{s}(k,\ell,t) = s[k - (dk-\overline{dk}), \ell - (d\ell-\overline{d\ell}),t]$$

$$+ q(k-\overline{dk},\ell-\overline{d\ell},t-T) \tag{6.10}$$

where $q(.,.,.)$ is the quantization noise added to the previous frame, which will be omitted in our computation for the sake of simplicity, and $\overline{dk}$ and $\overline{d\ell}$ are the components of the motion displacement estimate.

From (6.9) and (6.10) (ignoring the quantization noise) we have

$$e(k,\ell,t) = s(k,\ell,t) - s[k-(dk-\overline{dk}), \ell-(d\ell-\overline{d\ell}),t] . \tag{6.11}$$

Taking the Fourier Transform of (6.11) in the spatial coordinates

$$E(\omega_0, \omega_1, t) = S(\omega_0, \omega_1, t) \left[1 - e^{j(\omega_0 \bar{t}k + \omega_1 \bar{t}\ell)} \right] \tag{6.12}$$

where

$$\bar{t}k = (dk - \bar{d}k) \cdot T_J$$
$$\bar{t}\ell = (d\ell - \bar{d}\ell) \cdot T_L . \tag{6.13}$$

Inside the DPCM loop (see Figure 6.3) the MC error signal is split into $N = n^2$ narrow bands. For a basic 2-D separable quadrature mirror filter (QMF) n is the decimation factor and can take value of the power of two.

$$E_{ij}(\omega_0, \omega_1, t) = \frac{1}{N} \sum_{x=0}^{n-1} \sum_{y=0}^{n-1} H_{ij} \left(\frac{\omega_0 + 2x\pi}{n}, \frac{\omega_1 + 2y\pi}{n} \right)$$

$$\times E \left(\frac{\omega_0 + 2x\pi}{n}, \frac{\omega_1 + 2y\pi}{n} \right) . \tag{6.14}$$

The decomposed error bands $E_{ij}(.,.)$ are each quantized independently. Assuming that the quantization noise is purely additive, the quantized subbands can be shown as

$$L_{ij}(\omega_0, \omega_1, t) = E_{ij}(\omega_0, \omega_1, t) + Q_{ij}(\omega_0, \omega_1, t) . \tag{6.15}$$

Substituting $E_{ij}(\omega_0, \omega_1)$ from (6.14) in the above,

$$L_{ij}(\omega_0, \omega_1, t) = \frac{1}{N} \sum_{x=0}^{n-1} \sum_{y=0}^{n-1} H_{ij} \left(\frac{\omega_0 + 2x\pi}{n}, \frac{\omega_1 + 2y\pi}{n} \right)$$

$$\times S \left(\frac{\omega_0 + 2x\pi}{n}, \frac{\omega_1 + 2y\pi}{n} \right) \left[1 - e^{j(\omega_0 \bar{t}k/n + \omega_1 \bar{t}\ell/n)} \right]$$

$$+ Q_{ij}(\omega_0, \omega_1, t) . \tag{6.16}$$

The quantized bands are entropy coded, multiplexed and transmitted to the receiver. In the DPCM feedback loop the quantized subbands, after remapping, are sent to the interpolation QMF filters. The reconstructed output is given by

$$\bar{L}(\omega_0, \omega_1, t) = \sum_{x=0}^{n-1} \sum_{y=0}^{n-1} S(\omega_0 + \frac{2x\pi}{n}, \omega_1 + \frac{2y\pi}{n}) \left[1 - e^{j(\omega_0 \bar{t}_k/n + \omega_1 \bar{t}_\ell/n)} \right]$$

$$\times \sum_{i=0}^{n-1} \sum_{j=0}^{n-1} (-1)^{i+j} H_{ij}(\omega_0, \omega_1) H_{ij}(\omega_0 + \frac{2x\pi}{n}, \omega_1 + \frac{2y\pi}{n})$$

$$+ N \sum_{i=0}^{n-1} \sum_{j=0}^{n-1} (-1)^{i+j} H_{ij}(\omega_0, \omega_1) Q_{ij}(n\omega_0, n\omega_1, t) . \qquad (6.17)$$

For a perfect reconstruction separable 2-D QMF (6.17) can be simplified as,

$$\bar{L}(\omega_0, \omega_1, t) = S(\omega_0, \omega_1, t) \left[1 - e^{j(\omega_0 \bar{t}_k + \omega_1 \bar{t}_\ell)} \right]$$

$$+ N \sum_{i=0}^{n-1} \sum_{j=0}^{n-1} (-1)^{i+j} H_{ij}(\omega_o, \omega_1) Q_{ij}(n\omega_0, n\omega_1, t) . \qquad (6.18)$$

In the DPCM feedback loop the reconstructed signal is added to the MC estimate to obtain the decoded signal as

$$\hat{S}(\omega_0, \omega_1, t) = S(\omega_0, \omega_1, t) + N \sum_i \sum_j (-1)^{i+j} H_{ij}(\omega_0, \omega_1) Q_{ij}(n\omega_0, n\omega_1, t) . \quad (6.19)$$

This signal is the same as the decoded signal at the receiver side in the absence of channel noise.

6.2.2.2 Model II

In this model [22,36], as shown in Figure 6.4, the input video is first decomposed into smaller bands by means of 2-D QMF banks:

$$S_{ij}(\omega_0, \omega_1, t) = \frac{1}{N} \sum_{x=0}^{n-1} \sum_{y=1}^{n-1} H_{ij} \left(\frac{\omega_0 + 2x\pi}{n}, \frac{\omega_1 + 2y\pi}{n} \right) \times$$

$$S \left(\frac{\omega_0 + 2x\pi}{2}, \frac{\omega_1 + 2y\pi}{2} \right) \quad i, j = 0, 1 \ldots, n-1 . \qquad (6.20)$$

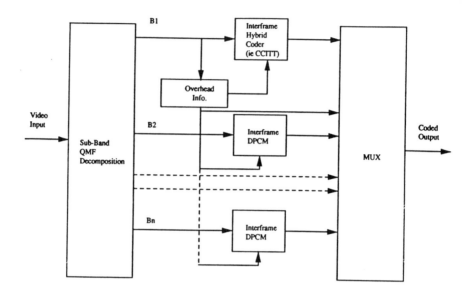

Figure 6.4 - Basic Structure of Subband Model II

The decomposed bands are each individuallypplied to   DPCM loops where they can then be subtracted from their corrending MC estimates. These estimates can be obtained by using two different hardware structures. In the first structure only one reference frame memory is used to store the decoded frame after QMF reconstruction. In this case, the stored frame needs to undergo further decomposition for the next inter-sub-frame prediction. In the second structure, however, the decoded subbands (before QMF reconstruction) are each stored in local frame memories of the coder and decoder. As a result, these sub-frames are readily available for the next sub-frame prediction. This structure has the advantage that it can forego further decomposition for the inter-sub-frame prediction. In our case, where we have assumed that a perfect reconstruction QMF is in place, both structures can provide identical results. Thus, in either case the interframe estimate of Model II can be shown to be,

$$\bar{S}'_{ij}(\omega_0,\omega_1,t) = \frac{1}{N} \sum_x \sum_y H_{ij}\left(\frac{\omega_0 + 2x\pi}{n}, \frac{\omega_1 + 2y\pi}{n}\right)$$

$$(6.21)$$

$$\times \ S\left(\frac{\omega_0 + 2x\pi}{n}, \frac{\omega_1 + 2y\pi}{n}\right) e^{j(\omega_0 \hat{t}k_{ij} + \omega_1 \hat{t}\ell_{ij})} .$$

In the above equation the quantization noise added to the previous frame has been ignored. Thus, the sub-frame prediction error can be obtained as

$$E'_{ij}(\omega_0,\omega_1,t) = \frac{1}{4} \sum_x \sum_y H_{ij}\left(\frac{\omega_0 + 2x\pi}{n}, \frac{\omega_1 + 2y\pi}{n}\right) \times$$

$$S\left(\frac{\omega_0 + 2x\pi}{n}, \frac{\omega_1 + 2y\pi}{n}\right) \left[1 - e^{j(\omega_0 \hat{t}k_{ij} + \omega_1 \hat{t}\ell_{ij})}\right] , \qquad (6.22)$$

where $\hat{t}k_{ij}$ and $\hat{t}\ell_{ij}$ are the time delays corresponding to motion displacements $\hat{d}k_{ij}$ and $\hat{d}\ell_{ij}$, which are independently estimated for each decomposed band and are given as

$$\hat{t}k_{ij} = \left(\frac{dk}{n} - \hat{d}k_{ij}\right) \cdot T_J$$

$$\hat{t}\ell_{ij} = \left(\frac{d\ell}{n} - \hat{d}\ell_{ij}\right) \cdot T_L \qquad\qquad i,j = 0,1,\ldots, n-1 , \qquad (6.23)$$

where n is the subsampling rate and its value for N band uniform decomposition is $\sqrt{N}$ (in the case of 2-D separable QMF $n = 2$).

In each DPCM loop the prediction errors are independently quantized. Assuming that the quantization noise is purely additive we have

$$L'_{ij}(\omega_0,\omega_1,t) = \frac{1}{N} \sum_x \sum_y H_{ij}\left[\frac{\omega_0 + 2x\pi}{n}, \frac{\omega_1 + 2y\pi}{n}\right]$$

$$\times \ S\left[\frac{\omega_0 + 2x\pi}{n}, \frac{\omega_1 + 2y\pi}{n}\right]$$

$$\times \left[1 - e^{j(\omega_0 \hat{t}k_{ij} + \omega_1 \hat{t}\ell_{ij})}\right] + Q'_{ij}(\omega_0,\omega_1,t) . \quad (6.24)$$

The quantized subframes are subsequently entropy coded, multiplexed and transmitted to the receiver. In the DPCM local decoders the quantized prediction errors, after remapping, are each added to their corresponding interframe estimates to form the decoded subframes. The DPCM decoded sub-frames are finally sent to the interpolation filter banks. Assuming perfect reconstruction QMF, the reconstructed frame is given as

$$\hat{S}'(\omega_0,\omega_1,t) = S(\omega_0,\omega_1,t)$$

$$+ N\sum_i\sum_j(-1)^{i+j}H_{ij}(\omega_0,\omega_1)Q'_{ij}(n\omega_0,n\omega_1)$$

$$i,j = 0,1,\ldots,n-1 \ . \tag{6.25}$$

6.2.3 Model Comparisons

To compare the two models we look at the quantized prediction error signals given by (6.16) and (6.24). The two models can perform equally if

$$\begin{cases} \overline{dk} = n\cdot\hat{dk}_{ij} \\ \overline{d\ell} = n\cdot\hat{d\ell}_{ij} \end{cases} \tag{6.26}$$

and

$$Q_{ij} = Q'_{ij} \ . \tag{6.27}$$

The first condition also indicates that in Model II each individual band should produce the same displacement estimate. Such a requirement cannot be easily met in practice. This is due to the fact that the nature of each decomposed band differs from one to another. It can therefore be expected that the horizontal, vertical, and diagonal bands, which contain mainly the vertical, horizontal and diagonal edges, could produce different displacement estimates.

Imposing condition (6.26) however, has inspired us to simplify Model II. Accordingly, two approaches have been considered.

In the first approach, no motion estimation is carried out except for the lowest band. Motion vectors estimated for this band can be then applied to all the bands for MC prediction. As will be discussed later, this approach has been adapted for our Model II based coding.

In the second approach, the motion estimation is performed on a full frame basis (before subband decomposition). The estimated components of the motion displacements, after being divided by n, are equally applied to each decomposed sub-frame for MC prediction. This arrangement, which can be considered as a compromise between the two models, is still more complex than Model I. Bear in mind that Model I has the advantage of requiring only one MC prediction DPCM loop. We should, however, point out that although Model II requires a series of DPCM loops, they can operate at the speed which is reduced by the factor of n^2. This can be regarded as a significant property for cases such as high quality, high compression inter-frame video coding where speed is important. Consequently, this model can be easily utilized to improve the service quality of the existing video codecs such as the newly proposed CCITT video codec [3].

6.2.4 Compatibility Consideration for High Quality Video Coding

The CCITT SGXV has recently drafted a coding scheme for video-phone and video-conferencing services. This codec is designed to operate at very low speeds, i.e., $64 \times n$ ($n = 1,2$) kb/s for video-phone and $384 \times m$ kb/s ($m = 1$ to 6) for video-conferencing applications. For video-conferencing the spatial and temporal resolution of the input is set at 360 (pixels) $\times$ 288 (lines) and 29.97 frames/sec, respectively (CIF). For video-phone, where the transmission rate is lower (e.g., 64 kb/s), the spatial resolution is reduced to 1/4 CIF (i.e., 180×144). Consequently, at these speeds and resolutions the codec is unable to produce video with the quality that can support the needs of the business community. The proposed standard, however, is expected to provide worldwide availability of the CCITT codec at a reasonably low cost. Thus, for any new system the cost can be reduced if a codec is CCITT compatible or at least utilizes hardware blocks that can also be used in the CCITT codec.

Furthermore, the installation of optical fibers and the trend towards BISDN is expected to encourage the proliferation of higher quality video services such as HDTV conferencing. To maximize the integration of all such services, it is necessary to design a system which can provide an unlimited range of video qualities. The Model II based system, due to its hierarchical video representation, is considered here to be most suitable for the integration of video conferencing services. As an example, the following presents a Model II based CCITT compatible coding algorithm.

6.2.5 CCITT Compatible Coding Algorithm

As mentioned earlier, it would be beneficial to design a video codec that can be compatible with the current CCITT proposal. Figure 6.5 displays the

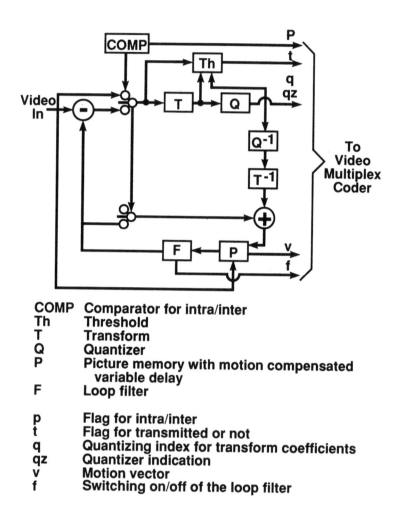

COMP Comparator for intra/inter
Th   Threshold
T   Transform
Q   Quantizer
P   Picture memory with motion compensated
            variable delay
F   Loop filter

p   Flag for intra/inter
t   Flag for transmitted or not
q   Quantizing index for transform coefficients
qz   Quantizer indication
v   Motion vector
f   Switching on/off of the loop filter

Figure 6.5 - CCITT Source Coder

block diagram of the CCITT codec. The operation of this codec is self explanatory and the details can be found in [3]. This codec is used as a part of our proposed codec and its block diagram is shown in Figure 6.6. The scheme is a modification of the subband Model II coder and is described as follows:

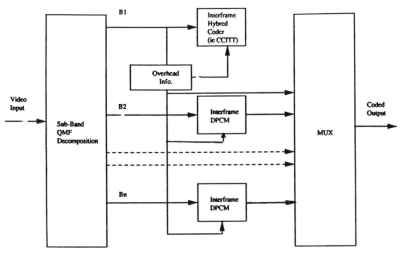

Figure 6.6 - Model II Based CCITT Compatible Video Encoder

The input video is first decomposed into N narrow subframes in such a way that the spatial resolution of the lowest band is reduced to 1/4 CIF. If the input video signal is based on CCIR Rec. 601 (625-line system) which is equivalent to $4 \times$ CIF (720×576 European standard), the number of bands would be sixteen for uniform and seven for non-uniform [7,9] decompositions respectively. For HDTV signals which have a different aspect ratio than CIF, special treatment is needed to convert the lowest band into 1/4 CIF. Here however, we concentrate only on the CCIR 601 where the spatial conversion to 1/4 CIF is more straightforward. As shown in Figure 6.6, the lowest band is coded by the CCITT hybrid DCT/DPCM for compatibility reasons. The higher band signals, however, can either be coded by direct intraframe PCM or MC interframe DPCM. Bear in mind that the higher band signals contain mostly the edges and interframe DPCM can be very effective particularly for video sequences with busy backgrounds. In addition, if the MC prediction is employed it should be capable of aligning moving edges of the consecutive sub-frames very accurately. This would, therefore, minimize the interframe edge difference which can then be efficiently coded.

This has been investigated by considering the following cases; 1) MC interframe prediction where each band uses its own independently estimated motion displacements, 2) MC interframe prediction where higher bands utilize the motion displacements estimated for the lowest band, 3) interframe without motion compensation prediction, and 4) intraframe PCM.

To evaluate each of the above cases, which were presented in decreasing order of coding complexity, we carried out a series of simulation tests. In these experiments, two successive video frames were first decomposed by a separable four band 2-D QMF. Subsequently, blockmatching motion estimation was performed on each individual band. The results, in terms of signal to prediction noise ratio and the entropy of the prediction error for each individual band, are given in Table 6.2. In this table, by comparing the entropy results, we conclude that while there is a significant advantage in the interframe prediction for band 1, there is none whatsoever for band 4. This band, which corresponds to the diagonal variation in the coding frame, should therefore be directly coded by intraframe PCM. With regards to the other two bands (2 and 3), the entropy reduction is not as significant as for the first band (this, of course, depends a lot on the busy-ness of the image background). The simplest approach would be to encode these bands directly by intraframe PCM. In our experiments however, we observed that for achieving the highest degree of compression, it is still advantageous to encode these bands by MC interframe prediction. We have therefore, adopted Case 2 for bands 2 and 3 as a compromising solution between the complexity and the coding efficiency. In addition, these bands also receive information such as block classification which is provided by the lowest band.

Table 6.2 Comparison of Various Coding Strategies

Case	Band 1		Band 2		Band 3		Band 4	
	Entropy	SNR	Entropy	SNR	Entropy	SNR	Entropy	SNR
1	3.26	26.1	2.46	35.27	2.14	39.3	1.92	42.33
2	3.26	26.1	2.48	34.0	2.18	38.4	2.04	41.2
3	3.32	22.08	2.63	31.74	2.34	37.5	2.06	40.4
4	7.64	N/A	2.81	N/A	2.53	N/A	1.86	N/A

Case 1: Each band uses its own displacement estimate
Case 2: All bands use motion displacements estimated for the lowest band
Case 3: Direct Difference
Case 4: Intraframe

For the case of 7-band non-uniform decomposition, similar experiments were carried out. Subsequently the arrangement depicted in Figure 6.7 was adopted. By looking at this figure we can observe that i) the four lowest frequency bands are coded in a same manner as before, ii) the diagonal bands are always coded by direct PCM and iii) the higher frequency vertical and horizontal bands are coded by interframe DPCM without motion compensation.

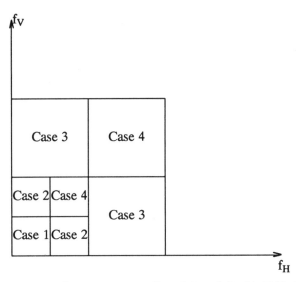

Figure 6.7 - Coding Arrangement. Cases 1-4 are defined in Table 6.1

6.3 Motion Estimation

In interframe predictive coding the temporal redundancy can be removed more efficiently by taking into consideration the displacements of moving objects in the coding process.

There are two distinct approaches for motion compensated prediction [1]: pixel recursive and block matching motion estimation. In pixel recursive motion estimation, the motion parameters are recursively estimated to minimize the motion compensated error signal at each pixel. The important feature of this approach is that it does not generate a blocking effect. In addition, the rotary and translatory motion, which is the nature of the real world scene, can be dealt with. In this chapter we concentrate only on block matching motion estimation, although in our investigations we have also considered pixel recursive motion estimation. Block matching motion estimation, due to its lesser implementation complexity, has been considered for most block coding methods in video conferencing applications [3,23] and some of them are described as follows.

6.3.1 Blockmatching Motion Estimation

In blockmatching motion estimation the coding frame is partitioned into blocks of size m-by-n by assuming that all the pixels within each non-overlapping block have the same motion displacement vector. The motion vector for each block is estimated by searching through a larger block (search window), centered at the same location on the previous coding frame, for the best match. *Let (k, ℓ) be pixel's location.*

Let's assume $S_f(k, \ell)$ to be the intensity of the pixel located on the upper most left in the block of the current frame, f, and $S_{f-1}(k+i, \ell+j)$ to be the intensity of the pixel on the previous coding frame, f-1, shifted by the i pixels and j lines (in the scanning directions). The accuracy of the estimate depends on the matching criteria applied in the search. For a matching criterion the main objective would be to minimize the mean distortion function. This has been defined by Jain and Jain [23] as

$$D(i,j) = \frac{1}{mn} \sum_k \sum_\ell g\left\{ S_f(k,\ell) - S_{f-1}(k+i,\ell+j) \right\} \qquad (6.28)$$

where g{x} is a positive distortion function of x. For a given g{x} the best match corresponds to a block with the spatial displacement $\bar{d}_h$ and $\bar{d}_v$, such that $D(\bar{d}_h, \bar{d}_v)$ is minimum. The criteria, which are based on the first and second order distortion functions, are known as mean absolute difference (MAD) and mean square difference (MSD). Consequently, for g{x} = |x| and g{x} = x^2 we have:

$$MAD(i,j) = \frac{1}{mn} \sum_k \sum_\ell |S_f(k,\ell) - S_{f-1}(k+i,\ell+j)| \qquad (6.29)$$

and

$$MSD(i,j) = \frac{1}{mn} \sum_k \sum_\ell \left[S_f(k,\ell) - S_{f-1}(k+i,\ell+j) \right]^2. \qquad (6.30)$$

Currently, blockmatching motion estimation with the mean absolute difference (MAD) [23] (see Equation 6.29) criterion is considered a good candidate for low bit rate video applications. This is mainly due to its relative ease of hardware implementation, although it may suffer from poor performance. More efficient criteria such as mean squared difference (MSD) [24] are too complex and their hardware realization seem far from being feasible.

For our application however, we have considered a different criterion which is based on minimum cost function [25]. This criterion has been shown to be extremely efficient and yet simple for implementation [26] and described as follows.

In this method, which is called pixel difference classification (PDC), each pixel in the block is classified as one of two:

1. Matching pixel

2. Mismatching pixel .

A threshold t is selected to perform the above classification.

$$T(k,\ell,i,j) = 1 \quad \text{if } |S_f(k,\ell) - S_{f-1}(k+i,\ell+j)| \leq t \quad (6.31)$$

$$= 0 \quad \text{otherwise} .$$

$T(k,\ell,i,j)$ is the binary representation of pixel difference and its value being one or zero corresponds to a matching or mismatching pixel respectively.

The matching role is then given by G (i,j) which operates as follows:

$$G(i,j) = \sum_k \sum_\ell T(k,\ell,i,j) \quad (6.32)$$

The value of $G(i,j)$ represents the number of matching pixels which exist between the current block and the block on the previous reference frame shifted by the i pixels and the j lines. Throughout the search (i.e., i = 0, ± 1 .... ± x and j = 0, ± 1 .... ± y) the largest $G(i,j)$ represent the best match. Thus,

$$G(\bar{d}_h, \bar{d}_v) = \max_{i,j} \{G(i,j)\} \quad (6.33)$$

where $\bar{d}_h$ and $\bar{d}_v$ are the horizontal and vertical components of the displacement.

Figure 6.8 displays the SNR (peak-to-peak signal to rms noise ratio) versus the size of search and for MAD, MSD, and PDC. In this figure the block size is fixed at 16× 16 using a sequence known as "Salesman". According to this figure all three schemes appear to be very close initially but as the search window becomes larger, the gap between MAD and the other two schemes becomes increasingly wider.

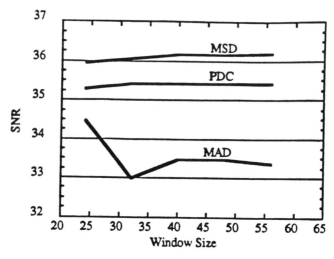

Figure 6.8 Peak-to-peak signal to rms prediction noise ratios
versus search window size -

These results are also presented subjectively in Figure 6.9. This figure
displays the motion compensated prediction error images for the "Trevor"

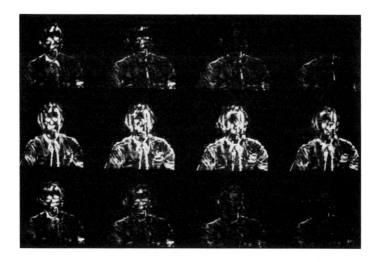

Figure 6.9 - MC Prediction error frame for "Trevor" sequence

sequence only. The first, second, and third rows of images correspond to PDC, MAD, and MSD respectively, whereas the first, second, third, and fourth columns represents the search area of sizes; 24×24, 32×32, 40×40, and 48×48. This subjective evaluation verifies the earlier SNR results with a clear indication of the deteriorating performance of MAD as the search area becomes larger. It should be noted that a larger search area should only increase the ability of the motion compensated predictor to track the fast moving object more efficiently and this conflicting behavior is a clear indication of a major deficiency of MAD.

Based on the above comparison, the PDC has been chosen for our MC interframe subband coding due to its simplicity as well its superior performance over MAD. In addition the PDC, due to its inherent use of thresholding does not require motion detection. Thus, in our experiments, the threshold is selected adaptively in accordance with buffer fullness.

6.4 Coding

The coding is comprised of two distinct operations; quantization and entropy coding. The function of the quantizer is to gracefully reduce the number of PCM levels of the input signal. The quantized levels are then coded noiselessly by the entropy coder.

In this section we present the quantization scheme which we have considered for our subband coding schemes, particularly the higher frequency bands. This is then followed by a description of the entropy coder which introduces an efficient runlength coding scheme for the transmission of quantized signals.

6.4.1 Quantization

In both interframe subband coding models the decomposed prediction error signals are first quantized before being entropy coded. An optimal approach, in the minimum mean square sense, is to design a non-uniform quantizer which matches the probability density function of the input signal (i.e., Laplacian). We have observed, however, that such a quantizer is not always suitable (in a subjective sense) for encoding the decomposed bands. This is mainly due to the existence of picture noise (e.g., camera noise) which manifests itself as a low level signal within these bands, and would result in a fine quantization of the noise (while the contour points which fall within the coarser range of the quantizer are encoded less accurately).

This observation led us to design a symmetric quantizer with the following characteristics [7] (see Figure 6.10): a) quantizer has a center dead zone, d, to eliminate the picture noise; b) quantizer range is fixed by the lower and upper limit thresholds, ± t, to cover a moderate change in the

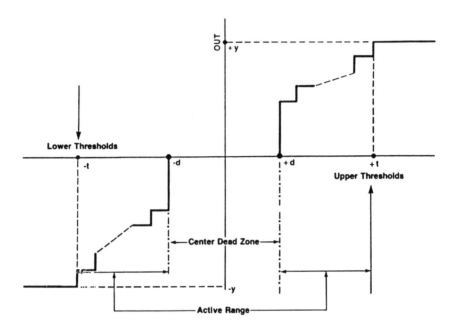

Figure 6.10 - Quantizer For High Frequency Bands

signal c) the signal falling within active range is uniformly quantized into L levels; and d) signals above threshold t or below threshold − t are mapped to saturation values +y and −y, respectively. Thus, the total number of quantization levels is L+ 3.

The decision in selecting the parameters d, t, y and L is based on a compromise between fidelity and bit-rate requirements. These parameters, however, change according to the buffer fullness. For example, to prevent buffer overflow, larger values for dead zone, d, (particularly for higher frequency bands) are selected.

6.4.2 Entropy Coding

The entropy coder consists of a variable word length (VWL) coder to encode the non-zero quantized values and a run length coder to encode their corresponding locations [7,9]. This is achieved by considering each scan line as a consecutive black and white run. The black run corresponds to zero and white runs to non-zero values. In this way only non-zero values need to be transmitted. The VWL code set is designed according to the average statistics of all bands so that only one VWL code set can be used for all bands, at the expense of a slight loss of efficiency. It should be noted however, that based on our earlier experiments [7,9], we observed that VWL coding of non-zero PCM does not significantly improve the coding performance (particularly, in the case of intraframe). Therefore, it may be more convenient to select the number of quantization levels within the active range (see Figure 6.10) in such a way that fixed length codes can be efficiently applied.

In the above approach a good deal of overall bits go toward the run-length coded positional information, therefore, any attempt to further improve the coding performance should be directed at increasing the efficiency in which the positional information is coded. Although this approach can effectively exploit the correlation in the horizontal direction, it does not take into consideration the 2-D spatial dependencies in the image. There are however, a number of schemes which have been developed for two level image compression (facsimile) and can be utilized for coding the positional information. These include edge difference coding [28], ordering techniques [29], contour coding [30], etc. Although these schemes are currently under consideration, in this chapter we present a simple 2-D run length coding strategy which provides a signal with long bit runs. The coding approach is described as follows.

In this approach the individual sub-images are first partitioned into non-overlapping blocks. The scanning is performed on a block-by-block basis starting from the first block on the upper most left and continuing in the horizontal direction until the last block (upper most right) is scanned. The resulting bit stream is then runlength coded and transmitted together with non-zero PCM coded values. This process continues in the same manner until the last strip of blocks in each sub-image is scanned and coded.

For the block size of m× n, the above arrangement can be viewed as transforming M× N image into P× Q where

$$P = M \times n$$

$$Q = \frac{N}{n} \quad .$$

In our discussion so far we have not mentioned the manner in which each block is scanned. There are normally three scanning techniques which can be considered here. These are horizontal (HS), vertical (VS), and diagonal (DS) scanning which are shown in Figure 6.11.

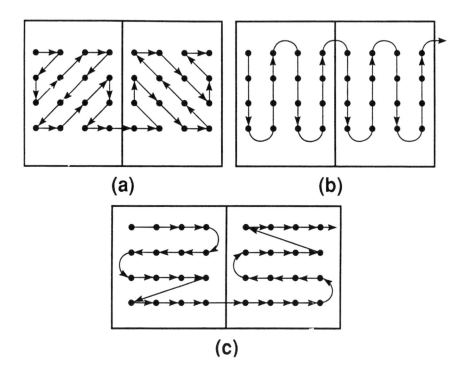

(a) Zig-zag Scanning for Diagonal Band
(b) Vertical Scanning for Horizontal Band
(c) Horizontal Scanning for Vertical Band

Figure 6.11 - Procedures for Block-by-Block Scanning of Address Information

To find out which of the above scanning techniques is more suitable for our application, let's consider the nature of each decomposed sub-image with respect to its position in the 2-D frequency spectrum. Figure 6.12 shows the sub-images which are the result of a basic 2-D decomposition of an image.

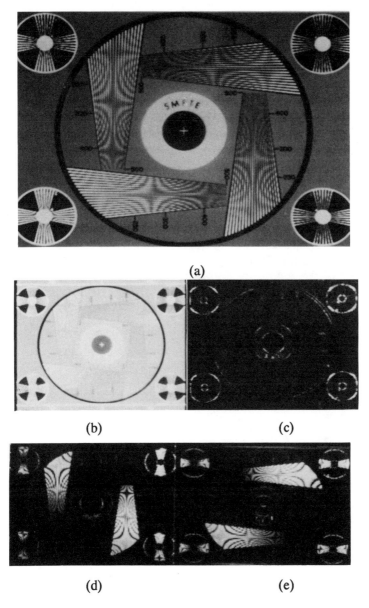

(a) Test Image, (b) Lowest Band
(c) Diagonal Band, (d) Horizontal Band, (e) Vertical Band

Figure 6.12 - Decomposition of Image into 4-Band

In this figure sub-images b, c, d, and e correspond to the lowest band (LB), diagonal band (DB), vertical band (VB) and horizontal band (HB), respectively. For better clarification, the position of each individual band in a 2-D frequency spectrum is shown in Figure 6.13. From this figure it can be easily deduced that the horizontal, vertical, and diagonal bands mostly contain vertical, horizontal, and diagonal edges, respectively. For interframe case, the decomposed prediction error bands in both models contain information which corresponds to the temporal variation of the video signal. These variations are normally distributed in the horizontal, vertical, and diagonal bands depending on the spatial movements of the objects. Thus, to increase the average black and white runs which could consequently improve the runlength coding efficiency, an appropriate scanning technique should be adapted in this respect. As a result, horizontal, vertical, and diagonal scanning are each assigned to the vertical, horizontal, and diagonal bands, respectively. In addition, to help the continuity of the black or white runs, the scanning from one block to the next is performed in the manner shown in Figure 6.11. Note that with this strategy the length of maximum run is now increased by the factor of n (i.e., $P = nM$). This can help to improve the effectiveness of the end of line (EOL) and end of image (EOI) termination process which is described as follows.

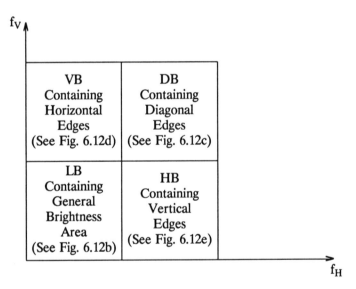

Figure 6.13 - Coding Arrangement. Cases 1-4 are defined in Table 6.1

6.4.2.1 EOL and EOI

As mentioned earlier, the subbands contain the information which results from the mismatches of moving objects in the consecutive frames. In addition, the energy distribution of the frequency spectrum depends on the contrast in brightness between background and moving objects. In video conferencing, where the variation of the movement is mostly concentrated in the center of the scene, it is beneficial to terminate the run-length coding process once the last non-zero value has been reached on the coding line. The same argument can also be applied once the last non-zero value is reached in the image. Termination of the runlength coding process along the line in the image can be achieved by assigning unique codewords to signify the end of line (EOL) or end of image (EOI). The usefulness of EOL and EOI, when applied to higher band sub-images, depends on how the image boundaries are treated by QMF. In a normal convolution approach these boundaries are viewed by QMF as sharp transitions. The sharpness of these transitions depends on the difference between the filters initialization value and the pixel intensity at the image boundaries. As a result, the decomposed higher band images can be expected to be surrounded, to a large extent, by the non-zero quantized values. This could consequently cause the deterioration of the coding performance. More importantly, the employment of EOL and EOI with the intention of improving the runlength coding efficiency, becomes no longer beneficial. To overcome this, an interesting solution would be to use the symmetric extension approach which has been suggested by Smith and Eddins [31] (see also Chapter 3, section 3.2). This technique, which is based on the appropriate QMF filter initialization, can limit the region of support so that the total number of pixels in the subbands would be the same as in the original image. However, the main contribution of this approach, in conjunction with our coding method, is the elimination of the unnecessary non-zero values near the image boundaries. Consequently, the utilization of EOL and EOI becomes advantageous in improving the runlength coding efficiency.

6.4.2.2 Runlength Coding Design

The address information is runlength coded by considering each scan line as a sequence of black and white runs where the white runs correspond to the length of the contour points. The optimal solution is to design two sets of codewords which can match the statistics of black and white runs. It is well known that for a given set of message probabilities, the most efficient code is the Huffman code [32]. Huffman codes, however, generally lack structure and their implementation can be complicated when the number of messages is large. As a trade-off between compression and ease of implementation, we have chosen a class of 1-D codes called B_1 codes [33]. These codes are most efficient for negative-power distributions (i.e., the codeword length increases approximately logarithmically with the run length) and are found to be suboptimal over a wide range of runlength coding applications [33]. In addition, they are also easily implementable requiring essentially only a counter.

In general, B codes of block length (n,m) are constructed as follows: Codewords corresponding to a given runlength consist of a sequence of blocks of length $(n+1)$ and $(m+1)$ bits for black and white runs respectively. The first bit of each block, called the color bit, is used to indicate the color of the coded run (i.e., black and white runs). Subsequent blocks with the same color bit form a codeword. In the decoding process, the color bit is used to examine whether a new block belongs to the codeword being decoded or is the start of a new codeword. Readers are referred to Table 6.3 for an example of B_1 (i.e., $n=m=1$) codewords corresponding to both black and white runs with lengths 1 to 4.

Due to the structure of the B_1 code (see Table 6.3), accommodation of a unique codeword to signal EOL is not straightforward. As a result, we have modified the B_1 code in the manner shown in Table 6.4. As can be seen from this table, the EOL codeword is given by the codeword C1C1

Table 6.3 Example of B1 Codes		Table 6.4 Modified B1 Codes	
RUN	CODE	RUN	CODE
1	C0	1	C0
2	C1	2	C1
3	C0C0	3	C0C0
4	C0C1	4	C0C1
5	C1C0	5	C1C0
6	C1C1	EOF,EOL	C1C1
7	C0C0C0	6	C0C0C0
8	C0C0C1	7	C0C0C1
9	C0C1C0	8	C0C1C0
10	C0C1C1	9	C0C1C1
11	C1C0C0	10	C1C0C0
12	C1C0C1	11	C1C0C1
13	C1C1C0	12	C1C1C0
14	C1C1C1	13	C1C1C1
	etc.		etc.

C is the color of the run.   C is the color of the run.
                                               EOF End of Frame.
                                               EOL End of Line.

which was originally assigned to the run of length 6 (see Table 6.3). The EOL codeword requires another bit to signify the uniqueness of EOL; e.g., if for the End of Line code, we select the codeword for the black-run of 6 (i.e.,

0101), it is essential that this codeword be followed by an additional bit of white color. This is to guarantee the termination of the codeword and thus isolate the 5-bit EOL codeword "01011". This arrangement can, of course, leave us with another unique codeword "11110" which is used as an end of frame code EOI. The EOL codeword, with differing lengths, can also be defined under the same conditions. In our applications we observed that 5-bit EOL is the most efficient length.

6.5 Intraframe Coding

Intraframe coding is applied whenever the previously coded frame is not available (i.e., first frame) or, if available, not suitable for use (i.e., scene cut). In this situation both subband models operate on an open loop basis. This would consequently result in a unified intraframe model which has been considered in the past for single image coding [7,11]. The coding is performed in such a way that the lowest band is coded by DPCM, whereas the other by direct PCM. In the DPCM encoder used here, a linear predictor constructs its predicted pixel as a weighted summation of previous pixels. The pixel configuration for the 2-D prediction is shown in Figure 6.14.

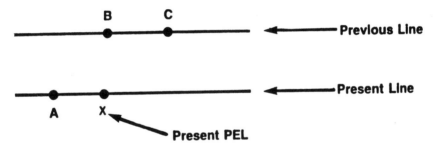

Figure 6.14 - Configuration of Pels Used for Prediction

$$x = 0.5A + 0.25B + 0.25C$$

where A, B, C are previously reconstructed pixels and x is the prediction of the present pixel. The prediction error signal is then quantized by symmetric non-uniform quantizers.

The higher band signals are coded in the same manner as in the interframe case and that is the combination of PCM quantization and entropy coding.

6.6 Subband Coding of Color Components

The CCIR 601 and CIF provide the color information as two color difference components C_r and C_b. C_r and C_b are weighted values of R-Y and B-Y, respectively. The spatial resolutions of these components are lower than luminance Y and are shown in Table 6.1, Section 6.2. These signals are basically coded in the same manner as Y except;

    i)   No motion estimation is performed on the color difference frames

    ii) The color components can be motion compensated using the motion vectors derived for the luminance component

    iii) The color subbands are quantized with lower accuracy (i.e., larger quantizer dead zone).

To meet the above item (ii) some arrangement is needed with regard to each interframe subband model. This is mainly to handle the lower spatial resolution of color components as compared with luminance components. For example, in Model I the motion estimation is performed on a full frame size. Thus, the components of the estimated motion displacement of the luminance needs to be divided by an appropriate factor so that they can be applied to the MC prediction of color components. This factor is related to the ratio of the spatial resolution between the luminance and each color component (in each direction).

For Model II however, where the motion estimation is performed on individual decomposed bands, a different arrangement is needed. In this case the number of decompositions is decided on the basis that the lowest band of luminance and the lowest band of each chrominance have the same spatial resolution. Under this condition the estimated motion displacements of the luminance bands can also be applied to the decomposed bands of each color difference frame. In this chapter, however, the color frames are coded without motion compensation. This is simply to avoid the above complexity as well as taking into consideration the fact that the color components have a much lower dynamic range than luminance signals.

6.7 Results and Discussion

This section contains the simulation results of the intra/interframe subband coding schemes developed in this chapter. First, we present the results of intraframe coding which is based on a single image input. In addition, the evaluation of the block runlength coding technique (see Section 6.4) as well as the effect of QMF filter impulse response on the overall intraframe coding performance are also discussed. This is followed by the results of interframe subband coding schemes developed in Section 6.2. Two video sequences with different spatial resolutions are considered. One is based on the CIF format which has a spatial resolution of 352 × 288 and is suitable for low bit rate video applications (64 to 384 kb/s). This sequence is known as "Salesman" and its first coding frame is shown in Figure 6.15. The other sequence has a spatial resolution of 720 × 480 (CCIR-601, 525 lines standard) which is generated for higher quality video conferencing applications (i.e., 1.5 Mb/s). This sequence is particularly suitable for Model II based CCITT compatible video coding (Section 6.2.5). The sequence is named "Diane and Rich" (see Figure 6.16) and consists of a moderately close up view of two people sitting behind a desk and involved in active conversation.

Figure 6.15 - Salesman

6.7.1 Intraframe

Computer simulations were carried out using the QMF system bank discussed in Section 6.1. Similar to our earlier evaluation, we employed 8-tap and 16-tap near perfect reconstruction filters designated as type A by Johnston [14]. In these experiments the bit accuracy at the output of each 1-D QMF was truncated to 12-bit. In the simulation the first frame of each sequence (shown in Figures 6.15 and 6.16) was coded using the intraframe coding technique described in Section 6.5. Accordingly the lowest band was coded by a 2-D DPCM. The prediction was performed using one previous pixel along the line with the weighing factor of 1/2 and the two nearest pixels on the previous line. One is exactly above and the other on its right hand

Figure 6.16 - Diane and Rich

side; both with equal weighing factors of 1/4. A 31-level non-uniform quantizer was used to quantize the prediction error signal. The quantized signal was then variable word-length (VWL) coded before being multiplexed with higher frequency bands.

The higher band images were coded as follows as described in Section 6.4.1. The quantization parameters were selected to the region where each sub-image is located in the 2-D frequency spectrum. For example, higher band images are quantized using larger values for d. The quantized images were subsequently binary coded.

The positional information was runlength coded on a block-by-block basis. The block size was selected according to the decimation factor as well as the original image resolution. In this application the block sizes of 8×8 and 4×4 were used. As described in Section 6.4 three scanning methods were applied. These were horizontal, vertical, and diagonal (zig-zag) scanning which are noted by HS, VS, and DS, respectively. The assignment of the scanning method to each sub-image was based on the location of the band on the 2-D frequency spectrum. Such an arrangement is shown in Figure 6.17. This figure shows the allocation of different scanning to each non-uniformly decomposed band. The block by block (from left to right) scanned pixels were then runlength coded using the modified B_1 code (see Table 6.4).

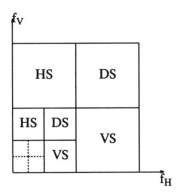

DS: Diagonal Scanning
HS: Horizontal Scanning
VS: Vertical Scanning

Figure 6.17 Assignment of the Block Scanning to each Band
in a Non-Uniform Decomposition

The modified B_1 code was designed to accommodate the end of line (EOL) and end of image (EOI) codewords.

To verify the effectiveness of the adaptive scanning we have applied all three methods, one by one, to each individual band. The resulting total bits per band for four-band decomposition are shown in Table 6.5. These results clearly indicate that VS, HS, and DS are best suited to be applied to horizontal, vertical, and diagonal bands, respectively.

Table 6.5   Average   bits/band   corresponding   to   the runlength   coded   positional   information   using different scanning method for each band

Sub-Band	Horizontal Scanning (Avg. Bits)	Vertical Scanning (Avg. Bits)	Diagonal Scanning (Avg. Bits)
Vertical-Band	1622*	2558	2054
Horizontal-Band	5586	2958*	4784
Diagonal-Band	324	338	262*

* Selected Scanning method

In our experiments we also evaluated the effect of filter length on the coding performance. In addition to 8-tap and 16-tap, we have also considered a simple FIR 2-tap QMF [15] which can satisfy the perfect reconstruction requirement expressed by (6.2) and (6.4). The results of this evaluation, in terms of average bits/pixel with their respective signal to noise ratios, SNR (peak to peak to rms noise) for the pictures shown in Figures 6.15 and 6.16, are presented in Tables 6.6 and 6.7, respectively. As can be seen from these tables for both seven and ten bands

Table 6.6. Results for the Picture of "Salesman"

Results	7-Band Decomposition			10-Band Decomposition		
	2-tap	8-tap	16-tap	2-tap	8-tap	16-tap
SNR	32.12	32.21	32.35	32.01	31.8	32.6
Bits/pel without block scanning	0.64	0.53	0.52	0.61	0.53	0.50
Bits/pel with block scanning	0.58	0.48	0.46	0.54	0.46	0.42

Table 6.7. Results for the Picture of "Diane and Rich"

Results	7-Band Decomposition			10-Band Decomposition		
	2-tap	8-tap	16-tap	2-tap	8-tap	16-tap
SNR	34.17	33.43	34.82	34.17	33.9	34.69
Bits/pel without block scanning	0.40	0.38	0.33	0.38	0.36	0.31
Bits/pel with block scanning	0.35	0.32	0.27	0.32	0.29	0.24

decompositions, the increase in the filter length can have a significant effect on improving the coding performance. Improvement can also be obtained with higher number bands (i.e., 7 to 10). For example, the 16-tap QMF using 10-band decomposition requires an average bits/pixel of 0.24 for the picture of Diane and Rich. This figure can go up to 0.32 bits/pixel for 2-tap and 0.27 bits/pixel for 16-tap with 7-band decomposition.

Furthermore, we have compared the results of our coding scheme with the 1-D runlength coding scheme [9,11]. These results are also given in Tables 6.6 and 6.7 for the same pictures which verifies the effectiveness of block runlength coding. As a result, a good quality image can be reconstructed at very low average bits/pixel. As an example, Figures 6.18

and 6.19 are the processed pictures coded at 0.42 bits/pixel and 0.24 bits/pixel with the SNR of 32.6 dB and 34.69 dB. These results clearly

Figure 6.18 - Processed "Salesman" Coded at 0.42 bits/pixel

Figure 6.19 - Processed "Diane and Rich" Coded at 0.24 bits/pixel

indicate that the overall performance of the subband coding relies heavily on how the higher frequency bands are coded. In addition, the filter impulse which guarantees the narrowness of the decomposed bands, can contribute significantly in further improving the coding efficiency.

6.7.2 Interframe Model I Based

Simulations were performed to assess the performance of the subband Model I based coding technique described in Section 6.2. In the simulation, the first frame of each sequence was coded using intraframe prediction. For the second frame and thereafter, motion compensated interframe prediction was applied. The input data was first organized into a block of 16×16 pixels to perform the block matching motion estimation. The size of the search window selected on the previous frame was considered to be 46×46. This could handle the tracking range of ± 15 pixels in both vertical and horizontal directions. The motion vector for each block was coded by VWL and then transmitted at the beginning of each block even though the actual data was not coded on a block by block basis. The input signal, after subtracting the motion compensated prediction signal, was then applied to the 10-band non-uniform QMF bank system on a pixel by pixel basis. Decomposed subbands were subsequently quantized by symmetric uniform quantizers deadzone d. The value of d was selected according to the buffer fullness. The buffer fullness was set at five threshold levels; 10%, 30%, 50%, 70%, and 90%. The block diagram of the overall encoder is shown in Figure 6.20.

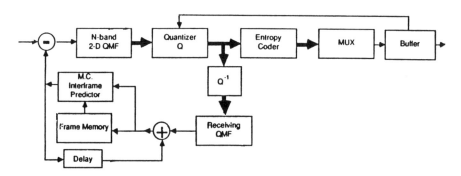

Figure 6.20 - Motion Compensated Interframe Subband Coder (Model I Based)

In the coding process a set of VWL was designed to encoded the non-zero quantized values. For positional information, as described in Section 6.4, the block runlength coding scheme was applied. Different scanning was assigned to different bands in accordance with the location of each band in a 2-D frequency spectrum of prediction error signals. In our experiments we observed that such an arrangement was significantly effective in the interframe case; particularly in conjunction with the modified B_1 code which utilizes the EOL and EOI codewords. The simulation was performed with a transmission rate of 64 kb/s. Similar to CCITT, at this rate the spatial resolution of the input sequence was reduced to quarter CIF (176× 144).

Four consecutive frames of the processed "Salesman" coded at the rate of 64 kb/s with a frame rate of 10 f/s are shown in Figure 6.21. The performance

Figure 6.21 - Four Consecutive Frames of Coded "Salesman"

of the proposed scheme was also evaluated by viewing the quality of the processed video sequences in real time. The reference for this evaluation was based on the results of the adaptive hybrid DPCM/DCT scheme presented in our earlier publications [34]. We observed that the predictive subband coder under the same conditions can produce a better quality video at its output than the hybrid DPCM/DCT scheme. The two methods were also compared in terms of SNR (Peak to peak signal to rms noise ratios) and the results are shown in Table 6.8. These results support the subjective evaluation and

Table 6.8 Average SNR at the Rate 64 kb/s

Sequence	Hybrid ref [34] (dB)	Subband Model I (dB)	Frame Rate (f/s)
Salesman	33.2 dB	35.3 dB	10

clearly verify the effectiveness of the proposed coding schemes. Moreover, this scheme has an advantage over the hybrid technique of being free of annoying block distortion.

6.7.3 Model II Based CCITT Compatible Coding

This model was mainly considered for a relatively high quality video conferencing applications (i.e., at a transmission rate of 1.54 Mb/s). The "Diane and Rich" video sequence consisting of 150 frames (2:1 interlaced) of 720 pixels × 480 lines (CCIR-601, 525 lines standard) was therefore, considered for such an application. The horizontal resolution was reduced to 704 so that it can have a 2:1 relation with a reduced CIF resolution of 352. For the sake of compatibility with CIF, the coding was performed on a frame-by-frame basis (interframe) although there are some hardware advantages by using the interfield prediction [34]. In the simulation basic 4 bands and non-uniform 7 bands decompositions were considered. In each case the individual bands were coded according to the arrangements described in Section 6.2.5 (see also Figure 6.7).

The lowest band was coded by the hybrid DPCM/DCT coding technique presented in our earlier publication [33]. This scheme, which is based on a sub-block approach, can produce more or less the same video quality as the CCITT hybrid coder [3]. A block size of 16× 16 was used for the full search blockmatching motion estimation (see Section 6.3). The search area was selected so that both horizontal and vertical components of the motion vectors had a motion tracing range of ± 15. The estimated motion vectors were then made available to some of the higher frequency bands. As discussed earlier, these bands were coded by a combination of the MC interframe using the motion vectors estimated for the lowest band, interframe prediction without motion compensation and direct intraframe PCM. It should be noted, however, that a simpler and less efficient approach would be to encode these bands directly by PCM (intraframe). Both PCM and DPCM employed a symmetric uniform quantizer with a center dead zone, d. The quantized signal was then entropy coded using a combination of variable word length (VWL) for non-zero quantized values and runlength coding for their corresponding positional information. Similar to the previous cases, the block runlength coding scheme with different scanning procedures was applied to encode the positional information. At the receiver each subframe, after being decoded, was sent to the QMF interpolation filter banks for reconstruction.

The performance of hierarchical subband coding was compared with the hybrid scheme and then modified to operate on a full frame basis. For 4-band decomposition, at an average transmission rate of 1.54 Mb/s, this coding scheme has shown a noticeable subjective improvement with an SNR gain of 1.2 dB over the hybrid scheme. This improvement has shown to be more significant at a higher number of decompositions (see Table 6.9).

Table 6.9 Average SNR at the Rate of 1.5 Mb/s

Sequence	Hybrid ref [34] (dB)	Subband Model II	
		4-band (dB)	7-band (dB)
Diane and Rich	34.6	35.8	37.02

Nevertheless, the main motivation for considering the hierarchical subband is not so much its relative coding efficiency but more its multi-layer video representation feature. This important feature should be considered as a basis for offering a wide range of video services including the current CCITT proposed scheme for videophone and videoconferencing applications.

6.8 Conclusion

This Chapter was mainly concerned with developing coding schemes for high compression video applications. Two distinct models for interframe prediction were presented. Both models were compared and show that they perform equally under certain conditions. The first model, due to its lesser hardware complexity, was considered for video-conferencing at lower speeds. The second model however, due to its generic structure, was found to be suitable for a wide range of video applications. As an example, a CCITT compatible video coding scheme was developed. For the best performance, the higher frequency bands were coded with a combination of interframe DPCM and direct intraframe PCM. In addition, to preserve its hierarchical structure, each band was coded independently of the higher frequency bands. However, they retained the ability to share information with the lower bands. In the coding process each band was quantized independently. To transmit only the non-zero quantized values, we initially applied a 1-D runlength coding scheme to encode their locations. To further improve the coding performance, we considered a block runlength coding method to exploit the 2-D correlation of the signal. In addition, the coding scheme takes into consideration the position of each individual band in the 2-D frequency spectrum of the original image. It was shown that such a scheme can significantly improve the coding performance.

Finally, simulation results for the Model I and Model II coders with transmission rates of 64 kb/s and 1.54 Mb/s, respectively, were performed. These results were also compared with the hybrid DCT/DPCM coding scheme under the same environments.

Acknowledgments

I would like to thank Jules A. Bellisio for his support in completing the work presented in this chapter. I also wish to express my sincere gratitude to Marilyn Welsh for her tireless efforts in typing and putting this chapter together.

Bibliography

[1] H. M. Musmann, P. Pirsh, and H. J. Gravoert, "Advances in picture coding," *Proc. IEEE*, vol. 73, April 1985, pp. 523-548.

[2] A. Habibi, "An adaptive strategy for hybrid image coding," *IEEE Trans. Comm.*, vol. COM-29, pp. 1736-1740, Dec. 1981.

[3] "Video Codec for Audiovisual Services at P×64 kb/s" draft revision of the CCITT recommendation H.261, WP XV/1 report, Part II, Dec. 1, 1989.

[4] H. Gharavi, "Low-bit radio video transmission for ISDN application," *IEEE Trans. on Circuits and Systems*, vol. CAS-35, No. 2, February 1988, pp. 258-261.

[5] M. Vetterli, "Multi-dimensional subband coding: some theory and algorithms," *Signal Processing*, vol. 6, April 1984, pp. 97-112.

[6] J. W. Woods and S. O'Neil "Subband coding of images," *Proc. ICASSP'86*, April 1986, pp. 1005-1008.

[7] H. Gharavi and A. Tabatabai, "Subband coding of digital images using two-dimensional quadrature mirror filtering," *Proc. SPIE*, vol. 707, September 1986, pp. 51-61.

[8] J. W. Woods and S. D. O'Neil, "Subband coding of images," *IEEE Trans. ASSP*, vol. 34 October 1986, pp. 1278-1288.

[9] H. Gharavi and A. Tabatabai, "Subband coding of monochrome and color images," *IEEE Trans. on Circuits and Systems*, vol. 35, February 1988, pp. 207-214.

[10] H. Gharavi and A. Tabatabai, "Applications of quadrature mirror filtering to the coding of monochrome and color images," *Proc. ICASSP '87*, vol. 4, pp. 2384-2387.

[11] H. Gharavi, "Subband coding of color images using differential vector quantization techniques," *Picture Coding Symposium*, PCS-87, June 1987, Sweden.

[12] P. H. Westerink, J. Biemond, and D. E. Boekee, "Subband coding of digital images using predictive vector quantization," *Proc. ICASSP '87*, vol. 3, pp. 1378-1381.

[13] P. H. Westerink, J. Biemond, D. E. Boekee, and J. W. Woods, "Subband coding of images using vector quantization," *IEEE Trans. Comm.*, vol. 36, No. 6, June 1988, pp. 713-719.

[14] D. LeGall and A. Tabatabai, "Subband coding of digital images using symmetric short kernel filters and arithmetic coding techniques," *Proc. ICASSP '87*, vol. 3, pp. 1378-1381.

[15] W. F. Schreiber, C. F. Knapp, and N. D. Kay, "Synthetic highs, an experimental TV bandwidth reduction system," *Journal of the SMPTE*, vol. 68, August 1959, pp. 525-537.

[16] R. E. Crochiere, S. A. Webber, and J. L. Flanagan, "Digital coding of speech in subbands," *Bell System Technical Journal, BSTJ*, vol. 55 October 1976, pp. 1069-1085.

[17] R. E. Crochiere, "On the design of subband codes for low bit rate speech communications," *BSTJ*, vol. 56 May-June 1977, pp. 747-770.

[18] A. Croisier, D. Esteban, and C. Galand, "Perfect channel splitting by use of interpolation, decimation, tree decomposition techniques," *1976 Int. Conf. on Information Sciences/Systems*, Patras, Greece.

[19] D. Esteban, and C. Galand, "Application of quadrature mirror filters to split band voice coding schemes," *Proc. 1977 ICASSP*, May 1977.

[20] J. D. Johnston, "A filter family designed for use in quadrature mirror filter banks," *Proc. ICASSP '80*, pp. 291-294.

[21] H. Gharavi, "Differential subband coding of video signals," *Proc. ICASSP-89*, pp. 1819-1821.

[22] J. W. Woods and T. Naveen, "Subband encoding of video sequence," *Proc. SPIE*, Nov. 1989.

[23] J. R. Jain and A. K. Jain, "Displacement measurement and its application in interframe image coding," *IEEE Comm. Trans.*, vol. COM-29, December 1981, pp. 1799-1808.

[24] S. Kappagantula and K. R. Rao, "Motion compensated predictive coding," *SPIE, 27th Proc. 432*, pp. G4-70, 1983.

[25] C. D. Bowling and R. A. Jones, "Motion compensated image coding with a combined maximum A posteriori and regression algorithm," *IEEE Trans. Comm.*, vol. COM-33, Aug. 1985, pp. 844-857.

[26] H. Gharavi and M. Mills, "Blockmatching motion estimation algorithms – new results," *IEEE Trans. on Circuits and Systems*, May 1989, pp. 649-651.

[27] T. C. Chen and P. E. Fleischer, "Subband coding for ATV signals based on spatial domain considerations," *SPIE Visual Comm. and Image Proc. 89*, Philadelphia, PA, vol. 1199, paper 54.

[28] R. Hunter and A. H. Robinson, "International digital facsimile coding standards," *Proc. IEEE*, vol. 68, July 1980, pp. 830-46.

[29] A. N. Netravali, ed., special issue on Digital Encoding of Graphics, *Proc. IEEE*, vol. 68, July 1980.

[30] W. F. Schreiber, T. S. Huang, and O. J. Tretiak, "Contour coding of images," *Conference Record, IRE Western Electronics Conference*, August 1968.

[31] M. J. T. Smith and S. L. Eddins, "Subband coding of images with octave band tree structures," *ICASSP Proc.*, Dallas, Texas, April 1987, pp. 1382-1385.

[32] D. A. Huffman, "A method for the reconstruction of minimum redundancy codes," *Proc. IRE*, vol. 40, No. 9, Sept. 1952, pp. 1098-1101.

[33] T. S. Huang, "Easily implementable suboptimum runlength codes," *Proc. IEEE Trans. Comm.*, vol. 1, June 1975, pp. 7.8-7.11.

[34] H. Gharavi, "Low bit rate video for ISDN," *IEEE Trans. on Circuits and Systems*, vol. CAS-35, Feb. 1988, pp. 1382-1385.

[35] F. A. Kamangar and K. R. Rao, "Interfield hybrid coding of component color television signals," *IEEE Trans. Comm.*, vol. COM-29, Dec. 1981, pp. 1740-1753.

[36] H. Gharavi, "Subband-based CCITT compatible video coding," *Proc. IEEE GLOBECOM'90*, San Diego, California, Dec. 1990.

Chapter 7

Advanced Television Coding Using Exact Reconstruction Filter Banks

by: Rashid Ansari and Didier Le Gall[#]
      Bellcore
      Morristown, NJ 07960

7.1 Introduction

Advanced Television (ATV) is a very promising new development in visual communications on account of its high picture quality and the variety of future services that it can offer. Viewers in the United States currently watch television pictures with a resolution specified by the NTSC (National Television Standards Committee) standard. The monochrome NTSC standard was promulgated in 1941, before consumer electronics witnessed the tremendous advances that were set in motion by semiconductor technology. In 1953 the standard was modified to incorporate color but in a way that would not make the black-and-white TV receivers obsolete. The NTSC signal is transmitted in the analog format over a 6 MHz channel.

Television technology has made enormous strides since the NTSC standard was adopted. We have witnessed significant progress in display, device and recording technologies. These developments have produced television equipment that demonstrates the possibility of providing a vastly improved picture with a new generation of television systems. Such systems are referred to by the generic label of *Advanced*

[#] Now with C-Cube Microsystems, San Jose, CA 95131

Television (ATV) systems. A *High Definition Television* (HDTV) system is an ATV system that provides roughly double the vertical and horizontal resolution of the NTSC picture, and in a wide-screen format with about thirty percent larger width-to-height ratio. The quality of HDTV pictures is suitable for demanding medical, industrial, and other imaging applications.

An HDTV system calls for a proper transport infrastructure to be in place for signal delivery. In this regard there are many issues pertaining to the choice of transmission medium, signal format, bandwidth allocation, formulation of standards that need to be addressed and will be discussed in the next Section. Initially television signal delivery was confined to terrestrial broadcast. But now, in addition to terrestrial broadcast, there exist alternative transmission media for television transmission such as electrical cable, direct satellite broadcasting and optical fiber. In particular optical fiber holds tremendous promise for the transport of a digital HDTV signal.

A major influencing factor in the development of ATV systems has been the emergence of flexible, efficient and cost-effective techniques of high-speed digital signal processing that have provided attractive options in the methods of compressing and representing the ATV signal. Digital signal processing techniques have had an impact on both analog and digital television signals. In analog television the manipulation of the digitized version of the signal opens up many avenues to exploiting the redundancies in the video signal. This is the basis of several new techniques of ATV signal representation that have been recently developed. Typically the procedures involve the use of signal processing techniques with multiple sampling rates and the exploitation of the frequency domain distribution of the signal energy. The ATV signal is finally converted into a suitable analog or digital format for transport. The analog ATV signal is subject to considerable transmission impairments in the presence of noise, interference and multipath. The digital television signal is significantly more robust and lends itself to intricate techniques of data compression. It is well-suited for the anticipated development of the fiber-based broadband integrated services digital network (B-ISDN). In this chapter we will study a class of compression techniques for digital television.

We begin by briefly reviewing analog and digital advanced television and describe some issues in the transmission of ATV signals over a broadband network. Then we focus on the main subject of the Chapter: *the compression and representation of digital ATV signals using subband coding.*

The emphasis in this chapter is on issues in subband coding when it is applied to high quality coding of HDTV signals. The signal formats, the desired bit rates and compression requirements are discussed in the following sections. As a prelude to the description of coding schemes, we shall briefly review two-dimensional separable and non-separable filter banks for application in intra-field HDTV data compression. Following this, methods of representing the subband signals with a given budget of bit rate are described, and issues and trade-offs in assigning the bit rates to the different subband signals are discussed. Simulation results using these coding techniques are then presented. A short description of the use of subband coding in the temporal direction is also provided.

7.2 Background on ATV Systems

In this Section we will look at the NTSC format and the attendant artifacts. Then we examine some options in providing an improved signal or an advanced television signal. A brief description of the role of the future broadband network in the transmission of digital HDTV is presented. Finally the framework and objectives for the digital coding of ATV signals are spelled out. For further information on these topics, the reader is referred to [1-4].

7.2.1 NTSC Standard and Its Limitations

The picture in the black-and-white (monochrome) NTSC standard consists of 525 lines/frame, 30 frames/sec with a 2:1 interlaced scan and a 4:3 aspect ratio. The video (luminance) signal is modulated using vestigial sideband amplitude modulation at 1.25 MHz above the lower band-edge of the 6 MHz channel. The sound information is mounted on a carrier at a frequency that is 4.5 MHz above the picture carrier frequency.

When the monochrome standard was modified to include color information, the parameters of the monochrome standard were unaltered except for a change in the frame rate to 29.97 Hz. This change was designed to limit the mutual interference between the color subcarrier and the sound carrier signal. The color information was added without increasing the bandwidth assigned to each channel. This was accomplished by limiting the high frequency information in the chrominance signals to a significant extent. The color subcarrier was modulated in quadrature by the two color-difference signals and the resulting signal was added to the luminance component. The color subcarrier frequency was carefully chosen to limit possible artifacts. However the composite form in which the luminance and chrominance information are assembled causes mutual interference artifacts when the video data contains certain types of details. The interference artifacts are referred to as cross-color and cross-luminance effects that manifest as spurious color and dot patterns. Thus even within the limitation of resolution due to the 525-line interlaced signal, the details can be distorted by the NTSC signal format. Also the color bandwidth is considerably narrow relative to the luminance bandwidth and produces a visible loss of resolution for highly saturated color transitions. It is these limitations that researchers have sought to overcome in designing ATV systems.

7.2.2 Types of Advanced Television Systems

In designing an ATV system, one option is to build a system that is compatible with the NTSC standard. The improvement is provided as auxiliary information on the signal details. However one can make a clean break from the composite NTSC format and instead directly code and multiplex the luminance and color-difference signals.

The extent of the improvement over the NTSC signal and the manner in which it is rendered can vary greatly among ATV systems. The systems are classified [5] into the following categories:

- Improved Definition TV (IDTV) -
  The improvements here remain within the confines of NTSC emission standards and they would require little, if any, Federal Communications Commission (FCC) action. The source signal and the receiver may be modified to provide improvements in encoding and processing. However all embellishments conform to compatibility with NTSC standard.

- Enhanced Definition TV (EDTV) -
  This system provides an NTSC receiver compatible signal with some IDTV-like improvements but with added features such as wider aspect ratio and improved resolution, though less than twice that of NTSC. The improvements require modifications in NTSC emission standards.

- Extended Quality TV (EQTV) -
  The quality may be at par with EDTV but the restriction of NTSC receiver compatibility is removed. Alternate media such as fiber,

coaxial cable or satellite may be used. A component-based signal rather than a composite signal is typically used.

* High Definition TV (HDTV) -
The HDTV signal is characterized by a doubling (roughly) of the horizontal and vertical resolution of the NTSC signal. Features of the above three systems may be included.

HDTV will provide a picture that rivals 35mm film in quality and sound that is comparable to compact disc output. Movies presented on HDTV will show a dramatic improvement over the NTSC format. The picture will appear in its original wide-screen format without the cropping that is necessary for NTSC presentation. There is immense potential for using HDTV in the film industry. The flexibility offered by videotape and the manipulation that it lends itself to can be a major cost-saver in the tasks of editing and composing. Other beneficiaries of the development of HDTV are the computer graphics industry, the computer-aided design (CAD) and computer-aided manufacturing (CAM) industry, electronic publishing and medical imaging (cf. Chapter 8).

7.2.3 Analog ATV systems

An important consideration in the design of ATV systems is the choice between analog and digital ATV. The choice is affected by the medium of signal transmission. For transport on optical fiber, digital ATV is the more suitable choice. However, in the case of terrestrial broadcast, spectrum limitation suggests the use of analog ATV, although a recent proposal [6] by General Instruments Corporation describes a digital transmission system for the 6 MHz channel. This system is based on the use of a hybrid discrete cosine transform (DCT)/ differential pulse code modulation (DPCM) technique together with quadrature amplitude modulation (16 QAM). A dominant consideration in terrestrial broadcast is the fact that the introduction of ATV should not immediately make the existing NTSC receivers obsolete. Options toward this end include the use of an augmentation channel to carry the information on the signal details or the use of "simulcasting" or simultaneously broadcasting the ATV signal on a different channel. The simulcast approach provides freedom from the use of a composite signal. The features of the signal that are typically exploited in analog ATV are the large amount of spatio-temporal correlation in a video sequence and the nonuniform distribution of energy in the three-dimensional spectrum. The perceptual significance of the different portions of the spectrum is not uniform and can be utilized in further compressing the bandwidth of the signal. These are also the features that are at the crux of subband coding of digital ATV signals.

Terrestrial broadcast of ATV in the US is constrained by the limited bandwidth that is available in the TV broadcast spectrum regulated by the FCC. As a result the television system for terrestrial broadcast is required to be limited to a bandwidth of 6 MHz or slightly more for the improved signal. There is an additional restriction that the existing receivers should not become obsolete. Under these constraints, three approaches have been suggested for ushering in ATV broadcasts. One approach is to ensure compatibility with existing NTSC receivers by squeezing the extra information into an enhancement component that can be included within the 6 MHz channel. The second approach is to provide the enhancement component as an augmentation signal in a 3 or 6 MHz channel in addition to the NTSC channel so that suitably designed receivers can synthesize the improved signal. The third approach is to simulcast the NTSC and ATV signals on separate 6 MHz channels. This approach is attractive for a variety of reasons such as the elimination of the the NTSC signal as an ATV signal component and allowing greater flexibility in signal representation. In the simulcast approach for terrestrial broadcast, the ATV signal initially has a bandwidth of approximately 15 to 30 MHz (for the luminance) and it has to be transformed into a signal that is suitable for transmission in 6 MHz channel. Several methods for representing ATV signals have been proposed according to FCC guidelines [7].

Early work on HDTV development and compression was done by NHK in Japan. A bandwidth compression scheme for HDTV developed by NHK is called the MUSE system where MUSE stands for Multiple Sub-Nyquist Sampling and Encoding [8]. The MUSE System was developed for the transmission of HDTV via a direct broadcast satellite, and has been used for a variety of experimental transmissions. There are several versions of MUSE and the basic principles for compression are described here. The essential idea is that only a quarter of the signal samples are retained in every field using a nonrectangular grid, and these samples are used to interpolate the samples on the original grid. The system uses a periodically repeated sampling pattern over a four field sequence together with motion dependent interframe interpolation. Stationary areas of the picture are reconstructed by temporal interpolation using samples from four fields. In areas of motion the picture is reconstructed by using samples within a single field to avoid the appearance of a multi-line blur in the reconstructed picture. Subsampling on a nonrectangular grid is a key feature of the compression procedure.

Versions of MUSE have been proposed for terrestrial broadcast in North America [9]. The 1125/60 HDTV studio standard adopted by SMPTE serves as the basis for generating different grades of quality in the MUSE family of signals. At the high end in quality of the hierarchy of MUSE is the signal referred to as MUSE-T followed by the signal labeled MUSE. The MUSE-T signal is intended for transmission of HDTV programs for *contribution* purposes, while the MUSE signal was developed for satellite *distribution*. A contribution signal is one that is amenable to high-quality editing and post-production processing after reception. A distribution signal is not likely to be subjected to high-quality processing after reception except for measures to combat degradations in transmission. The main concern is to obtain the highest subjective picture quality within given bandwidth limitations. The systems below MUSE are labeled ADTV and are designed for terrestrial ATV broadcasting using 6 MHz channels. These ADTV systems are referred to as MUSE-6, MUSE-9 and Narrow-MUSE. Of these systems MUSE-6 and MUSE-9 are NTSC compatible systems, where MUSE-9 has a larger bandwidth and requires an augmentation channel. The Narrow-MUSE signal has a bandwidth of 6MHz and it is not NTSC-compatible. It is based on a modification of the MUSE signal which has a baseband width of 8.1 MHz. The reduction for fitting into the 6 MHz channel is accomplished by using 750 scanning lines that are derived from a 1125 line signal. Other proposals [1],[7] for terrestrial broadcast in North America include those by Massachusetts Institute of Technology (MIT), North American Philips, David Sarnoff Research Center and Zenith. A discussion of all these systems is beyond the scope of this Chapter but the principles of a system proposed by MIT will be discussed in Section 7.6. It should be pointed the format of ATV signal representation proposed by MIT is based on 3-D *subband coding* that is suitable both for analog and digital ATV transmission.

A family of ATV transmission systems has been evolving in Europe which is based on the Multiplexed Analog Component (MAC) format. The MAC transmission systems originally were intended for satellite transmission of television signals. The signal consists of time-multiplexing of compressed luminance, chrominance, sound and data. HDTV signals can also be represented by building upon the principles of a MAC system. Again, a key feature of MAC systems for HDTV transmission is the use of spatio-temporal subsampling [3].

7.2.4 Digital HDTV Transmission and Broadband Networks

Digital transmission is a promising method for the delivery of high definition television signals, especially when a high quality signal is desired at the receiver. This is particularly attractive for the future broadband network where different services such as voice, data and video will share resources in a common network. It is assumed that the network will be fiber-based and will offer wide bandwidth capability. However the cost of the supporting electronics can limit the extent to which the bandwidth is utilized. In view of this it is desirable, and it turns out to be feasible, to significantly compress the bandwidth of the raw HDTV signal while maintaining high quality.

We need to examine the network considerations that set the objectives for the coding work. Toward this purpose we will briefly examine the role of the future communication network in providing HDTV services. The matter is briefly discussed, and the reader is referred to [4] for further information.

As mentioned before, HDTV will have a significant potential impact on future visual communication services. HDTV is expected to be one of the many different services provided by the communication network of the future. Such a network will represent a culmination of the current evolution in which different services will be offered with integrated access, transport, switching and network management. The Broadband Integrated Services Digital Network (BISDN) is a concept that is motivated by the need to establish a unified mechanism for the provision of all services that have traditionally been provided by segregated networks. Voice, data, and television communication are provided by a variety of networks typically using circuit switching, packet switching, and broadcast techniques respectively. New services including interactive visual communication will be increasingly in demand. Segregated networks will not meet these demands in a cost-effective and efficient manner. The Broadband Integrated Network is expected to meet these needs with an efficient and flexible use of the communication resources. It will facilitate easy introduction of new services by providing integrated access. The large bandwidth capability of the fiber-based network will allow integrated transport of many different services.

Two technologies are well suited to help usher in the fully integrated network. These are the Synchronous Optical Network (SONET) and Asynchronous Transfer Mode (ATM). SONET [10] is a standard for the transport of a family of signals on an optical fiber network. It provides the physical interface to the network and facilitates the cross-connection of large granularity signals. ATM [11] is the target technology for BISDN that will provide the desired service flexibility. It captures the useful features of both circuit and packet switching.

In the SONET layer the basic unit is referred to as the Synchronous Transport Level 1 (STS-1) module with a transmission rate of 51.84 Megabits per second (Mbps). Three STS-1 modules are combined to get the STS-3c signal with a transmission rate of 155.52 Mbps. The information is organized as a frame consisting of 9 rows by 270 columns of octets (bytes of information) as shown in Figure 7.1.

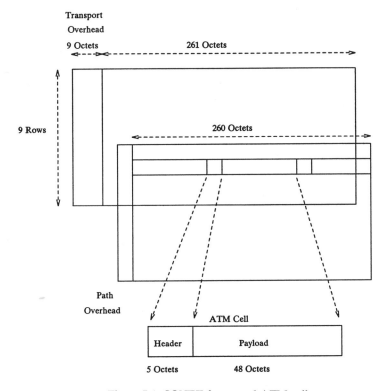

Figure 7.1: SONET frame and ATM cell

The frame period is 125 μsecs. The first nine columns of octets are allocated for transport overhead. Within the 261 columns of the payload the first column is assigned for path overhead. This leaves a net capacity of 149.76 Mbps available to carry ATM information units or other payloads. The basic ATM information unit has a fixed length. It is called a cell and has a header and an information field. The size of the ATM cell was chosen to be 53 octets by the CCITT of which 48 octets belong to the information field. As a result the available information payload capacity in an STS-3c signal is 135.6 Mbps. This rate seems to be an appropriate upper bound for the transmission of HDTV.

7.2.5 Objectives of HDTV Coding

While a full bandwidth PCM component HDTV signal requires as much as 1.2 Gbits/s, digital video coding makes compression to about a tenth of this bit rate feasible using only intrafield techniques. Rates roughly in the range of 120-140 Mbps are attractive from the viewpoint of the broadband network features. As mentioned before the information payload capacity in an STS-3c signal of approximately 135 Mbps appears appropriate for the transmission of high definition television. The main objective of the coding techniques described here is to fit the HDTV signal within this rate of 135 Mbps. This rate is also within the bounds of the fourth level (139.264 Mbps) of the European transmission hierarchy. The feasibility of fitting HDTV signals within 140 Mbps was demonstrated in [12-13]. Since the cost of implementation of a motion compensated high quality inter-frame coder remains fairly high for HDTV, it is important to investigate efficient intrafield coding techniques. Much of this Chapter is devoted to the study of intrafield coding techniques using a subband decomposition. A brief discussion of an inter-frame subband procedures will be presented in Section 7.6.

An important consideration in coding is the frame size and frame rate of the the signal that is to be coded. There is no agreed upon picture definition for the raw HDTV signal. A key parameter of the signal format is the number of lines per frame. Under the likely conditions of home viewing, a satisfactory picture with little perceptible line structure from a distance of three times picture height requires 1000 lines or more in a 2:1 interlaced format. Typical numbers proposed for HDTV signals are therefore in the vicinity of 1000 active lines/frame. The original NHK system had 1125 lines, 2:1 interlace, 60 fields/sec, and a 5:3 aspect ratio. In a movie theater a wider aspect ratio is attractive. In order to allow HDTV production equipment to also be conveniently used in production of theater movies, a slightly wider ratio of 16:9 is more suitable and has been used in several proposed formats. The horizontal sampling rate is suggested by the CCIR Recommendation 601 with a doubling of the density. The CCIR 601 format has 720 pixels/ line. Taking the aspect ratio into account, one would arrive at a figure of 1920 pixels (luminance samples)/ line for doubling the horizontal sampling density. Since studies suggest that a luminance to color-difference bandwidth ratio of 2:1 is suitable, this leads to 960 samples/active line for color-difference signals. As far as vertical sampling goes, figures such as 1050, 1125 and 1250 lines/frame with 2:1 interlace have been proposed. If we consider a figure of 1035 active lines/frame, with 30 frames/sec, and 8 bit/sample then the resulting rate is approximately 950 Mbps. The sampling rate for the luminance component of this signal is around 75 MHz (note that in computing the raw bit rate at this sampling rate one has to take into account blanking periods in the picture). It should be pointed out that this is just one possible signal format. With 50-60 MHz sampling rates, the raw bit rate is roughly between 640 to 780 Mbps.

As discussed earlier one objective of the coding work that is described here is to produce an ATV signal with a coded bit rate which is within 135 Mbps for the STS-3c channel. Within this restriction one makes allowances for overhead information. The raw signal is represented with 16 bits/pixel assuming that color-difference signals are sampled at half the luminance rate. A target coded bit rate of 2.5 bits/pixel was assumed in the work on coding that is described here.

(a)

(b)

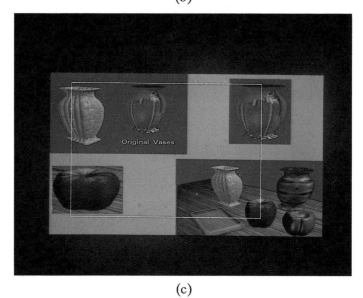

(c)

Figure 7.2: Test material for simulation

Only intrafield techniques were considered in the coding and the decoded picture quality is required to be very high. Subband coding was found to be attractive for this purpose and the procedure and results of coding are described in the next few Sections.

Simulation of the intrafield coding procedures described here were carried out on test material consisting of the following:

- *Kiel Harbor*, an artificially generated sequence obtained from the Heinrich Hertz Institute, depicting a very busy harbor scene.

- *MIT*, an artificially generated sequence obtained from Massachusetts Institute of Technology, depicting a woman surrounded by a large collection of objects.

- *Vases*, an artificially generated sequence depicting still life objects.

In Figure 7.2 three frames are shown, including a marked area indicating the portions that were used for detailed and critical examination. The bit rates in the coding are computed using entropy estimates. Using fixed Huffman tables the actual bit rates were found to be within five percent of the entropy estimates.

7.3 Issues in HDTV Subband Coding

Here the focus is on intra-field coding and the exploitation of temporal correlation will be briefly discussed later. The use of intra-field coding is motivated by the goal to keep the decoder cost low. The implementation is relatively simple since frame stores and motion detection/estimation hardware is not required. Each field in the video sequence is represented by using two-dimensional subband coding [14]. As we have seen in earlier chapters, subband coding refers to the class of techniques where, by the application of the signal to a set of filters, the input signal is decomposed into several subband signals that are decimated and encoded separately.

Let us review the basic ideas in subband coding of two-dimensional signals. Subband coding is based on the observation that for a typical natural scene, the signal energy is nonuniformly distributed in the frequency domain and the signal components from the different regions of the spectrum have different tolerance to quantization errors from a perceptual point of view. We take advantage of this observation to partition the signal into subband components and suitably allocate the bits according to their perceptual significance.

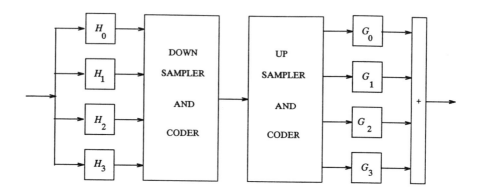

Figure 7.3: Subband coding procedure

As shown in Figure 7.3 each field in a video sequence is decomposed into several components (four in the figure) using a bank of filters called the analysis filter bank. The lowest frequency component is a lower resolution version of the original scene, and the high frequency bands mostly carry information about the contours, edges and other finer details. These filtered signals are then downsampled to yield the subband signals, which are coded using procedures appropriate to the information content. With a prescribed bit rate, we have to apportion the bits among the different subband signals. One makes a judicious allocation of a given budget of bits/pixel by choosing quantizers to produce an overall good reconstruction. With roughly equal bandwidths for the different analysis filters, usually the lowest frequency subband signal ends up with the largest slice of the budgeted bits.

At the receiver the subband signals are decoded, upsampled and merged by the synthesis filter bank to yield a close approximation to the input scene. In the coding of ATV signals, keeping the artifacts at a minimum and preserving consistently high quality is of paramount importance.

We will briefly describe the main issues in subband coding of ATV signals and we will consider methods for addressing these issues. One important consideration is the selection of the analysis and synthesis filter banks. Since system complexity at the high processing rates is a concern, we wish to to use filters that allow simple implementation. We also require that in the absence of coding and quantization errors a back-to-back connection of analysis and synthesis filters should provide exact reconstruction of the input signal. And in the presence of coding distortions we wish to limit the observable artifacts. The choice of filter banks is tied to the related issue of the selection of the downsampling grid that is used following the analysis filter bank.

A second important consideration is the procedure used to code the different subband signals. The coded bit-stream has a variable rate and there is a need to control the fluctuations to provide a smoothed out stream at the prescribed rate. This brings us to the third major issue, which is the choice of the mechanism for bit rate control. There are some options here which in turn are related to the first two issues described. The focus is primarily on these three issues in this chapter.

7.3.1 Analysis and Synthesis Filter Banks

We will briefly look at a back-to-back connection of analysis and synthesis filter banks. For detailed development refer to Chapters 2 through 4.

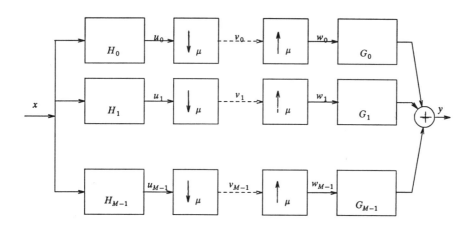

Figure 7.4: M-channel analysis/synthesis filter bank

In Figure 7.4 we see an M-channel filter bank with analysis filter frequency responses denoted by $H_k(\omega)$, $k = 0,1,...,M-1$ and $\omega = [\omega_1, \omega_2]^T$. The outputs are downsampled on a grid defined by the matrix μ [15-16]. The matrix μ is a 2×2 nonsingular matrix with integer entries. The output w_k, $k = 0,1,...,M-1$, of a downsampler is expressed in terms of the input u_k, $k = 0,1,...,M-1$, as:

$$w_k(\mathbf{n}) = u_k(\mu\mathbf{n}), \quad \mathbf{n}\epsilon\mathbf{Z}^2 \tag{7.1}$$

where $\mathbf{Z}$ denotes the set of integers. Note that the downsampling factor is assumed to be M, equal to the number of channels in the subband decomposition. The absolute value of the determinant of matrix μ is equal to the downsampling factor M. An example of μ for a downsampling factor of $M = 8$ is given by

$$\mu = \begin{bmatrix} 2 & 2 \\ -2 & 2 \end{bmatrix} \tag{7.2}$$

According to (7.1) only samples at locations defined by linear combinations (with integer coefficients) of the column vectors of μ are retained after downsampling. It is convenient to associate the following index set $P(\mu)$ with a matrix μ [17]:

$$P(\mu) = \{\mathbf{k}\epsilon\mathbf{Z}^2 : \mu^{-1}\mathbf{k}\epsilon[0,1)^2\} \tag{7.3}$$

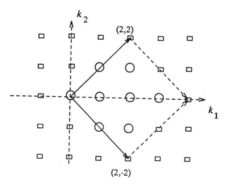

Figure 7.5: Index set $P(\mu)$ for μ in (7.2)

This set consists of sample locations in $\mathbf{Z}^2$ contained in the parallelogram defined by the two column vectors of μ except the samples on the two outer boundaries. For the matrix in (7.2) the set $P(\mu)$ is indicated in Figure 7.5 by the circles at sample locations.

The set $P(\mu)$ is useful in writing down the relation between the input and the output of back-to-back connected analysis and synthesis filter banks. The upsampled signal v_k can be expressed in terms of the signal w_k as:

$$
v_k(\mathbf{n}) = \begin{cases} w_k(\mu^{-1}\mathbf{n}) & \mu^{-1}\mathbf{n}\epsilon\mathbf{Z}^2 \\ 0 & \textit{otherwise} \end{cases}
\tag{7.4}
$$

Let $U_k(\omega)$ and $V_k(\omega)$ denote the Fourier Transforms of the signals u_k and v_k in Figure 7.4. It can be shown that they are related by

$$
V_k(\omega) = \frac{1}{M}\sum_{\mathbf{m}\epsilon P(\mu^T)} U_k(\omega - (\mu^T)^{-1}\mathbf{m}2\pi).
\tag{7.5}
$$

Using (7.5), the output Fourier Transform in Figure 7.4 can be written as:

$$
Y(\omega) = \frac{1}{M}\sum_{k=0}^{M-1} G_k(\omega)\sum_{\mathbf{m}\epsilon P(\mu^T)} H_k(\omega - (\mu^T)^{-1}\mathbf{m}2\pi)
$$
$$
X(\omega - (\mu^T)^{-1}\mathbf{m}2\pi).
\tag{7.6}
$$

From (7.6) the conditions on the filters for exact reconstruction can be derived in a straightforward manner. In the work described here tree-structured filter banks with $M = 2$ are used. Figure 7.6 shows a single-stage four-channel filter bank and a two-stage tree-structured filter bank. Consider the downsampling matrix μ in Figure 7.6(a) to be given by:

$$
\mu = \begin{bmatrix} 2 & 0 \\ 0 & 2 \end{bmatrix}
\tag{7.7}
$$

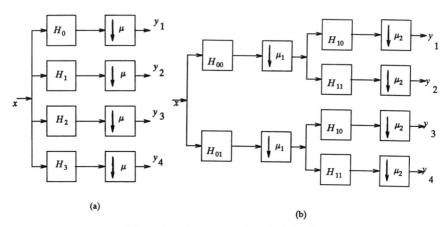

(a)

(b)

Figure 7.6: Four-channel analysis bank.
(a) Single-stage (b) Two-stage

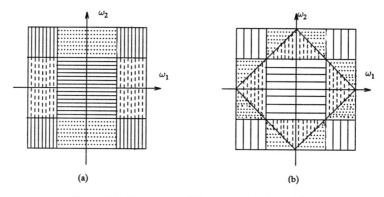

(a)   (b)

Figure 7.7: Two cases of frequency band partition

with downsampling factor $M = |det \mu| = 4$. In Figure 7.6(b) the
matrices μ_0 and μ_1 are obtained by factorizing μ with integer matrix
elements and with determinants equal to two. Two possible
factorizations are given below:

$$\mu = \begin{bmatrix} 2 & 0 \\ 0 & 1 \end{bmatrix} \begin{bmatrix} 1 & 0 \\ 0 & 2 \end{bmatrix} \qquad (7.8a)$$

and

$$\mu = \begin{bmatrix} 1 & 1 \\ -1 & 1 \end{bmatrix} \begin{bmatrix} 1 & -1 \\ 1 & 1 \end{bmatrix}. \qquad (7.8b)$$

In the work that is the focus of this chapter, only two stages of tree-structuring are used, as a result of which four subband signals are generated. Note that the first factor in (7.8a) corresponds to horizontal downsampling by two and the second factor corresponds to vertical downsampling by two. In (7.8b) the two matrices correspond to the dropping of every other diagonal in the data array. There is no unique subband partition associated with either of the two factorizations in (7.8a) and (7.8b). However if we require that one of the subband signals carry the lowest frequency information, then it is natural to choose the ideal four-band partitions shown in Figure 7.7(a) and (b) for the factorizations in (7.8a) and (7.8b) respectively. The rectangular partition of Figure 7.7(a) is used with horizontal and vertical downsampling. The processing can be carried out in a separable manner. The partition in Figure 7.7(b) is based on a diamond-shaped sectioning and is naturally suited for the factorization in (7.8b). These are the two cases that are considered in detail in this chapter.

7.3.2 Coding of Subband Signals

The downsampling following the analysis filter bank reduces the sampling rate of each of the M subband signals to $1/M$ times the input signal sampling rate. The subband signal carrying the lowest frequency information usually has a high degree of spatial correlation that is suitable for coding using either predictive techniques such as Differential Pulse Code Modulation (DPCM) or transform coding using Discrete Cosine Transform (DCT) [18-19]. As has been mentioned earlier, in the coding of HDTV signals the distortion in the signal should be kept low. Therefore the compression ratio is low for the lowest frequency subband. The high frequency subband signals can be directly quantized in the spatial domain. A large fraction of the quantized data in the high frequency band signals consists of zero samples and these signals are therefore well suited for runlength encoding [20].

7.3.3 Bit Rate Control

The output of the subband coders is a variable rate bit stream which can have large fluctuations depending upon the input signal. The coders are equipped with a buffer that smooths out the fluctuations, but if the data is too busy and the coder is fixed then it is possible for the buffer to overflow. In order to avoid this problem we need a mechanism to adjust the coded bit rate to maintain it constant at the prescribed bit rate. This can be accomplished by providing suitable buffer control mechanisms. Two control options available in subband coding are shown in Figure 7.8. These are:

- The prefiltering of the input signal prior to subband splitting. This would render the signal less demanding of bits, especially in the higher frequency subbands as the prefilter reduces the high frequency content of the input signal. And/or

- The adjustment of the subband quantizers. The lowest subband carries critical information and the adjustment of the low subband quantizers should be the last option in bit rate control. Selection of one of a fixed number of settings of the high subband quantizers is an effective means of adjusting the bit rate.

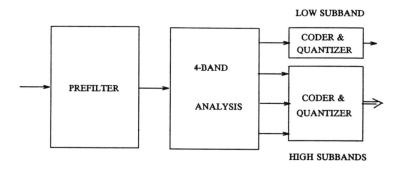

Figure 7.8: Options for bit rate control

We will now take a detailed look at the subband decomposition and coding for the separable horizontal/vertical case described by (7.8a). Following this we will consider the case of ATV coding using filter banks with diamond-shaped passband or stopband. In this case the downsampling is described by (7.8b).

7.4 Subband Coding Using Separable Filter Banks

The procedure for coding and decoding is shown in Figure 7.9 where the input signal x stands for either the luminance (Y), or the color-difference signals (U, or V). The Y, U, and V signals of each field are applied to analysis filter banks. The signal is decomposed into four subbands: low-low (LL), low-high (LH), high-low (HL), and high-high (HH). In this notation the first letter L/H denotes that the subband signal contains the horizontal low/high frequency content of the original signal, while the second L/H denotes the vertical content. The use of only four subbands avoids the added complexity of multiplexing several streams of coded data. The use of DCT on the lowest frequency band provides highly efficient coding. If the lowest band is coded using DCT, then there is no significant gain in coding efficiency when more than four subbands are used, especially in the case of coding demanding signals such as HDTV.

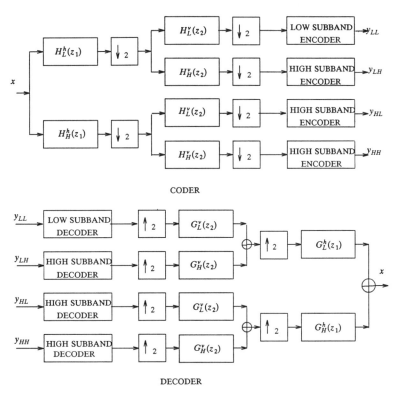

Figure 7.9: Separable analysis/synthesis filter banks

The subband decomposition is performed in two stages using one-dimensional filters that process the data along the rows and columns of the data array. The input signal x is first applied to horizontal filters $H_L^h(z_1)$ (lowpass) and $H_H^h(z_1)$ (highpass), and horizontally down sampled to get the signals x_L and x_H respectively. In the second stage of the decomposition each of the signals x_L and x_H is applied to the two vertical filters $H_L^v(z_2)$ (lowpass) and $H_H^v(z_2)$ and vertically down sampled to get the signals x_{LL}, x_{LH}, x_{HL}, and x_{HH}. For the purpose of reconstruction the signals are merged by upsampling and filtering using the synthesis filters. The vertical synthesis filters are denoted by $G_L^v(z_2)$ (lowpass) and $G_H^v(z_2)$ and the horizontal synthesis filters are denoted by $G_L^h(z_1)$ (lowpass) and $G_H^h(z_1)$ (highpass).

Since the processing of HDTV signals involves high sampling rates, it is very desirable to keep the filter bank complexity low. All the filters considered here are implementable using a small number of shift and add operations. The use of general multipliers is avoided. Both Finite-duration Impulse Response (FIR) and Infinite-duration Impulse Response (IIR) filters can be used in coding HDTV signals [13], [21] and the filters used will be described now.

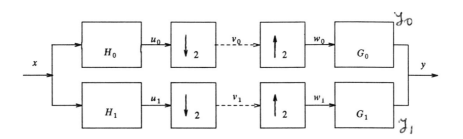

Figure 7.10: Two-channel analysis/synthesis filter banks

Since we are interested in tree-structured exact reconstruction filters, we need to focus only on the back-to-back connection of a two-channel one-dimensional partition shown in Figure 7.10. The analysis filter bank splits the input signal into two-channel signals by processing it with a lowpass filter $H_0(z)$ in one path and with a highpass filter $H_1(z)$ in the other. The filtered signals u_0 and u_1 are downsampled by a factor of two to obtain the subband signals. For convenience a scaling factor of two is introduced in each path. In a back-to-back connection, these signals are upsampled and processed by filters with impulse response g_0 and g_1, and transfer functions $G_0(z)$ and $G_1(z)$. The output signals y_0 and y_1 are added to yield the approximation y to the original input signal x. The signals u_k and the upsampled signals v_k, $k = 0,1$, are related by

downsampled

$$v_k(n) = \begin{cases} 2u_k(n) & n \text{ even} \\ 0 & n \text{ odd} \end{cases}$$

$$= (1 + (-1)^n)\, u_k(n). \tag{7.9}$$

The corresponding z-transforms are related by

$$V_k(z) = [U_k(z) + U_k(-z)]. \tag{7.10}$$

Therefore from Figure 7.10 and equation (7.10), the z-Transform of the output y can be expressed as

$$Y(z) = [G_0(z)\, H_0(z) + G_1(z)\, H_1(z)]\, X(z)$$

$$+ [G_0(z)\, H_0(-z) + G_1(z)\, H_1(-z)]\, X(-z). \tag{7.11}$$

For exact reconstruction we require that the first term on the right hand side of (7.11) be exactly equal to $X(z)$ with a possible delay while the second term should vanish in order that the aliasing be cancelled. We assume that the filters can be noncausal and the system delay can be suitably absorbed in the filter transfer functions. Taking this fact into account, the conditions for exact reconstruction can be expressed as

$$G_0(z)\, H_0(z) + G_1(z)\, H_1(z) = 1 \tag{7.12}$$

and

$$G_0(z)\, H_0(-z) + G_1(z)\, H_1(-z) = 0. \tag{7.13}$$

We wish to choose filters that satisfy the above conditions (7.12) and (7.13). A technique of deriving them from halfband filters [22] will be used. However we do not confine the impulse response of the halfband filter to be of finite duration. Let $G_{HB}(z)$ be the transfer function of a (zero-phase) symmetric halfband filter. The impulse response $g_{HB}(n)$ of the halfband filter satisfies the following constraints:

$$g_{HB}(n) = \begin{cases} \tfrac{1}{2} & n = 0 \\ 0 & n \text{ even} \\ g_{HB}(-n) & n \text{ odd.} \end{cases} \tag{7.14}$$

We can factorize the transfer function $G_{HB}(z)$ as

$$G_{HB}(z) = H_0(z)\, G_0(z). \tag{7.15}$$

Now we choose the transfer function $H_1(z)$ as

$$H_1(z) = z^{-1}\, G_0(-z) \tag{7.16}$$

and $G_1(z)$ as

$$G_1(z) = z\, H_0(-z). \tag{7.17}$$

Then

$$(G_0(z)\, H_0(z) + G_1(z)\, H_1(z)) = G_{HB}(z) + G_{HB}(-z)$$

$$= 2g_{HB}(0) = 1. \tag{7.18}$$

Also

$$G_0(z) \, H_0(-z) + G_1(z) \, H_1(-z) =$$

$$G_0(z) \, H_0(-z) - z \, H_0(-z) \, z^{-1} \, G_0(z) = 0 \qquad (7.19)$$

so that the aliasing cancellation condition (7.13) is satisfied. There are many possible filters that can be derived as described above. We will choose those filters that satisfy the additional requirement of simple coefficients. We will first look at FIR filters and later at IIR filters that have the above properties.

7.4.1 FIR Filter Bank

We are interested in deriving FIR filter banks by starting with halfband filters that have simple coefficients. An example of an FIR halfband filter with simple coefficients is:

$$G_{HB}(z) = \frac{1}{2} + \frac{9}{32}(z + z^{-1}) - \frac{1}{32}(z^3 + z^{-3}) \qquad (7.20)$$

One useful factorization of $G_{HB}(z)$ in (7.20), with the notation in (7.15) will be presented now. The lowpass analysis filter $H_0(z)$ and the lowpass synthesis filter $G_0(z)$ are given by

$$H_0(z) = \frac{1}{8}(-z^2 + 2z + 6 + 2z^{-1} - z^{-2}), \qquad (7.21)$$

and

$$G_0(z) = \frac{1}{4}(z^{-1} + 2 + z). \qquad (7.22)$$

The filters in (7.21) and (7.22) are used to generate the analysis/synthesis filter banks in [23]. The highpass filters needed to provide exact reconstruction are given by

$$H_1(z) = \frac{1}{4}(-z^{-2} + 2z^{-1} - 1) \qquad (7.23)$$

and

$$G_1(z) = \frac{z}{8}(-z^2 - 2z + 6 - 2z^{-1} - z^{-2}). \qquad (7.24)$$

In the absence of coding and quantization errors the filters given by (7.21)-(7.24) yield exact reconstruction. However coding distortions will always occur in practice. But in the case of HDTV signals these coding distortions should be kept small.

7.4.2 IIR Filter Bank [24]

Here we will consider the following symmetric IIR halfband filter transfer function:

$$G_{HB}(z) = \frac{1}{2} + \frac{1}{4}[zT(z^2) + z^{-1}T(z^{-2})] \qquad (7.25)$$

where $T(z)$ is allpass and can be expressed as the product of first order sections. $T(z)$ is given by

$$T(z) = \prod_{k=1}^{L} \frac{a_k^{(L)}z + 1}{z + a_k^{(L)}}. \qquad (7.26)$$

Some of the coefficients $a_k^{(L)}$ may be greater than zero. The coefficients $a_k^{(L)}$ are all positive so that the poles of $G_{HB}(z)$ in (7.25) are all on the imaginary axis in the z-plane. We assume that $T(z)$ is the transfer function of a stable, possibly noncausal, linear shift-invariant system. Also the phase response is approximately linear. Noting that

$$T(z)\, T(z^{-1}) = 1 \qquad (7.27)$$

we see that $G_{HB}(z)$ can be written as

$$G_{HB}(z) = \frac{1}{2}[1 + zT(z^2)] \cdot \frac{1}{2}[1 + z^{-1}T(z^{-2})]. \qquad (7.28)$$

We therefore pick $H_0(z)$ and $G_0(z)$ as

$$H_0(z) = \frac{1}{2}[1 + zT(z^2)] \tag{7.29}$$

and

$$G_0(z) = \frac{1}{2}[1 + z^{-1}T(z^{-2})]. \tag{7.30}$$

We can now choose the highpass filters according to (7.16) and (7.17). But an alternate choice allows a simpler implementation, where the analysis filters $H_0(z)$ and $H_1(z)$ can be obtained by a sum-and-difference operation of common intermediate outputs. This can be seen by comparing (7.29) and (7.30) with the transfer functions of $H_1(z)$ and $G_1(z)$ given by:

$$H_1(z) = \frac{1}{2}[1 - zT(z^2)] \tag{7.31}$$

and

$$G_1(z) = \frac{1}{2}[1 - z^{-1}T(z^{-2})]. \tag{7.32}$$

The filters $H_0(z)$ and $H_1(z)$ can be designed to have Butterworth or elliptic magnitude response. In our work we used filters with Butterworth magnitude response but with possibly a noncausal impulse response. The filters have approximately linear phase response. The filter coefficients are quantized so that multiplication with the coefficients can be accomplished with a small number of shift and add operations. The performance of filters with different values of L in (7.26) was examined. For $L = 0$ the filters reduce to two-tap FIR filters. The results of coding with $L = 0$, 1, 2 were compared. With $L = 2$ the filter impulse response has a large spread and gives rise to prominent ringing artifacts when coarse high frequency band quantizers are used. On the other hand the filter with $L = 0$ does not provide adequate smoothing in the interpolation when coarse high frequency band quantizers are used. It was found that $L = 1$ offers a good compromise with $a_1^{(1)} = 1/4$. This filter was used in the simulations.

Since the filter $T(z^{-2})$ is noncausal and requires a large memory for implementation, an FIR synthesis filter is used instead. The FIR approximation $T'(z^{-1})$, given by

$$T'(z^{-1}) = 1/8 + (1 - 1/64)(z - z^2/8 + z^3/64) \tag{7.33}$$

is found to yield no noticeable distortion.

7.4.3 Coding of the Subband Signals

For reasons mentioned earlier we will consider the case in which subband decomposition is used to create four subband signals which are labeled LL, LH, HL and HH as described in the last section. The results presented here are based on the work described in [13], [21].

Encoding of the Lowest Frequency Band

As noted earlier most of the perceptually significant information of the HDTV signal resides in the baseband signal. This signal forms an image with half the number of samples vertically and horizontally. Since HDTV has roughly twice the vertical and horizontal resolution of 525 (625) lines systems, this baseband (or a section of it) can be viewed as an embedded 525 lines sub-signal. Unlike digital video generated within the frequency constraints of CCIR Recommendation 601, the baseband produced by an exact reconstruction filter bank has a significant amount of information all the way to half the sampling frequency. This high frequency content of the baseband places heavy demands on the budget of bits. In order to achieve a sufficiently high compression, DCT is applied to the baseband of the luminance and the color-difference components with subsequent entropy coding of the transform coefficients.

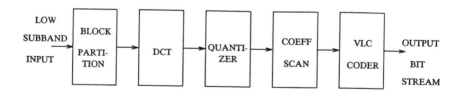

Figure 7.11: Discrete cosine transform coding

The algorithm considered for the LL-band image compression is a conventional DCT transform coding technique illustrated in Figure

7.11. The input image is first divided into disjoint blocks of size $N_1 \times N_2$. The DCT is computed for each block of LL-band image data as follows:

$$X_{LL}(k_1,k_2) = \frac{4}{N_1 N_2} c(k_1)c(k_2)$$

$$\sum_{n_1=0}^{N_1-1}\sum_{n_2=0}^{N_2-1} x_{LL}(n_1,n_2)\cos\frac{(2n_1+1)k_1\pi}{2N_1}\cos\frac{(2n_2+1)k_2\pi}{2N_2}$$

$$k_i = 0,1,...., N_i - 1, \quad i = 1,2. \tag{7.34}$$

In (7.34) the function $c(.)$ is given by

$$c(k) = \begin{cases} \frac{1}{\sqrt{2}} & \text{for } k = 0 \\ \\ 1 & \text{otherwise.} \end{cases} \tag{7.35}$$

The DCT coefficients $X_{LL}(k_1,k_2)$ are applied to a uniform quantizer to get the quantized coefficients $X_{LL}^Q(k_1,k_2)$ given by

$$X_{LL}^Q(k_1,k_2) =$$

$$\text{SZ} \times \text{NINT}\left\{\frac{X_{LL}(k_1,k_2)}{\text{SZ}}\right\} \tag{7.36}$$

where NINT{·} stands for the nearest integer and SZ denotes the quantizer step size. The 2-D data is then arranged into a 1-D sequence by zigzag scanning [25]. Finally, the nonzero amplitudes and run-length of zeros are entropy-coded, and sent to the channel coder for transmission or storage. Since the interpolation process at the reconstruction end in the subband coding scheme increases the visibility of the artifacts, a high image quality (or low distortion due to compression) for the LL-band reconstruction should be maintained. For this reason, the block size has been restricted to 4×4 and 8×4 for the luminance (which, in the reconstructed image, correspond to 8×8 and 16×8 blocks respectively). In coding the LL-band block of size 8×4 for the DCT with a coefficient step size SZ=1.0 was found to be satisfactory. For the more demanding signals, the payoff in assigning any additional bits to the low subband signals was less significant than that in assigning the bits to the high subband signals while maintaining the same overall bit rate.

There are other possibilities for coding the lowest frequency subband signal of an HDTV signal. One option is to use differential pulse code modulation (DPCM) [26], and a method based on this is described in Section 7.4.5.

Encoding the Higher Bands

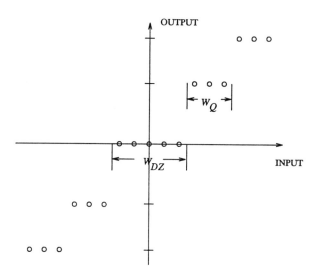

Figure 7.12: High frequency band quantizer

The high frequency band signals are not suited for transform coding. Instead these signals are directly quantized in the spatial domain using a variation of the technique in [20]. The quantizer for the high subband signal is shown in Figure 7.12. It has a deadzone centered at zero and uniformly spaced outer quantization intervals. A nonuniform quantization scheme for lowering the mean squared error was examined. This results in considerably wider intervals for outer quantization values. As a result the quantization error was large in high contrast areas. Though the occurrence of such large sample difference is sparse this causes noticeable artifacts due to ringing caused by the synthesis filters. The penalty in bit rate is not very large if the quantizers are kept uniform. With the value of the outer quantization interval w_Q equal to 4 or 5 and the width of the deadzone w_{DZ} set to one of the three values: 5, 7, 9, the reconstruction is of very satisfactory quality, with some softening of detail for larger deadzones noticeable on close observation. After the quantization, the high frequency band signals are then coded using runlength encoding of blocks of data within a large window. Following a study of different block and window sizes, and scanning procedures, a choice of 2×1 blocks within 360×1 window was found to be satisfactory. This can be implemented with a table lookup procedure.

Decoding Procedure and Simulation Results

The decoded subband signals y_{LL}, y_{LH}, y_{HL}, and y_{HH} are applied to the synthesis filter banks, as shown in Figure 7.9, to obtain the reconstructed image. The inverse DCT is computed as

$$x_{LL}^Q(n_1,n_2) = \sum_{k_1=0}^{N_1-1}\sum_{k_2=0}^{N_2-1} c(k_1)c(k_2)\, X_{LL}^Q(k_1,k_2)$$

$$\cos\frac{(2n_1+1)k_1\pi}{2N_1}\cos\frac{(2n_2+1)k_2\pi}{2N_2}$$

$$n_i = 0,1,....,\ N_i-1,\quad i = 1,2. \qquad (7.37)$$

The coding and decoding procedure were applied to some available test material. One of the test sequences used was *Kiel Harbor* and was provided by the Heinrich Hertz Institute (HHI), Berlin, Germany. This sequence is a synthetic zoom and pan generated from a very high resolution still picture and contains a full 27 MHz of very busy information with extremely critical high frequency details. Despite the fact that this sequence does not have a 16:9 aspect ratio it is very useful test material for the algorithm simulation in view of its high spatial and temporal frequency content. Other test material considered included the *MIT* sequence, computer generated stills, and stills derived from analog HDTV.

The test data were in RGB format with equal bandwidth in the three components. Before any encoding the signals were converted to

luminance and color-difference components. The color difference
signals were filtered and subsampled by a factor of two.

Table 7.1 (a) Bit rate (in bpp) for *Kiel Harbor* Field #1.

Component	LL	$H(W_{DZ} = 7)$	$H(W_{DZ} = 9)$
Y	0.944	1.196	1.065
U	0.147	0.022	0.013
V	0.174	0.044	0.026
Total		2.527	2.369

Table 7.1 (b) Bit rate (in bpp) for *MIT* Field #1.

Component	LL BAND	HIGH($W_{DZ} = 7$)	HIGH($W_{DZ} = 5$)
Y	0.840	0.725	0.840
U	0.196	0.085	0.119
V	0.204	0.094	0.139
Total		2.144	2.338

The results of simulation showed that the coding procedure yields
pleasing results both spatially and temporally. In the simulations slightly
different procedures were used in the FIR and IIR approach and the
details of the bit rates can be found in [13], [21]. Here we present the
bit rates and signal-to-noise ratio (SNR) for the IIR approach using
frames from the *Kiel Harbor* and *MIT* sequences. With the procedures
and parameters described above, the bit rates are given in Table 7.1(a)
and (b) in bits per pixel (bpp) for *Kiel Harbor* and *MIT* respectively.
For the *Kiel Harbor* data the peak SNR for the Y-signal was 39.2 dB
using w_Q equal to 5 and w_{DZ} equal to 7. The DCT on the LL-band
was computed for blocks of size 8×4 and SZ = 1.0. The U and V
signals had peak SNR better than 40 dB. With the same parameter
settings for the *MIT* picture a lower entropy rate of about 2.144
bits/pixel produced comparable SNR values. The differences from the
original are observable on high quality monitors under very critical
viewing conditions. These differences are not visible on 35 mm prints
and slides. There is a slight amount of ringing that is observable around

the extremely high-contrast edges. These artifacts are noticeable when the background is uniform and the image is viewed from a very short distance.

7.4.4 Channel Rate Adaptation strategy

Two options in providing a mechanism for controlling the bit rate were shown in Figure 7.8. The first option is an adjustable prefilter while the second is the adaptation of the subband quantizers. A rate buffer strategy is suggested by the simulation results. If the prefilter is used as the sole mechanism for bit rate control, then for signals that are highly demanding of bits the picture can be considerably softened. In the case of subband quantizers one can adjust the parameters of either the lowband or the high band quantizers. In view of the critical nature of the low frequency band signal, the adjustment of the low band quantizers should be exercised as the last option. The high band quantizer can be varied among a few preselected settings. However since an excessively coarse quantizer leads to objectionable artifacts, a suitable balance should be struck between the two options of an adjustable prefilter and the settings of the high band quantizer.

The low frequency band is quantized relatively finely. Therefore most of the signal error is contributed by the high frequency bands in situations where the high band quantizer is coarse. This error is due quantization in the spatial domain. When the subband signals are applied to the synthesis filters then the high band quantization error is spread out by the convolution operation. For coarse quantizers this results in noticeable ringing artifacts. The extent of the ringing artifacts depends on the spread of the impulse response. For higher order filters this spread is considerably larger. In order to limit the visibility of ringing artifacts it is preferable to choose filters with a small spread of the impulse response within the restriction of exact reconstruction. However if the impulse response spread is too narrow then the filter will not provide adequate smoothing in the process of interpolation. In practice a filter with intermediate extent of the impulse response spread offers a good compromise. The difference shows up when demanding signals are coded. In this case there is no significant difference in bit rate when compared with filters with a larger spread of impulse response.

The filters described in Section 7.4.1 and 7.4.2 do offer the compromise discussed above. With such a choice of filters, a small number of settings of fine and coarse quantizers are suitable for controlling the bit rate. Additional control can be exercised through a prefilter with variable coefficients. The coefficient values and hence the effective bandwidth of the prefilter is controlled by one adjustable parameter. The parameter varies the filter frequency response from one extreme of allpass to the extreme of a relatively narrow diamond-shaped response. In the case of very demanding signals the coarse quantizers should be used in conjunction with the diamond prefilter.

7.4.5 A Low Complexity Subband Coder

A low complexity subband coder is described in [26]. The coding is based on the use of simple two-tap sum and difference filters at each stage in the filter banks. This approach allows for a considerable reduction in coder complexity. A six-band decomposition is used. Differential coding is applied to the lowest frequency band. In this subband coding scheme the use of a prefilter is suitable for controlling the bit-rate. The coarseness of the quantizers cannot be used to the same extent as in the methods described earlier as the filters do not provide adequate smoothing to remove the blockiness due to coarse quantizers. In addition to the six-band decomposition, the coder uses a statistical coder with an option of coding tables.

7.5 ATV Coding Using Nonrectangular Subband Decomposition

In the previous Section we considered rectangular subband coding of ATV signals for achieving an intrafield compression from 16 bits per pixel to 2.5 bits per pixel. In some situations the original signal format may require a greater reduction, down to 2 bits per pixel or less. In this case an attractive option is to prefilter and downsample the signal in a way that retains as much as possible of the significant information. The prefilter can be chosen from perceptual consideration, and depending on the filtering the downsampling grid pattern may not be rectangular. Image and video signals can be sampled on a variety of grids, the choice of which can be made based on the nature of the significant information content in the frequency domain [15-16]. It is known that in most naturally occurring scenes the image information is largely confined to a diamond-shaped frequency band and information in this band is efficiently packed in the digital signal by choosing a nonrectangular sampling grid, referred to as a line quincunx sampling pattern [16]. If a given rectangularly sampled signal is required to be decimated onto a line quincunx pattern, the signal should first be processed by a filter with a diamond-shaped passband with the ideal response shown in Figure 7.13(a). This is followed by 2:1 downsampling on a line quincunx pattern shown in Figure 7.13(b).

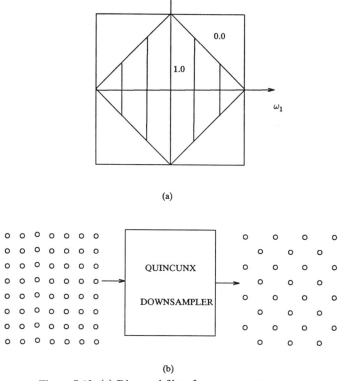

(a)

(b)

Figure 7.13: (a) Diamond filter frequency response
(b) Quincunx sampling grid

In quincunx downsampling every other pixel is dropped in each line, but the downsampling is staggered by one pixel in adjacent lines. In effect samples on every other diagonal are dropped. This type of decimation is well-known and has been used in data compression applications including the coding of HDTV signals at rates below 140 Mbps [12]. Diamond filters can also be used in the construction of filter banks for carrying out the subband decomposition of images for the purpose of data compression [27-30].

Here we will describe the application of the diamond filters in the data compression of ATV signals based on an intrafield subband decomposition using nonrectangular filter banks [28]. A diamond filter is used to prefilter the given signal before quincunx downsampling. Following this a pair of filters, one with a diamond-shaped passband and the other with a diamond-shaped stopband, are used to perform a subband partition of the signal. These subband signals are then coded. If we process frames of rectangularly sampled ATV signals with 2 million pixels at 30 frames/sec, then by limiting the coding to a budget of 2 bits/pixel one can code the signal at rates below 135 Mbps. Note that the rate of 2 bits/pixel is with reference to the rectangularly sampled data but implies a rate of 4 bits/pixel for the downsampled data if both luminance and chrominance are decimated by a factor of two. This approach is more suitable for progressively scanned data though it works satisfactorily with interlaced signals.

We will first consider the design of the filters for (a) the prefiltering operation prior to downsampling and (b) the analysis and synthesis operation in subband coding and decoding.

7.5.1 Diamond Prefilter

The ideal frequency response of a diamond lowpass filter is shown in Figure 7.13(a). A procedure for approximating this response using a 1-D prototype filter [28] will now be described. We note that the filter in Figure 7.13(a) can be obtained from the filter in Figure 7.14(a) by a transformation of variables. If $F_a(z_1,z_2)$ is the transfer function of the filter approximating the response in Figure 7.14(a), then the diamond filter transfer function $D(z_1,z_2)$, given by

$$D(z_1,z_2) = F_a(z_1^{1/2}z_2^{1/2}, z_1^{-1/2}z_2^{1/2}), \qquad (7.38)$$

will approximate the response in Figure 7.13(a). The response in Figure 7.14(a) can be expressed as the sum of the two responses shown in Figures 7.14(b) and (c). We observe that if $F_b(z_1,z_2)$ is the transfer function of the filter approximating the response in Figure 7.14(b) then

$$F_c(z_1,z_2) = F_b(-z_1,-z_2) \qquad (7.39)$$

will approximate the response in Figure 7.14(c). This is due to the fact that negating the arguments shifts the (periodic) frequency response of

F_b by (π,π).

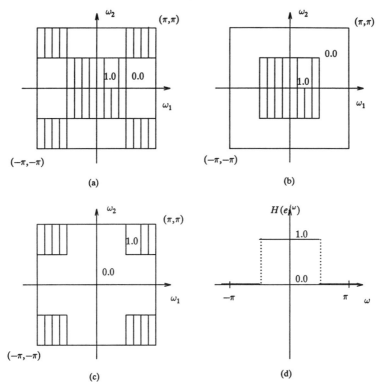

Figure 7.14: Ideal frequency response of filters
(a) $F_a(z_1,z_2)$ (b) $F_b(z_1,z_2)$ (c) $F_c(z_1,z_2)$ (d) $H(z)$

The response in Figure 7.14(b) can be expressed as the product of two ideal 1-D lowpass filters, one horizontal and one vertical, which have the response ideal shown in Figure 7.14(d). This 1-D frequency response can be approximated with either FIR or IIR filters. One possible choice of the 1-D prototype filter is a halfband filter. This will produce a response in which the transition band straddles both sides of the cutoff frequency boundaries in Figure 7.13(a). If we wish to constrain the transition band to lie within the boundaries of the diamond-shaped region in Figure 7.13(a), then we should choose a 1-D filter whose stopband interval is $(\pi/2,\pi)$.

Let $H(z)$ be the transfer function of the prototype 1-D lowpass filter approximating the response in Figure 7.14(d) with a suitably chosen transition boundary. The transfer function $H(z)$ can always be expressed as

$$H(z) = T_1(z^2) + zT_2(z^2). \tag{7.40}$$

The transfer function F_a is given by

$$F_a(z_1, z_2) = H(z_1) H(z_2) + H(-z_1) H(-z_2). \tag{7.41}$$

Combining (7.38), (7.40) and (7.41) we get

$$D(z_1, z_2) = 2T_1(z_1 z_2)T_1(z_1^{-1} z_2) + 2z_2 T_2(z_1 z_2)T_2(z_1^{-1} z_2). \tag{7.42}$$

As mentioned before $H(z)$ can be chosen to be a halfband filter with

$$T_1(z^2) = 0.5. \tag{7.43}$$

In our work the diamond filters used as prefilters prior to the 2:1 quincunx downsampling were derived from 1-D FIR filters. Both halfband FIR filters and filters with narrower passbands were used.

In constructing the filter banks for subband coding, there are procedures for obtaining exact reconstruction filters by applying transformations to one-dimensional filters [27], [30]. Here we will consider the use of IIR filters. In this case halfband IIR filters can be used with $T_2(z^2)$ chosen to be an allpass filter. For this choice of analysis filters one can design synthesis filters that allow exact reconstruction in the absence of coding and quantization. In our work filters with simple coefficients were used in the simulation due to the relative ease with which they can be implemented.

7.5.2 Coding of Subband Signals and Simulation Results

We now describe a scheme in which a signal sampled on a rectangular grid is first decimated to get a quincunx sampled signal. In effect the information in the frequency band outside a diamond shaped region is discarded in this scheme. The prefilter in our scheme has to be designed carefully in order to prevent distortion due to aliasing while not rendering the picture too soft by excessive filtering. Two types of

FIR 1-D prototypes, one of which has a monotone passband response with a stopband interval of $(\pi/2,\pi)$ and the other a Lagrange halfband filter were used in the processing [28]. The corresponding diamond filters are labeled as F1 and F2.

The signal to be coded is assumed to be in the Y,U,V format with the U and V signals decimated horizontally by a factor of two. The Y signal of each field data is applied to a prefilter with a diamond-shaped passband followed by a downsampling by a factor of two on a line quincunx grid. The same processing can be used for the U and V signals. But since the high frequency information in these signals is typically low, one can use rectangular prefiltering and direct downsampling of the data. The first step, consisting of diamond prefiltering and quincunx downsampling, is used in the HDTV compression scheme of [12] following which the quincunx sampled signal is applied to a DPCM coder. We will consider the application of DCT coding to the data following the first step. Toward this purpose we will generate a subsignal that is available on a rectangular sampling grid. This is accomplished by decomposing the quincunx sampled signal into two rectangularly sampled subband signals by the analysis portion of an exact reconstruction pair of nonrectangular filter bank. The original signal information is split into three bands in the frequency domain, ideally as shown in Figure 7.15.

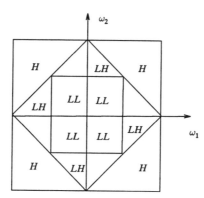

Figure 7.15: Frequency band partition using diamond filters

The information in the high frequency band (H in Figure 7.15) is discarded. This is motivated by the fact that natural scenes often contain very little information in these bands and also studies suggest that the human eye is less sensitive to resolution along the diagonals.

The coding procedure for the low (L) frequency band quincunx
sampled signal is shown in Figure 7.16. The details are discussed later.

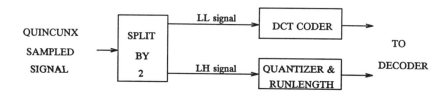

Figure 7.16: Coding procedure using diamond filters

As mentioned before the analysis filter splits the quincunx
sampled signal into the low-high(LH) and low-low(LL) components.
The two-band analysis-synthesis system is shown in Figure 7.17.

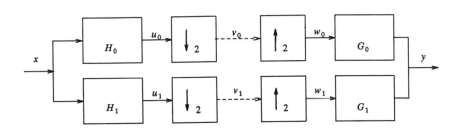

Figure 7.17: Two-channel diamond filter banks

The analysis filters are denoted by H_i, $i = 0,1$, where H_0 is a diamond
lowpass filter and H_1 is a diamond highpass filter. The quincunx
downsampling by the factor 2 is done on a grid that is already on a
quincunx grid. This yields rectangular sampling grids for the decimated
signal. The transfer function $H_i(z_1, z_2)$, $i = 0,1$ is given by

$$H_i(z_1, z_2) = 1/2 + (-1)^i \ 2z_2 \ T_2(z_1 z_2) \ T_2(z_1^{-1} z_2) \qquad (7.44)$$

where

$$T_2(z) = \tfrac{1}{2} \prod_{k=1}^{L} \frac{a_k z + 1}{z + a_k}. \tag{7.45}$$

Note that H_0 is the same as $D(z_1, z_2)$ in (7.42) with T_1 as defined in (7.43) and T_2 is allpass. In the synthesis filter bank if we choose

$$G_i(z_1, z_2) = 2 H_i(z_1^{-1}, z_2^{-1}), \quad i = 0,1 \tag{7.46}$$

then the conditions for exact reconstruction are satisfied [27]. In our decomposition we used quantized coefficients derived from Butterworth filters with $L=1$ and $L=2$. With $L=1$ we examined $a_1 = 1/4$. With $L=2$ we used quantized coefficients of the noncausal version of a fifth order Butterworth filter. Two coefficient sets chosen were $a_1 = 32/3$ and $a_2 = 17/32$, and $a_1 = 8$ and $a_2 = 1/2$. For $L=2$ the section with coefficient a_1 is implemented in a noncausal manner. At the output of the analysis filter bank, the coding is carried out as shown in Figure 7.16. The LL band signal is represented by quantizing the coefficients of the discrete cosine transform of blocks of the data. The LH band signal is quantized using a deadband quantizer and the resulting data is coded using runlength and Huffman coding. Simulations were performed on HDTV data using frames of the *Kiel Harbor* and *MIT* sequences in interlaced format and *Vases* frame in a noninterlaced format. Results of the coding are shown in Table 7.2, with rates computed using entropy estimates in bits per pixel (bpp).

Table 7.2 Bit rate (in bpp) for Nonrectangular Subband Coding

Picture	Prefilter	Y	U	V	Total
Kiel	F1	1.06	0.145	0.175	1.380
Kiel	F2	1.24	0.145	0.175	1.560
Vases	F1	0.35	0.114	0.128	0.592
Vases	F2	0.39	0.114	0.128	0.632

In the simulation only the luminance signal was coded using nonrectangular subband coding. The color-difference data was decimated horizontally and vertically by a factor of 2:1 before coding. The reconstruction is satisfactory with no noticeable difference between the prefiltered signal and the reconstructed signal for normal viewing distances. However, in comparison with the unfiltered original we observe, at close viewing distances, some softening of diagonal edges due to diamond filtering.

We will briefly consider the implementation of the prefilter and the subband filters. The complexity is kept low by exploiting several features of the filters. We note that only filters with simple coefficients are used in all filtering operations so that a small number of shift and add operations are required to implement the filters. The relation between the lowpass and highpass filters is used to share operations. The fact that the filters are derived from 1-D filters is exploited in employing only 1-D processing which is carried out along the diagonal directions. All noncausal infinite impulse response filters are replaced by FIR filters or right-sided impulse response approximations to the synthesis filters.

7.6   HDTV   Subband   Coding   Exploiting Temporal Redundancy

The focus in this chapter has been on intrafield subband coding techniques for coding HDTV signals. In the future the costs of frame-stores are expected to drop and the exploitation of temporal redundancy will be a viable option even with constraints of low-cost decoders. Subband coding procedures that exploit temporal redundancy are attractive for this application. Some possible subband procedures using field/frame storages have been described in [31-32] and in Chapter 6. Here we will briefly examine the underlying principles of the scheme described in the MIT proposal [31].

A compression scheme based on three-dimensional subband coding has been proposed [31] which is suitable for both analog and digital transport. We will briefly describe the proposed channel-compatible system in which both analog and digital information is transmitted. The underlying subband procedure is suitable for an all-digital transmission as well. In the channel-compatible system the signal is not compatible with NTSC receivers, but it fits within the 6 MHz channel. The transmission bandwidth requirement of the signal depends on the number of subband signals transmitted. A significant reduction in this requirement is achieved by the exploitation of the spatial and temporal correlation in the signal. The saving in bandwidth accrues from the infrequent transmission of subbands containing diagonal spatio-temporal frequency content. Several possible signal formats and frequency domain partitions have been examined at MIT's

Research Laboratory for Electronics and one combination is described here.

Consider a video signal in RGB component form with 1280 lines/pixel, 720 lines/frame, and 60 fames per second. The system first converts RGB signals to luminance and color information signals labeled YIQ. These signals are then applied to an analysis filter bank using separable filters, with 16 taps along the spatial dimensions and three taps in the temporal dimension. In this manner the YIQ components of the ATV signal are decomposed into 8×8 spatial frequency partitions and three temporal partitions, generating a total of a 192 three-dimensional subband signals. For every three frames, each subband contains 160×90 signal values for each component.

Within the 6 MHz channel one can fit 10 M analog samples and 10 Mbps of digital data. The information is packed into the channel by using an adaptive selection of subband signals. For all three YIQ signals the four lowest frequency components (2×2×1) are always included and each sample is represented by a combination of bits (two for Y and one each for IQ) and one analog value. This representation takes 4.608 Mbps and 3.456 M analog values/s. The remaining 6.544 M analog values/s are assigned to the higher frequency band samples. The location information of these samples is sent digitally using 1.692 Mbps. The analog values are adaptively amplitude modulated using a 3-bit factor for 20 values on the average in order to improve resistance to channel impairments. This consumes 1.5 Mbps. Digital audio is allocated a total of 0.5 Mbps and auxiliary data for sending closed captions and other information is assigned 56 Kbps. In order to provide protection for the adaptive selection and adaptive modulation information, the remaining 1.644 Mbps is used for forward error correction. The analog values are scrambled for decorrelating coherent information.

For a purely digital transmission, the above techniques can be carried over except that scrambling is not applicable in this case. The signal will, in the case of digital transmission, lend itself to suitable statistical channel coding.

7.7 Summary

In this chapter we discuss issues in subband coding when it is applied to high quality coding of ATV signals. The main concern is to fit ATV signals into a prescribed bit rate by using only intrafield coding using rectangular and nonrectangular sampling grids. The Chapter begins with a review of analog and digital advanced television and proceeds to a discussion of some issues in the transmission of ATV signals over the broadband network. Methods of representing the subband signals with a given budget of bit rate are described, and issues and trade-offs in assigning the bit rates to the different subband signals are discussed. A large part of this discussion is based on the authors' work in [13], [21] and [28]. A short description of the use of subband coding in exploiting redundancy in the temporal direction is also presented.

The coding of ATV signals is a very active research area and hierarchical coding techniques for ATV have received a great deal of attention. Subband and pyramidal techniques play an important role here.

Acknowledgements

The authors wish to thank Dr. R. Schäfer of HHI for providing the *Kiel Harbor* data and Professor W. Schreiber of MIT for providing the *MIT* digital sequence used in the simulations. We are grateful to our colleagues who collaborated with us on some of the work described here.

Bibliography

[1] Special Issue, *IEEE Transactions on Consumer Electronics*, Vol. 34, No. 1, pp. 1-120, Feb. 1988.

[2] *IEEE Transactions on Consumer Electronics*, Vol. 35, No. 3, pp. 133-248, Aug. 1989.

[3] "EBU Studies on High-Definition Television," Reprinted from *EBU Review (Technical)*, No. 219, October 1986.

[4] R. C. Lau, Ed., "Research on Advanced Television for Broadband ISDN," *Bellcore Technical Document*, SR-ARH-001637, March 1990.

[5] *ATSC Executive Committee Report*, July 11, 1989.

[6] *Digicipher HDTV System Proposal*, submitted by General Instrument Corporation, June 1990.

[7] Seminar on *Advanced Television*, Columbia University, New York, NY, November 3, 1989.

[8] Y. Ninomiya, et al., "An HDTV Broadcast System Utilizing a Bandwidth Compression Technique - MUSE," *IEEE Transactions on Broadcasting*, Vol. BC-33,

pp. 130-155, Dec. 1987.

[9] Y. Kanatsugu, et. al, "Development of MUSE Family Systems," *IEEE Transactions on Consumer Electronics*, Vol. 35, No. 3, pp. 153-159, Aug. 1989.

[10] R. Ballart and Y. C. Ching, "SONET: Now It's the Standard Optical Network," *IEEE Communications Magazine*, March 1989.

[11] T1S1 Technical Subcommittee, *Broadband Aspects of ISDN Baseline Document*, May 1989.

[12] S. Matsumoto et al., "120/140 Mbps Intrafield DPCM Systems Transmission of HDTV Programs," *Proc. Second Int. Workshop Sig. Proc. of HDTV*, L'Aquila, Italy, Feb. 29 - March 2, 1988.

[13] D. J. Le Gall, H. P. Gaggioni and C. T. Chen, "Transmission of HDTV Signals under 140 Mbit/s using Subband Decomposition and Discrete Transform Coding," *Proc. Second Int. Workshop Sig. Proc. of HDTV*, L'Aquila, Italy, Feb. 29 - March 2, 1988.

[14] J. W. Woods and S. D. O'Neil, "Subband Coding of Images," *IEEE Trans. Acoust., Speech, Signal Processing*, vol. ASSP-34, pp. 1278-1288, Oct. 1986.

[15] R. M. Mersereau and T. C. Speake, "The Processing of Periodically Sampled Multidimensional Signals," *IEEE Trans. Acoust., Speech, Signal Processing*, vol. ASSP-31, pp. 188-194, Feb. 1983.

[16] E. Dubois, "The Sampling and Reconstruction of Time-Varying Imagery with Application in Video Systems," *Proc. IEEE*, vol. 73, pp. 502-522, April 1985.

[17] R. Ansari and S. H. Lee, "Two-dimensional Nonrectangular Interpolation, Decimation and Filter Banks," *Bellcore Technical Memorandum TM-ARH-010787*, Dec. 1987.

[18] A. K. Jain, *Fundamentals of Digital Signal Processing*, Prentice Hall, Englewood Cliffs, NJ, 1989.

[19] J. S. Lim, *Two-dimensional Signal and Image Processing*, Prentice Hall, Englewood Cliffs, NJ, 1990.

[20] H. Gharavi and A. Tabatabai, "Sub-band Coding of Monochrome and Color Images," *IEEE Trans. Circuits and Systems*, vol. CAS-35, pp. 207-214, Feb. 1988.

[21] R. Ansari, A. Fernandez and S. H. Lee, "HDTV Subband/DCT Coding Using IIR Filter Banks: Coding Strategies," *Proc. SPIE Conference on Visual Communications and Image Processing*, Philadelphia, PA, vol. 1199, 1291-1302, Nov. 1989.

[22] M. G. Bellanger, J. L. Daguet and G. P. Lepagnol, "Interpolation, Extrapolation and Reduction of Computation Speed in Digital Filters," *IEEE Trans. Acoust., Speech, Signal Processing*, vol. ASSP-22, pp. 232-235, Aug. 1974.

[23] D. J. Le Gall and A. Tabatabai, "Sub-band Coding of Digital Images Using Short

Kernel Filters and Arithmetic Coding Techniques," *Proc. IEEE Int. Conf. on Acoust., Speech, Signal Processing*, pp. 761-764, April 1988.

[24] M. J. T. Smith, R. M. Mersereau and T. P. Barnwell, III, "Exact Reconstruction Filter Banks for Subband Image Coding," Proc. Miami Technicon, pp. 121-124, Oct. 1987.

[25] W. Chen and W. K. Pratt, "Scene Adaptive Coder," *IEEE Trans. Commun.*, vol. COM-32, pp. 225-232, March 1984.

[26] T. C. Chen and P. E. Fleischer, "Sub-band Coding for ATV signals based on Spatial Domain Considerations," *Proc. SPIE Conference on Visual Communications and Image Processing*, Philadelphia, PA, vol. 1199, pp. 787-798, Nov. 1989.

[27] R. Ansari and C.-L. Lau, "Two-dimensional IIR Filters for Exact Reconstruction in Tree-structured Subband Decomposition," *Electronic letters*, vol. 23, No. 12, pp. 633-634, June 1987.

[28] R. Ansari, H. P. Gaggioni and D. J. Le Gall, "HDTV Coding using a Nonrectangular Subband Decomposition," *Proc. SPIE Conference on Visual Communications and Image Processing*, Boston, MA, vol. 1001, pp. 821-824, Nov. 1988.

[29] M. Vetterli, J. Kovacevic and D. J. Le Gall, "Perfect Reconstruction Filter Banks for HDTV Representation and Coding," *Proc. Third Int. Workshop on HDTV*, Aug. 1989.

[30] R. Ansari and C. Guillemot, "Exact Reconstruction Filter Banks using Diamond FIR Filters," *Proc. 1990 Bilkent Int. Conf.*, Ankara, Turkey, pp. 1412-1424, July 1990.

[31] *MIT Channel-Compatible System Proposal*, submitted by Advanced Research Television Program, Massachusetts Institute of Technology, Cambridge, MA, March 1990.

[32] G. Schamel, "Motion-Adaptive Subband Coding of HDTV signals with 140 MBit/s," *Proc. Picture Coding Symposium*, Cambridge, MA, pp. 1.7-1 - 1.7-3, March 1990.

Chapter 8

Medical Image Compression: Possible Applications of Subband Coding

by: Otto Rompelman
    Information Theory Laboratory
    Delft University of Technology
    2600 GA Delft, The Netherlands

Contemporary health care is inconceivable without diagnostic imaging techniques. These range from very basic X-ray images to highly advanced techniques such as MRI and PET (to be discussed in more detail later). Many images have to be kept for later inspection. As an example we may think of the comparison between two images made before and after a particular therapy. At present most image storage is in the form of photographic copies, even if the original imaging system was inherently digital. It seems feasible to expect that in the near future images will be stored in a digital format. Several reasons make this assumption a likely one. A few advantages of digital storage are insensitivity to aging, simple copying facility, multi-site inspection facilities and reliable archiving and retrieval. However, it should be mentioned that the photographic image has a number of advantages as well. The resolution of the photographic image outperforms any digital display. Furthermore the photographic image is both a storage, a communication and a display medium alike. Still at a number of places researchers are working towards what is called a picture archiving and communication system (PACS) or, as it has been called more recently: image administration and communication system (IMACS).

In this Chapter we will shortly review a number of medical imaging modalities. For more detailed information the reader is referred to e.g. [20]. The digital representation of the different images will be discussed in relation to their resolution. Some basic considerations of picture archiving systems will be introduced and image compression will appear to be an essential constituent part of these systems. A case study of subband coding using vector quantization is described as well as some results on subband coding using multidimensional scalar quantization. Finally we will discuss possible hardware implementations for fast subband coding of medical images.

8.1   Medical Imaging Modalities

This Section starts with an historical overview of medical imaging modalities, from conventional X-ray to MRI. This is followed by a discussion on image formats and the spatial and contrast resolution of the commonly used techniques.

8.1.1   Historical Overview

Up till the end of the 19th century the medical doctor had only one imaging system at his disposal i.e. his own pair of eyes observing the reflected light of the subject under study. This implies that observations were restricted to electromagnetic radiation with wavelengths in the range of 400 - 800 nm. In 1895 Wilhelm Konrad Röntgen discovered the X-rays which for the first time enabled the doctor to look at the inside of the intact living body. No one was really concerned with radiation hazards in those days. Major developments that improved the image quality and/or reduced the radiation dose were the rotating anode tube, high efficiency screens and the image intensifier.

At the recording side important improvements were reached in the field of both photographic materials such as very sensitive and low grain films. A very important invention was the plumbicon tube which led to the development of a highly sensitive video camera. This enabled the doctor to observe the radiological image at an arbitrary distance from the patient hence avoiding radiation hazards due to cumulative exposure. Nonwithstanding the developments mentioned, until the early fifties X-ray radiography remained the only imaging technique available. However in that period a number of new imaging techniques emerged. Radioactive isotope imaging triggered off the new field of nuclear medicine. Although first only one-spot measurements were done, later mechanical scanning devices were used that produced a matrix of spot measurements or in other words: an image. The introduction of the gamma camera in 1958 highly improved the method and from then on dynamical studies became possible. Ultra sound imaging appeared to be a feasible technique which was of particular interest in applications were doctors were very reluctant to expose their patients to any form of electromagnetic irradiation as is the case in pregnancy or in brain studies. The first application of ultra sound was the so called A-scan (line scan) showing the partial reflection of an ultrasonic pulse (frequency in the order of a few MHz) as a function of the penetrating depth in the patient. Later, however, scanning devices were developed composing an image from a number of line scans. The first ultrasound imaging systems employed mechanical scanning devices which introduced severe motion artifacts when observing moving objects. At present electronic sector scanning systems allow for real time inspection of e.g. cardiac movements or fetal behavior.

The imaging modalities mentioned so far were mainly based on analog techniques. The introduction of computer processing of data made it possible to develop a new range of imaging systems. The first digital imaging technique was the Computed Tomography (CT, from the Greek words $\tau o \mu o \varsigma$: slice and $\gamma \rho \alpha \varphi \epsilon \iota \nu$: to write), which was introduced in the early seventies. This technique made it possible to create images representing the radiation absorption in a slice of the body; this being in contrast with the integrated absorption over a specific area as was the case in traditional X-ray images. Also the X-ray imaging systems were improved by digital data processing. Therefore analog-to-digital image conversion methods had to be developed. As an example we mention the image receptor plates. In these plates the X-ray image is built up and consequently stored as a latent image. By means of a laser scanner the stored energy is released and converted to light in proportion to the stored energy. This light is detected with a photodetector, the output signal of which is then converted to a digital data stream thus creating a digital radiographic image. Digital radiography allowed for the introduction of Digital Subtraction Angiography (DSA). This technique is used to visualize parts of the arterial system. First an image is made of the organ of interest. Consequently a radio opaque contrast medium is injected whereafter a second image is made. Subtraction of both images shows the arterial system in great detail and contrast whereas other structures are highly suppressed.

A very promising new technique is Magnetic Resonance Imaging (MRI). This technique is based on the fact that the nuclei of particular elements (e.g. $^1H, ^{31}P, ^{13}C, ^{14}N$) have an intrinsic spin. This spin can be altered by applying a very strong magnetic field (in the order of a few Tesla). By means of a powerful radio transmitter pulses are fed into the tissue. The direction of the net magnetic moment is changed when a high frequency electromagnetic pulse is emitted. After the pulse has stopped the magnetic moment returns to its original value and a weak electromagnetic signal is emitted. An antenna placed around the patient detects this signal. By slowly modulating the magnetic field and measuring both frequency and amplitude of the returned signals it is possible to reconstruct a two-dimensional (2-D) distribution of the nuclei of interest. As in CT, an image of a slice is obtained which implies that MRI is a tomographic technique. Originally the method was called Nuclear Magnetic Resonance (NMR) imaging. However this name caused serious unrest since from the word "nuclear" it was related to radio-activity and hence blindly assumed to be extremely hazardous.

The latest development in medical imaging is Positron Emission Tomography (PET). This method is based on the detection of a distributed isotope as in nuclear medicine. The isotope emits a positron which at some place in the tissue is combined with an electron. This combination causes annihilation yielding two gamma photons of 511 keV each. These photons are emitted in opposite directions. With combinatorial logic the detector system only detects coinciding photons. The pictorial information is built up like in conventional CT-scanners employing either a rotating detection system or a multiple detector arrangement. The final image is created by a tomographic reconstruction procedure.

8.1.2   Image Formats; Some Remarks on "Resolution"

A digital image is a discrete 2-D representation of a set of measurements made over a spatial region. As an example, in the case of projection type imaging modalities (such as X-ray imaging) the pixel values in the image are related to the values of some physical quantity (e.g. X-ray absorption coefficient) integrated over a line (or better: strip) of tissue. Tomographic techniques are based on the reconstruction from a number of projections of a 2-D distribution of the physical quantity. Some modalities are inherently digital (such as CT and MRI) and the images are usually displayed on a CRT- or a video monitor. Analog imaging modalities, such as the conventional X-ray, still play an important role. The images are usually projected on a photographic medium and displayed after photographic development. Scanning devices are employed to convert the photographic image into a digital data stream. Ultra sound imaging systems are in some way hybrid in the sense that scanning takes place in a discrete number of directions whereas the reflections are measured in relation to continuous time which represents (continuous) depth in the tissue.

The number of pixels as well as the number of bits per pixel is a very important issue in medical imaging. We will give some basic considerations underlying this question however without going too much in detail. Let us shortly review some aspects of an imaging chain as depicted in Figure 8.1. We will assume that the imaging system is isotropic, linear, homogeneous

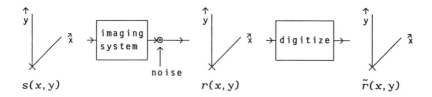

Figure 8.1: The imaging chain.

and space invariant. The object is denoted $s(x,y)$. As an example, in an X-ray system $s(x,y)$ represents the projection on a 2-D plane of the three-dimensional distribution of the X-ray attenuation. The imaging system creates an image $r(x,y)$ where G may be the blackness of the photographic image or the charge of the image receptor plate. Due to the point spread function (PSF) of the imaging system, $r(x,y)$ in fact represents a degraded version of $s(x,y)$. Furthermore noise will be introduced in the imaging chain. For the sake of simplicity we will assume here that the net result is the contamination of the output by wide band additive noise. The next step is the digitizing (sampling) of the image $r(x,y)$. The sampling distance Δx ($=\Delta y$) has now to be defined. It is obvious that Δx has to be chosen in relation to the PSF of the imaging system or, in other words, to the modulation transfer function (MTF).

Theoretically Δx could be chosen on the basis of the 2-D sampling theorem (e.g. [2]). The sampling theorem states that, if there is no power present above a spatial frequency f_0 lp/mm (line pairs per mm), the image can be perfectly reconstructed from the equidistant samples with inter sample distance $\Delta x = (f_0/2)$ mm. However some remarks should be made with respect to this theorem. Perfect reconstruction theoretically requires a sinc-function interpolation and, very important, reconstruction is only possible if the number of samples is very large (theoretically infinite). This can intuitively be understood if we realize that the sinc-function has significant values over an extreme long support. In image processing, however, we deal with a very limited number of samples. We may envisage a real image as being obtained by selecting a finite area from an infinite scene. This process is in fact the rectangular windowing procedure well known in signal analysis. This procedure introduces a leakage with infinite support. The consequent sampling will introduce unavoidable aliasing errors [26]. We may reduce the error by increasing the sampling density (oversampling, decrease of Δx and Δy). As a consequence the aliasing error becomes less severe. Furthermore we may employ interpolation functions (reconstruction filters) with a "nicer" PSF (shorter support), than the sinc-function approach. Another argument for oversampling is the fact that in practice we seldom deal with images for which the spectral power above f_0 is identical to zero.

The question now rises: how much oversampling will be necessary? This depends on three factors viz. *(a)* the 2-D power spectral density of the original image, *(b)* the MTF of the imaging system and, very important, *(c)* the final goal of the sampling procedure. As far as the original image and the imaging system are concerned it is obvious that if the original image comprises significant high frequency power and the imaging system allows for the passage of a significant portion of high spatial frequency power, the aliasing error will be large requiring a sampling density much in excess to the cut off frequency of the imaging system. As far as the goal of the digitizing procedure is concerned we can make the following remarks, referring to Figure 8.2.

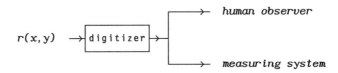

Figure 8.2: Two principally different destinations for image data.

If the image is to be observed by a human observer as is the case in image *processing*, we should indeed reconstruct the image since our eyes are analog instruments. In practice however the images are never reconstructed but digitally interpolated such that our eyes accept the upsampled digital image as an approximation of the reconstructed analog image. Due to the fact that our eyes are not perfect, we can allow for some reconstruction errors as long as they are not observable. We might go even further by saying that the reconstruction error should not lead to wrong interpretations (diagnostic errors), if the processed image is used for that purpose. We will come back to this point in Section 8.4. The conclusion is that due to the imperfectness of our visual system we may accept a certain distortion. Furthermore, our eyes will perform the final interpolation thanks to the limited bandwidth of our visual system.

At this point it is interesting to note that the term *resolution* in fact is ambiguous. At one hand the term is used for the ability of the system to separate two objects (usually two lines); we will refer to this ability as the *physical resolution* expressed in line pairs/mm. On the other hand resolution is often used for the number of pixels per line or column in the digitized image; this feature will be referred to as *representation resolution* expressed in pixels/mm. It is obvious that both resolutions in general will be different. As has been mentioned above the displayed digital image is an upsampled version of the digitized image. Therefore usually the representation resolution is larger than the physical resolution, assuming that 1 lp/mm corresponds to 2 pixels/mm. This latter assumption though not exactly correct, can be justified when taking the low pass filter characteristics of the human visual system into account. With the help of psychophysical evaluation techniques it is possible to find the minimum allowable sampling density in individual situations such that no statistically significant differences between the original and the reconstructed images are observed. It has been shown that for e.g. X-ray images the representation resolution should be in the order of about 1.2 to 1.5 times the physical resolution. As a result X-ray images should be displayed at a resolution of at least 2000*2000 [13]. Due to the fact that the relation between the physical and the representation resolution for the images is less than two, the images are usually sampled at a sampling density equal to the minimally necessary representation resolution.

If the digital image is used for measurements as is the case in image *analysis*, the error in the quantity to be assessed (e.g. the length of a curve) can be related to the sampling density. The sampling density then follows from the predescribed maximum allowable error. It should be emphasized that due to the previously given arguments this error will never be equal to zero since perfect reconstruction of the image is impossible. Furthermore, it should be noted that this error is also highly dependent on the algorithm used for the determination of the parameter looked after. For the given example of the length of a curve we refer to [3]. It can be concluded that we may arrive at two different values for the sampling density according to two different types of criteria viz. psychophysical and measurement criteria. The relation between these values has not been given much attention in the literature [26].

Finally some remarks on the number of bits per pixel or, as it is often called: *contrast resolution*. As was the case with spatial resolution the contrast resolution is dependent on *(a)* the properties of the image and the imaging system and *(b)* on the goal of the sampling procedure. The contrast resolution problem is identical to the scalar quantization problem in signal processing. The number of quantization levels is dependent on the number of levels that can be discriminated by the system and is mainly determined by the signal-to-noise ratio. As will be discussed different medical imaging systems show large differences as far as their ability to discriminate between two levels is concerned. As far as the human observation is concerned it is well known that our eyes can only discriminate between about 100 to 150 intensity levels. Therefore, usually 8 bits are used to represent visual images. If the discriminating power of the imaging system surpasses this figure we need to interactively select the relevant information. The latter technique is usually applied in CT- imaging and is known as *windowing* as illustrated in Figure 8.3. This means that by linear transformation a part of the range of variation of the pixel values is projected on the total range of visible values.

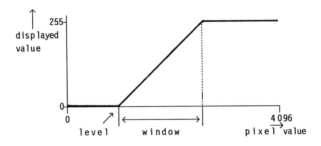

Figure 8.3: The window-level principle.

To conclude this section, in Table 8.1 a number of imaging systems are given with their respective (digital) image formats as well as their total number of bits/image. For the digital image systems these figures represent the

imaging system	*pixels*	*bits/pixel*	*bits*	
nuclear imaging	128*128	8	131	k
ultrasound	512*512	8	210	k
DSA	512*512	10	2.6	M
CT	512*512	12	3.2	M
MRI	512*512	12	3.2	M
X-ray	2048*1048	10	42	M

Table 8.1: Spatial and contrast resolution of some medical imaging systems

actual digital data produced by the system. For the analog imaging systems the number of pixels are in agreement with the minimally required spatial representation resolution.

8.2   Application Area: Picture Archiving and Communication Systems (PACS)

As has been mentioned in the Introduction, there is a tendency towards digitization of medical images. Apart from the inherently digital systems such as CT, MRI and DSA, also originally analog systems are now becoming digital. Recalling that an image is in fact a 2-D distributed set of measurements, this implies that the measured phenomena are directly obtained in the form of a matrix of either continuous or discrete elements, rather than in the form of a continuous 2-D distribution, which is digitized afterwards. If the matrix element are continuous, they are be converted to discrete values, usually by a scanning analog-to-digital conversion process. As an example we refer to the X-ray receptor plates as mentioned previously.

It is obvious that the digitization of medical imaging procedures will require a digital image archive and a network for storing and retrieval of the pictorial data. At present there is much research activity going on, possibly leading to an integrated picture archiving and communication system (PACS). We refer to the proceedings of the annual SPIE conferences on Medical Imaging. The tremendous amounts of data to be stored and transmitted impose heavy demands on such a system. A few aspects of a possible PACS structure will be discussed.

The amount of pictorial data to be stored in an average hospital of about 800 beds is estimated to be in the order of 2-10 Tbit. Modern magnetic disks have a capacity of about 700 Mbyte, hence about 900 disks would be needed per year. More promising are the optical disks (Digital Optical Recorder, DOR). Though DOR disks at present are so called WORM (Write Once, Read Many) devices, this is no problem for this is in fact the actual application at issue. DOR disks have a storage capacity of a few Gbyte and it may be expected that this capacity will increase in the near future. An archive is thought to be composed of "juke boxes" containing many DOR disks. Retrieval of large data files such as images from an optical disk is about five times slower than from a magnetic disk. As an example the retrieval of a 2000*2000 image of 8 bits/pixel from a modern magnetic disk takes about 2 sec whereas the same image will need about 10 sec to be retrieved from a DOR disk. Much more time will be needed for the mechanical device to select the proper disk in a juke box system. This leads to overall access times far in excess of acceptable waiting times. The waiting time between the retrieval command and the actual display of the image at the viewing station should be less than a few seconds.

A second bottle neck is the speed of communication between the archive and the viewing station. If we think of a local network with coaxial cables the transmission speed is limited to about 1 Mbyte/sec. In practice the actual throughput is much less. Optical fibers have a higher capacity but are still physically limited to about 10 Mbyte/sec.

Finally the display of images on a monitor will require a resolution corresponding to the representation resolution of the image. This implies that screens are needed with a resolution of at least 2048*2048 pixels. Furthermore these screens should have a contrast resolution corresponding to 8 bits/pixel in order to compete with the contemporary X-ray film resolution.

The problems concerning limited storage capacity and the reduced access times of mass storage devices as well as the limited speed of data transmission leads to the need of *(a)* a layered structure of the system with intermediate data buffers and *(b)* data compression.

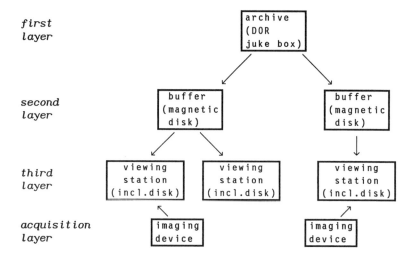

Figure 8.4: Layered structure of a picture archiving and communication system (PACS).

In Figure 8.4 a schematic illustration of a possible PACS is shown. The first layer consists of the DOR disk juke box. Here all pictorial data are collected. Due to this rather slow system images to be viewed at the viewing station have to be pre-ordered from the archive. This does not deviate from the contemporary praxis where images have to be collected from the photographic image archive. Doctors are used to the fact that they have to order these images a few hours to even one day before they are needed for inspection. In the proposed structure images are transported to and loaded into an intermediate buffer. This intermediate buffer should be much faster than the DOR disks but may have a much smaller storage capacity since only a very limited amount of images need to be stored here. Furthermore WORM devices are not applicable at this level. Hence magnetic disks (e.g. large Winchester drives) will be employed here. The third layer consists of the actual viewing stations.

It is envisaged to create three kinds of viewing stations viz:

- *Extended Work Station (EWS)*
  To be used by the radiologist for detailed investigation of images. At least one very high resolution display screen will be present. However more screens may be needed in order to carefully compare images. Image processing facilities will be present. Also a very fast access magnetic disk will be present for storage of a limited amount of images. Also processing software is stored on this disk. Interactive connection with image on site acquisition equipment for comparison of acquired images with archived images.

- *Simple Work Station (SWS)*
  Work stations located at offices for preview of a small series of images, possibly at a lower spatial resolution. A medium size disk will allow for the storage of a very limited amount of images. Some simple image processing capabilities may be present. As an example we might think of a conventional work station.

- *Output Only Station (OOS)*
  Very simple display system with low spatial resolution. May be located at various places within and outside the hospital. These systems could be personal computers equipped with a video screen or a simple conventional workstation

The location of possible image compression/expansion modules in this structure is still a matter of discussion. If compression/expansion modules are expensive and slow it is advisable to place them very near the first layer. This reduces the problem of the storage capacity and only the less critical communication time between the archive and the intermediate buffer is increased. The constraints concerning data transport times between the buffer and the viewing stations and vice versa however remain the same. It would therefore be much more attractive to place the image compression/expansion module near the third layer. An additional advantage could be that the processing power needed for expansion at a viewing station level could be employed for image processing facilities as well.

Finally, it would be most interesting if the compression technique to be applied is a so called *progressive coding technique*[1]. This implies that the data stream after coding comprises gradually more image details as time progresses. As a consequence when the expansion indeed takes place at a work station level the amount of data that is needed for compression will be less if this is just an OOS, whereas more data will be needed in more sophisticated work stations. Subband coding is a very feasible technique when taking these considerations into account.

8.3   Some Principles of Image Compression

In image compression (and in data compression in general) we may distinguish between two principles:

- *error-free (or lossless) compression*: removal of redundancy from the data: perfect reconstruction is possible,

- *irreversible compression*: removal of details from the data that may be neglected in accordance to a pre-established criterion: perfect reconstruction is not possible.

We will shortly review these two approaches. In error-free compression use is made of the internal structure of the data. Once this structure has been made explicit in statistical terms it is envisaged to remove or at least reduce the interdependencies of adjacent pixels (e.g. by DPCM). Furthermore the probability density functions (pdf's) of the pixelvalues themselves can be exploited (e.g. by applying Lloyd-Max quantizers and variable length coding). A combination of both approaches is possible by dividing the image into sub-images. The compression ratio to be gained with e.g. 512*512 CT-images (12 bits/pixel) is between 2 and 3, depending on the method chosen. Obviously, the advantage of error-free compression over irreversible compression is the fact that the original digital data can be perfectly reconstructed. The compression ratio however is too low in order to be of practical use for image archiving and communication purposes.

With irreversible compression methods those components of the data are removed that are supposed to be of no relevance for the actual destination of the data. In practice a more sophisticated approach is followed in the sense that the number of bits allocated to data components is tuned to the relevance of these components. In other words: the more important the data, the more

---

[1]Not to be confused with progressive versus interlace display as discussed in Chapter 2.

bits are used for that data. It might be argued that distortion is principally inadmissible in medical imaging. This means that "no relevance" does not apply here. Still the key question should be put forward: what is the actual meaning of "no relevance" ? This is completely dependent on the destination as has been discussed in Section 8.2. In the case at issue the destination of the data is the radiologist who uses the pictorial information for diagnostic purposes. This means that we may rephrase our previous question as "which components of the image may be left out without affecting the diagnostic quality of the image"? In order to answer this question we need a definition of "diagnostic quality". Though a detailed discussion concerning this question is beyond the scope of this chapter some points will be made when discussing a case study with CT-images.

8.4 Case Study with 512*512 CT Images

A case study with 512*512, 12 bits/pixel CT images has been carried out using subband decomposition and Vector Quantization. The images were obtained from the Radiology department of the Academic Hospital, Leiden University, the Netherlands. They were stored on floppy disks and transported to the Information Theory Laboratory of the Delft University of Technology. The actual goal was to find out to what extent the diagnostic quality was affected after compression/expansion with different compression ratios.

8.4.1 Subband Coding

The band-splitting as used in this study is based on a two stage four band spatial frequency filtering of the image. This is accomplished in the following way. First the image is split into four subbands. The location of these bands in the spatial frequency domain is as depicted in Figure 8.5.

The four output images of these filters are down-sampled, or decimated as it is usually called, by a factor two in each dimension. This down sampling procedure is allowed because of the fact that the filtering procedure reduces the spatial frequency range of the original image to half its original value. Subsequently, sub-sampled images are filtered and decimated once more by the same quadruple filter and decimation procedure. This results in 16 band-filtered versions of the image. The original image can be reconstructed by using an upsampling or interpolation procedure followed by a reconstruction or synthesis filter operation. The expansion of the image can be considered as the inverse procedure of the previously described filter decimation procedure.

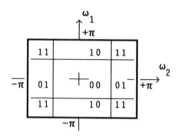

Figure 8.5: Subband splitting scheme in the frequency domain: first step.

It can be concluded that starting from a single 512*512 image we end up with 16 128*128 images as shown in Figure 8.6. Another way of putting this is that the original data consisted of 512*512 scalar values whereas the resultant data consists of 128*128 16-dimensional vectorial values.

This way of looking at the problem will be convenient when discussing the subsequent vector quantization procedure. The ideal filter characteristics as supposed to underly the spectral segmentation as depicted in Figure 8.5 cannot be realized in practice. Practical filters (e.g. FIR filters) deviate from this ideal filter. Due to both the filtering/decimation and the interpolation/filtering procedures aliasing errors are introduced. These errors are unacceptable. The actual band- splitting, however, is accomplished by a so called quadrature mirror filter (QMF) bank. With this type of filter aliasing errors at the output of the decimator are permitted but the synthesis or reconstruction filters are designed in such a way, that these errors are cancelled. This approach was introduced first for speech coding [4, 5] and later expanded for multi- dimensional signals [19]. In the latter paper it was also shown that separable QMF filters may be considered, which means that the multi- dimensional filter problem is reduced to a 1-D problem. Starting with a 1-D filter pair $H_0(\omega)$ and $H_1(\omega)$, to split off two subbands from a 1-D signal the reconstruction filters $G_0(\omega)$ and $G_1(\omega)$ in the QMF approach are chosen such that

$$G_k(\omega) = 2(-1)^k H_k(\omega) \qquad k = 0 \ or \ 1. \tag{8.1}$$

Although the aliasing errors are fully cancelled, the overall amplitude transfer function of the system exhibits a small ripple. This ripple can be minimized if some constraints are imposed on the choice of $H_k(\omega)$ [7]. For further details we refer to a useful review paper [16]. In the method using QMF as discussed above errors due to quantization of the sub bands have not been taken into account. This quantization will lead to an imperfect canceling of the aliasing errors. However, recent research has shown that these errors play a negligible

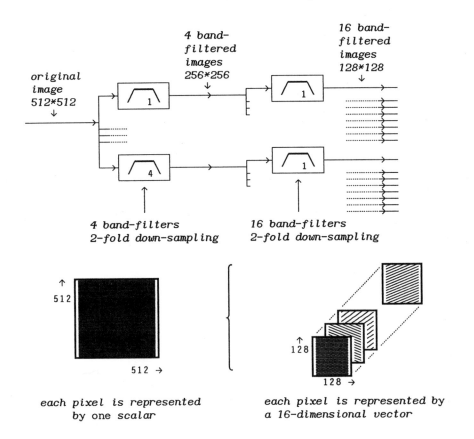

Figure 8.6: Complete subband splitting scheme: Transformation of scalar data into vectorial data.

role in the overall error introduced by quantization/coding procedure [23]. Following [22] and [25], in the coding scheme as used in the study reported here the coefficients of a 32-tap QMF designated as 32D in [7] were employed.

8.4.2 Vector Quantization

Vector Quantization (VQ) is a multidimensional generalization of an optimal scalar quantization. A scalar quantizer maps a finite or infinite range of continuous values on a finite set of discrete values. Perhaps the best known scalar uniform quantizer is the traditional analog-to-digital converter. A quantizer with a minimal mean squared error (MMSE) is the Lloyd-Max quantizer [11]. In this case the distribution of quantization levels is chosen in accordance with the pdf of the signal to be quantized. This implies that the underlying pdf is known. If not, this distribution has to be estimated on the basis of a representative set of test data. The vector quantizer can be considered as a multidimensional Lloyd-Max quantizer (if based on the MSE).

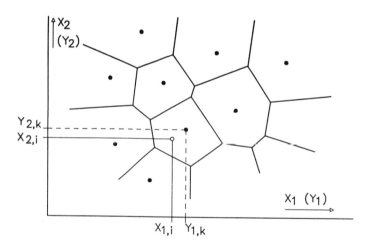

Figure 8.7: Principle of vector quantization.

The principle of the vector quantizer will be elucidated with the help of Figure 8.7. Assume that the data to be quantized consist of two-dimensional vectors (X_1, X_2). By means of some algorithm the X_1, X_2-plane is subdivided into a prescribed number of areas such that these areas optimally match the densities of the test data. The coordinates of the centers of the areas (viz.

(Y_1, Y_2) are used as the output values of the quantization procedure. These values are called the code vectors and the set of all code vectors is referred to as the codebook. When quantizing a new incoming data vector the algorithm determines (on the basis of some distance measure) to which code vector the actual vector has to be attributed. For more detailed information on VQ we refer to [6].

Codebook Generation

A codebook is generated using a set of images (e.g. four) called the training set. Using a particular algorithm (the LBG algorithm) a nearly optimal set of codebooks can be computed. The size of a codebook is given by its code-rate, R. The LBG algorithm computes a single codebook of rate R. Codebooks of successively increasing rates can be obtained by using a splitting technique. The LBG-algorithm was first proposed by Linde, Buzo and Gray (hence LBG) in 1980 [10]. Since then it has been widely used. It creates a codebook of rate $R=2$ (i.e. it contains 2 code-vectors) given an initial codebook. The latter is a "good guess" of what the codebook could look like before it is processed by the LBG algorithm. The codebook is then optimized by minimizing a distortion criteria. If this distortion is less than a pre-established threshold a, an optimal codebook of rate R has been found. A codebook of rate $R+1$ can only be computed if the codebook of rate R is known.

Actual Vector Quantization

Once the codebook has been obtained, VQ can be applied to the outputs of the subband filters to encode them. The vectors are formed by taking corresponding samples from each subband, resulting in 128*128 16-dimensional vectors. Using the codebook, which is basically a table containing a limited set of code-vectors, each image vector is replaced by the nearest code- vector. "Nearest" is taken to mean the closest code-vector to an image vector in a geometric sense. Therefore, in order to achieve vector quantization each image vector is used to compute a distance with respect to all the code-vectors. The code-vector with the minimum distance is taken as a replacement for the original image vector. In the algorithm, instead of computing the distance, another measure is used which is simply the square of the distance:

$$d_j = \sum_{i=1}^{16}(c_{ij} - v_i)^2 \qquad\qquad j = 1, ..., M \qquad\qquad (8.2)$$

with M the number of code-vectors, c element i of code vector j and v the data vector to be coded. This saves M square root operations. Of course, other distortion measures are possible and a variety of them have been proposed. However, the squared distance distortion measure is appealing from a mathematical point of view.

The next question that arises is how many vectors there should be in the codebook. Obviously, the larger the codebook the better the quality of the reconstructed image will be. However, a codebook that is too large will not result in meaningful compression ratios. Conversely, a small codebook will result in big compression ratios but will produce an image of unacceptable quality.

It is convenient to utilize codebooks which contain a number of code-vectors that is a power of two. Indeed, if there are 2^R code-vectors in the codebook each one can be identified by an R-bits address, which represents an index number. Thus, storage of a table which is $128*128*R$ bits long is enough to subsequently be able to decode and reconstruct an image. For a $512*512$ 12 bits/pixel images this results in a compression ratio C (i.e. the reduction factor for required memory space) according to:

$$C = \frac{512 * 512 * 12}{128 * 128 * R} = \frac{192}{R} \qquad (8.3)$$

For instance, if $R = 10$ (hence a codebook comprising 1024 code vectors), $C = 19.2$.

Complete Compression and Reconstruction Scheme

Combining subband filtering with vector quantization leads to a structure as depicted in Figure 8.8. First the original image is split into 16 subbands whereafter it is encoded using vector quantization. Then it is either stored or transmitted. Decoding an image simply involves fetching the appropriate code-vector from a codebook. Finally, using QMF's, the image can be reconstructed.

In summary, the entire coding/decoding procedure consists of the following steps:

1. Generation of a codebook (only once),

2. Coding:

    (a) splitting of the image in band filtered and down sampled (subband) images

    (b) vector quantization,

3. Decoding:

(a) decoding of the subband images (vector decoding)

(b) reconstruction of the image from the decoded subband images.

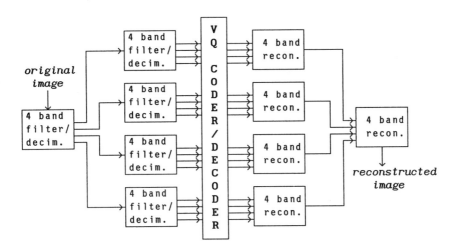

Figure 8.8: The complete SBC-VQ CODEC.

8.4.3   Results

The presentation first centers on processing times per image at a range od coding rates. The is then followed by a discussion of the corresponding resultant image qualities.

Processing Time

The technique discussed above has been applied to a number of CT-images. The images were of the 512*512 format, 12 bits/pixel. Our studies ran on a Digital Equipment VAX-11/750 (operating system: VMS) with an array processor (Analogic AP500). The processing times required for the procedures described above depend on a number of parameters. In terms of practical applications the processing time required for the different procedures involved as described above should decrease in the order A,B,C. From a test study insight was obtained about the processing time required in the different steps of the procedure [12].

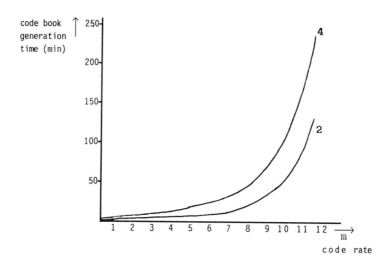

Figure 8.9: Codebook generation time in relation to codebook size.

The generation of the codebook is a very time consuming process. In Figure
8.9 the time required is plotted as a function of the code-rate for two values
of the size of the learning set. It should be noted however that this procedure
has to be carried out only once, whereafter in principle the code- book could
be used forever for the kind of images at issue (e.g. abdominal CT-scans). In
Table 8.2 a few results are shown. The time values given are average values
obtained during an experiment comprising about 40 images.

code-rate	compr. ratio	*coding*		*decoding*		total proc. time
		subband splitting *(2-a)*	vector quant. *(2-b)*	vector decod. *(3-a)*	image recons. *(3-c)*	
12	16	10.7	627	1.5	12	740 s
10	19	10.3	159	1.5	12	254 s
8	24	10.3	42	1.5	12	131 s
7	27	10.4	22	1.5	12	110 s

Table 8.2: Times (in sec.) required for compression and expansion of 512*512
CT-images with SBC-VQ

These values may vary somewhat due to the fact that the decoding time
was not logged separately but was estimated to be about 1-2 sec. It can be
concluded that indeed the procedures that from an applications point of view

are most time critical (i.e. decoding and reconstruction) require least time of the entire procedure.

Image Quality

Two different compression ratio's were tested viz. 16:1 and 19.2:1, which correspond to codebook sizes of 4096 and 1024 code vectors respectively. In Figure 8.10 a typical example of a result is shown. The image shows a cross section of the abdominal region.

Figure 8.10-*(a)* shows an original 512*512 CT-image whereas Figure 8.10-*(b)* and Figure 8.10-*(c)* show the compressed/expanded images with compression ratios of 16:1 and 19.2:1 respectively. The importance of the proper choice of the training set may become clear from an experiment in which the training set consisted of images with a larger size of the abdominal cross section than that of the images to be compressed. The codebook was generated on the basis of four large abdominal cross section images. The compressed/expanded image was a small abdominal cross section image. Figure 8.10-*(d)* shows the result of the compression/expansion procedure. It is obvious that this result is completely unacceptable as it is in sharp contrast with the images shown in Figures 8.10-*(b)* and - *(c)*.

The potential of the method for CT-image compression has been tested by means of a thorough psychophysical evaluation. In this study both checklist and ROC-analysis techniques were used. The results of this study are reported in more detail elsewhere [13, 16]. One of the striking results was that in terms of preservation of diagnostic value there was no statistically significant difference between the original images and the compressed/expanded images for both the 19.2:1 and 16:1 compressed images. Moreover, in a number of cases even the compressed/expanded images were called originals by the radiologists [17].

8.5 Some Results on SBC using Multidimensional Scalar Quantization (MDSQ)

As has been discussed VQ is a very time consuming process. Therefore it was decided to employ another form of prior information in order to arrive at a faster technique. We recall that VQ is based on prior statistical information of the class of images to be compressed. This information is extracted from a learning set. From a theoretical point of view VQ optimally exploits the both

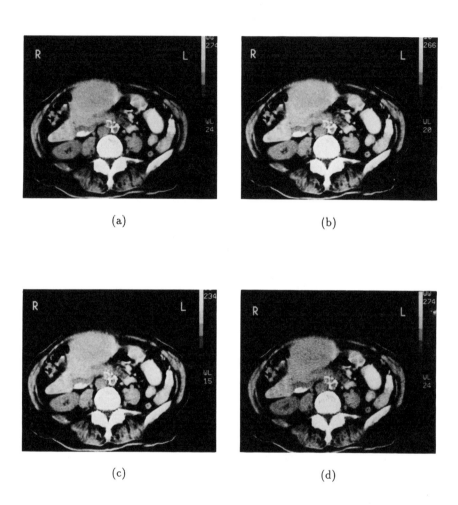

(a)   (b)

(c)   (d)

Figure 8.10: *(a)*: original CT-scan; *(b)* and *(c)*: CT-scans after compression/expansion with compression ratios of 16:1 and 19.2:1; *(d)*: result of image outside the family underlying the training set.

linear and nonlinear statistics of the vectors representing the subband data
[24]. A much simpler and faster technique obviously is the scalar quantization
(SQ) of the individual subband elements. Optimal quantization in the least
squared error sense is obtained by applying Lloyd-Max quantizers [11]. The
development of such quantizers requires knowledge about the probability density functions (pdf) of the data in the individual subbands. It has been shown
that the pdf's can be adequately described by the Generalized Gaussian pdf,
viz:

$$p(x) = ae^{-|bx|^{\gamma}} \tag{8.4}$$

with

$$a = \frac{b\gamma}{2\Gamma(1/\gamma)} \tag{8.5}$$

and

$$b = \frac{1}{\sigma_x}\sqrt{\frac{\Gamma(3/\gamma)}{\Gamma(1/\gamma)}} \tag{8.6}$$

The variances of the subband data are estimated and are used to normalize
the data so as to create unity variance pdf's. Consequently the data are
subjected to a standardized Lloyd-Max quantizer. For further details we refer
to Westerink et al's Chapter 5 in Section 5.2. The parameter γ has to be
chosen. For a large class of pictures it appeared that the histogram of the
subband data, apart from the lowest subband, is properly fitted to (8.4) if
$\gamma=0.5$. For the lowest subband, after subtraction of the mean, $\gamma=2$ appeared
to be a good value. Since the low pass filtered image is still significantly
correlated it makes sense to apply some form of predictive coding. Applying
DPCM to the lowest subband indeed leads to improvements. The applied
DPCM-algorithm is based on a three point quarter-plane model as shown in
Figure 8.11.

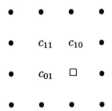

Figure 8.11: Three point quarter-plane model for DPCM.

The predicted pixel value $\hat{x}(m,n)$ follows from the previously estimated
values $\tilde{x}(i,j)$ of $x(i,j)$:

$$\hat{x}(m,n) = c_{01}\tilde{x}(m,n-1) + c_{11}\tilde{x}(m-1,n-1) + c_{10}\tilde{x}(m-1,n) \tag{8.7}$$

with $c_{01} = c_{10} = 0.86$ and $c_{11} = -(c_{01} * c_{10}) = -0.74$. Finally some improvements can be obtained by applying Huffman-coding to the resultant quantized data of the subbands. After DPCM the pdf of the resultant data could be adequately described by (8.4)-(8.6) with $\gamma=0.75$.

As an example we discuss the compression of an X-ray image of a human jaw. This kind of image is routinely made in an orthodontist practice. The image was converted into a 512*448 8 bits/pixel digital image using a video camera and a frame grabber. The subband filtering was carried out by a number of 16-tap QMF filters as discussed previously. The resultant frequency domain splitting was as depicted in Figure 8.12.

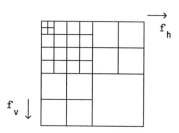

Figure 8.12: Frequency domain splitting scheme applied prior to MDSQ.

Consequently, three different coding methods were used:

1. Quantization of all subbands with quantizers designed according to the Generalized Gaussian pdf's as discussed above

2. Same as method 1, but with DPCM applied to the lowest subband

3. Same as method 1, but with post quantizing Huffman-coding applied to all subband data.

Bit allocation was performed as discussed in [21]. In Figure 8.13 an example of the results as well as the original image is shown.

In order to compare the results of the different approaches in terms of their efficiency, in Figure 8.14 the rate distortion curves are shown. As a measure of distortion we applied the mean-squared error (MSE), which for an $M * N$ image size is defined as

$$MSE = \frac{1}{M * N} \sum_{i=1}^{M} \sum_{j=1}^{N} [x(i,j) - \hat{x}(i,j)]^2 \qquad (8.8)$$

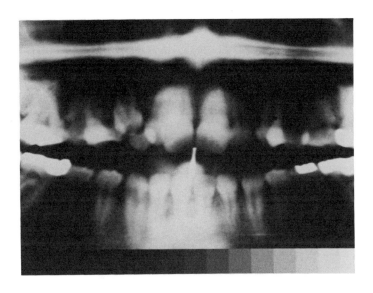

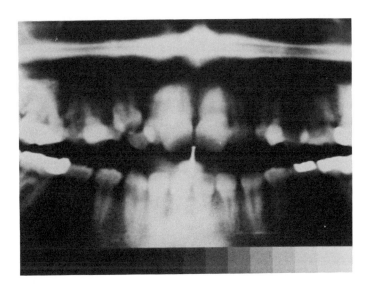

Figure 8.13: Original (*a*) and compressed/expanded X-ray image (*b*) of a human jaw: compression ratio 1:20.

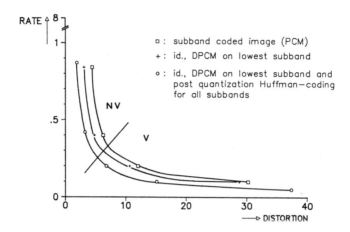

Figure 8.14: Rate-distortion curves for three different compression strategies; discrimination between *no visible distortion* (NV) and *distortion visible* (V) is indicated.

with $x(i,j)$ the original pixel values and $\hat{x}(i,j)$ the reconstructed pixel values.

The produced images were judged by the medical expert. They were classified into three categories: *no visible distortion, some minor distortion visible* and *significant visible distortion*. In Figure 8.14 the discrimination between the first two categories is indicated. From the results obtained thus far it may be concluded that indeed SBC with MDSQ is a promising technique for the compression of this type of images. As the compression ratio (or the final bit rate) is concerned it should be emphasized that this figure of merit may be highly flattered, if the original images are oversampled. In the case discussed above this was indeed the case. Still it may be concluded that a compression ratio of 10-20 may be obtained without noticeable reduction in diagnostic image quality.

8.6   Future prospects

It has been shown that SBC is a promising technique for medical image compression. As has been discussed previously PACS is an important field of application for medical image compression. We recall the main requirements for a compression system to be used in a PACS-environment:

- fast compression (order: a few seconds)

- very fast expansion (order: one second)

- compression/expansion hardware possibly to be located at workstation level, hence limited hardware costs required

- preferably not dependent on learning information

- preference for progressive coding

Though VQ has been applied to the subband filtered image, this method is highly dependent on prior information (learning sets). The theoretically less efficient MDSQ technique uses a more robust type of prior knowledge, i.e. the shape of the probability density functions of the subband pixel values, and seems to be an equally performing method as compared to VQ. Furthermore, the application of MDSQ leads to much faster running implementations. Finally, the SBC technique can be easily implemented in the form of a progressive coding technique. Hence, the MDSQ approach is to be favored, since it allows the low frequency subbands to be encoded and sent first. A software implementation (medium size mini-computer with an array processor) however, prevents it from practical application in a PACS environment. The problems concerning both computing time and power can be solved with special purpose hardware as will be discussed below. Furthermore, if the compression/expansion modules are to be located at the work station level, the special purpose hardware should be cost effective. Next we will discuss an experimental subband coder. In particular we will focus on the most time consuming part i.e. the subband filter.

8.6.1 Some Aspects of Hardware Implementation

It has been mentioned that SBC using MDSQ takes about 70 sec for compression and about 12 sec for expansion on a DEC VAX 11/750 with an Analogic AP500 array processor. A dedicated computing system with embedded parallelism is expected to lead to significant improvements in performance. This holds in particular for the computationally demanding filter operations.

The key problem in fast digital signal processing is the large amount of multiplications and accumulations (multiple vector multiplications). Many manufactures have tackled this problem by designing VLSI circuits specially suited for these tasks viz. DSP's (Digital Signal Processors) [8,9]. The core of these chips is a very fast multiplier/accumulator, one multiply- accumulation action (MAC) being performed in one or two clock cycles. Originally these chips comprised a 16*16 bit multiplier and a 32 bit accumulator. Later accumulator/buffer size was enlarged so as to allow for multiple (e.g. 256) multiplications/additions without overflow. Also the word size was increased to 24 bits in a number of cases. Though most devices handle fixed point numbers, recently floating point processors have been introduced such as the Motorola 96002, the Texas Instruments TMS 320C30, or the AT&T DSP32C.

Though these DSP's are very versatile as far as 1-D signal processing is concerned, their application to image processing is generally hampered by their limited address space for external RAM. As an example with a 16 bit address bus only 65 kwords can be addressed; a CT-image for comparison comprises 263 kwords. This implies the need for external address generators, which increases the complexity of the system and hence decreases performance. As has been discussed before, medical images have a dynamic range of 8 to maximally 12 bits per pixel. Hence fixed point processing will usually yield sufficient accuracy provided the pixel values are prescaled to 16 bits and the filters are of the moving average type. In order to get some idea of the total time required for an image to be split into subbands, the number of instructions needed to filter a $N*N$ image into 16 equally sized subbands will be addressed. The main feature of the DSP is the number of clock cycles necessary to perform one MAC as well as the clock speed. We recall that in a DSP multiply-accumulation and addition/subtraction operations each require the same number of instructions viz. one or two.

Assume that an $N*N$ image is to be split into subbands. First the pixel data are prescaled requiring a total number of $2*N$ instructions. Furthermore, it can be shown that, if separable filters of the polyphase structure are used the total number of instructions (MAC's, ADD's, etc.) to filter the image into 16 equally sized subbands is $n = 2 * N * (k + r)$, with k the number of filter taps and r an overhead factor due to the fact that sums and differences are to be calculated in the polyphase structure. If properly programmed r may be brought down to 4. Finally, as has been discussed previously it is necessary to calculate the variance of the subbands in order to carry out the bit allocation for the consequent quantizers. This will require an additional $2*N$ instructions. When applying a 16-tap filter (e.g. the the Johnston QMF16B [7]) the filtering and variance calculations (needed for the normalizing the data as to apply a fixed Lloyd-Max quantizer as discussed in Section 8.5) will require $44*N$ instructions. Since the instruction time of the applied DSP is 100 ns, the filtering of the image including variance calculations requires minimally 1.15 sec for a 512*512 image and 4.6 sec for a 1024*1024 image. In practice these figures will be somewhat larger due to interrupts for data transfer.

As has been mentioned before, the address space of most DSP's does not cope with the size of the images. Furthermore it is not feasible that the image memory and the filter processing units will be located on one board. This leads to separate boards for image memory and processing (with some buffering facilities) interconnected by some bus structure. The average time needed for address calculation and data transfer to the DSP-buffer should be of the same order as the time needed for filtering.

8.6.2   Example of a Hardware Subband Coder

As an example of a hardware system we will discuss the image compression/expansion system as it has been developed at the Delft University of Technology [18]. The system to be designed had to fulfill a number of requirements. A number of aspects of subband coding are still subject of investigation. Therefore, in order to allow for separate development and testing of the different functional units the system should be modular, the modules being a memory, a filter board and a compression/decompression module containing the quantizers. This lead to a bus type architecture.

The large amount of data involved in image processing requires a large address space. An image size of e.g. 1024*1024 pixels and 12 bit/pixel takes an address space of 1536 Kbytes (21 bit). Hence the address bus in the system must be at least 24 bit wide; the data bus should at least be 16 bit wide. Furthermore, reliable software (development tools) should be available for the host processor. The bus specifications should meet an open bus standard.

It was decided to develop a system based on a VME-bus with a Motorola 68020 as a host processor. The performance of such a system was supposed to be significantly improved by separating the control and program data stream from the actual image data stream. Therefore we apply an additional data bus, i.e. the VSB. The modules are functional units which communicate with each other by the busses. The modular structure is depicted in Figure 8.15.

A Motorola 68020 acts as a host processor. A hard disk is provided for storage of development tools, programs, images and numerical data necessary for filtering (coefficients) and compression/-decompression (tables). Images are loaded in the 2 Mbyte dual ported memory board and consequently transferred via the VSB to the subband filter module. This module contains a Motorola DSP56001 [1] and a separate address generator based on a Motorola 68020. This address generator was necessary because of the limited internal address space of the DSP56001. In Figure 8.16 the architecture of the DSP in relation to both the address generator and the bus is depicted.

Filtered data is (again via the VSB) transferred back to the memory board. Finally the filtered data is fed into the coder board and after coding (=com-

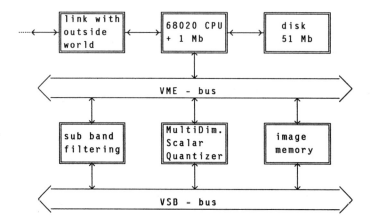

Figure 8.15: Structure of a bus-based subband compression/expansion system.

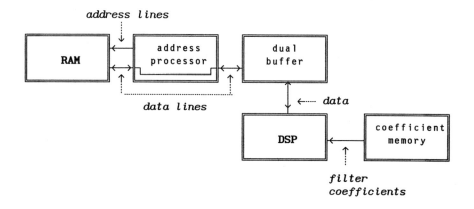

Figure 8.16: Configuration around the DSP.

pression) stored in the memory or on hard disk. Decoding takes place analogously to coding albeit in reverse order. In order to allow for separate development and testing of the individual boards, communication with the VAX-11/750 is provided through an ETHERNET interface. The operating system is OS9 whereas the ETHERNET communication with the VAX is based on the OS9-NET protocol.

8.6.3 Results

The performance of the described VME based image compression system has been compared to that of the VAX-11/750 + AP500 software solution. In Table 8.3 the results obtained with a 512*512 (8 bits/pixel) image are listed.

		VAX 11/750 + AP500	VME/VSB -system	
filter time	[1]	10.7 s	2.1 s	
coding time	[2]	18.6 s	0.6 s	[3]
decoding time		– [4]	–	[4]
reconstr. time		9.3 s	2.1 s	
S/N-ratio	[5]	60 dB	40 dB	

[1] - including calculation of variances
[2] - depends on contents of image and maximally allowable distortion
[3] - estimated value
[4] - not yet available; expected to be much faster than coding: only look-up's
[5] - without coding

Table 8.3: Comparison of SBC-VQ processing times for simulation (VAX + array processor) and dedicated hardware

At the time of writing only a prototype version of the coding module was available. Therefore the coding time results are preliminary. The processing time for image compression with the VME-system for this type of image is 2.7 sec. Apart from this, data transmission from the memory board to the filter board and back (via the VSB) takes about 1.5 sec. This latter figure should not be compared with the VAX-array processor implementation since data transmission times are highly dependent on the workload due to the multi-user environment.

The first results obviously show that the filter module is significantly faster than the simulation software running on the VAX-11/750 + array processor

AP500. For practical interactive applications the total coding and decoding times should be in the order of 1 sec. The net processing speed is mainly determined by the speed of the filter module and the data transmission rate.

The filter can be made faster by developing special purpose hardware e.g. based on bit slices. However this violates the desired flexibility. We think that a better performance may be reached by employing one of the recently announced floating point DSP's which are also characterized by a larger address space. As an example we refer to the Motorola DSP96002 [15]. The main advantage of this processor, apart from its higher speed, is its ability to generate 32-bit addresses. This will avoid the need of a separate address generator, which appeared to be the main time consuming element of the filter board.

The data transmission rate is highly dependent on the access time of the memory board and the VSB performance which factors are in fact mutually dependent. Still, if we would push a VSB/memory-combination to its limits it is to be expected that the 1 sec specification can not be met. This leads to the conclusion that for this application and maintaining the 1 sec requirement we have to refrain from this bus concept.

For the near future it is supposed that the performance of hardware image subband coders will be improved both by standard devices and by dedicated VLSI design. A subband coder based on standard devices will comprise transputers for data-routing and vector signal processors (e.g. as has been developed by Zoran Corporation) or FIR-chips for image filtering. If SBC will indeed prove to be an important technique for image compression, dedicated VLSI for such tasks as image filtering, bit-allocation and quantization are expected to be developed.

8.7 Concluding Remarks

Subband coding provides a useful technique for medical image compression. Possible artifacts will not show in false contours as is the case in block transform coding such as DCT-coding. Due to its hierarchical character the technique can be implemented using dedicated hardware. This feature makes application in a PACS environment most attractive. It has been shown that a significant data reduction can be obtained without affecting the diagnostic quality of the image. However, before jumping to conclusions thorough psychophysical studies need to be carried out before this irreversible compression method is introduced in a clinical environment.

Acknowledgements

This contribution could not have been produced without the support and the many stimulating discussions with Dr. Peter Westerink, Mr. Ad de Ridder and Mr. Peter van de Velde. The continuous support of Prof. Jan Biemond and Prof. Dick E. Boekee is highly appreciated. I am pleased to mention that Dr. Herman Kroon (University Hospital, Leiden, Netherlands) and Dr. Ted Bennett (orthodontist in Dordrecht, Netherlands) were so kind to supply us with the image material for this chapter.

Bibliography

[1] Anonyn., "Motorola's sizzling new signal processor", *Electronics*, vol. 10, pp. 30-33, 1986

[2] K.R. Castleman, *Digital Image Processing*, Prentice Hall, Inc. Englewood Cliffs, NJ, 1979

[3] L. Dorst and A. Smeulders, "Length estimators for digitized contours", *Computer Vision, Graphics and Image Proc.* , vol. 40, pp. 311- 333, 1987

[4] D. Esteban and C. Galand, "Application of quadrature mirror filters to split-band voice coding schemes", *Proc. of IEEE ICASSP*, pp. 191-195, April 1977

[5] C.R. Galand and H.I. Nussbaumer, "New quadrature mirror filter structures", *IEEE Trans. on ASSP*, vol. ASSP-32, pp. 522–531, 1984

[6] J.D. Johnston, "A filter family designed for use in quadrature mirror filter banks", *Proc. of IEEE ICASSP*, pp. 291-294, 1980

[7] E.A. Lee, "Programmable DSP architectures: Part I", *IEEE ASSP Magazine*, vol. 5, pp. 4-19, 1988

[8] E.A. Lee, "Programmable DSP architectures: Part II", *IEEE ASSP Magazine*, vol. 6, pp. 4-14, 1989

[9] Y. Linde, A. Buzo and R.M. Gray, "An algorithm for vector quantizer design", *IEEE Trans. on Information Theory*, vol. IT-28, pp. 227-232, 1982

[10] J. Max, "Quantizing for minimal distortion", *IRE Trans. on Information Theory*, vol. IT-6, pp. 7-12, 1960

[11] O. Rompelman, E. Voogt, M. Kupers, P.H. Westerink, G.W. Seeley, J.P.J. de Valk and H.M.J.A. Kroon, "CT-image compression by sub- band coding using vector quantization", *Proc. of the 7th Int. Conference on Medical Informatics*, MIE-87, Rome, September 21-25, 1987, vol. III, pp. 1384–1387, 1987

[12] G.W. Seeley, H.D. Fisher, M.O. Stempski, M. Borgstrom, J. Bjelland and M.P. Capp, "Total digital radiology department: spatial resolution requirements", *American Journal of Radiology*, vol. 148, pp. 421–426, 1987

[13] G.W. Seeley, J.P.J. de Valk, H.M.J.A. Kroon, O. Rompelman and A.R. Bakker, "Image compression evaluation: an example of a PACS component analysis chain using psychophysics", *Proc. of the SPIE-Conf. on Med. Imaging*, New Port Beach, January 31- February 5, pp. 792 – 798, 1988

[14] G.R.L. Sohie and K.L. Kloker, "A digital signal processor with IEEE floating-point arithmetic", *IEEE Micro*, pp. 49–67, 1988

[15] P.P. Vaidyanathan, "Quadrature mirror filter banks, M-band extensions and perfect reconstruction techniques", *IEEE ASSP Magazine*, vol. 4, pp. 4–20, 1987

[16] J.P.J. de Valk, H.M.J.A. Kroon, G.W. Seeley, P.R. Algra, R.B. Noordveld, O. Rompelman and A.R. Bakker, "First results using a PACS component evaluation chain based on psychophysics: image compression", *Journal of Medical Imaging*, vol. 1, pp. 325–330, 1987

[17] P. van de Velde, A.C. de Ridder, O. Rompelman, "VMEBus - Data compression system", *Proc. International Conference on VMEBus in Industry*, Paris, November 21-22, 1989

[18] M. Vetterli, "Multi-dimensional sub-band coding: some theory and algorithms", *Signal Processing*, vol. 6, pp. 97-112, 1984

[19] S. Webb (ed.), *The Physics of Medical Imaging*, Medical Science Series, Adam Hilger, Bristol, 1988

[20] P.H. Westerink, J. Biemond, D.E. Boekee, "An optimal bit allocation algorithm for sub-band coding", *Proc. Int. Conf. on Acoust., Speech and Signal Process. (ICASSP)*, New York, pp. 757-760, 1988

[21] P.H. Westerink, D.E. Boekee, J. Biemond and J.W. Woods, "Sub-band coding of images using vector quantization", *IEEE Trans. on Communication*, vol. COM-36, pp. 713-719, 1988

[22] P.H. Westerink, J. Biemond and D.E. Boekee, "Quantization error analysis of sub-image filter banks", *Proc. IEEE Int. Symp. on Circ. & Systems*, Helsinki, June 7-9, pp. 819-822, 1988

[23] J.W. Woods and S.D. O'Neil, "Sub-band coding of images", *IEEE Trans. on Acoust. Speech and Signal Process.*, vol. ASSP-34, pp. 1278- 1288, October 1986

[24] I.T. Young, "Sampling density and quantitative microscopy", *Anal. and Quant. Cytology and Histology*, vol. 10, pp. 269-275, 1988

Index

Video Tape of Subband Coding Results

produced by *Rensselaer Video*

This video tape will contain all the still monochrome and color image coding results from Chapters 3, 4, 5, and 8 as well as several coded video sequences to illustrate the video conferencing and advanced television results of Chapters 6 and 7. The tape will be 15 minutes in duration and will be made available in NTSC for both VHS and SVHS tape formats. The cost for a VHS copy is $14.95 plus $2.50 mailing (and state tax if NY resident). The SVHS copy is $19.95 plus $2.50 mailing (and state tax if NY resident). They may be purchased with a copy of the order form below.

....

ORDER FORM

Mail to: Rensselaer Video, P. O. Box 828, Clifton Park, NY 12065.

Name ....

Address...

....

....

☐ check enclosed
☐ purchase order attached
☐ COD (add $3.00)

☐ VHS copy
☐ SVHS copy

Total Enclosed: ....